Math Methods for ☺ sake!

A guide for VCE Mathematical Methods 3&4

2023-2027 VCAA Study Design

Georgia Gouros

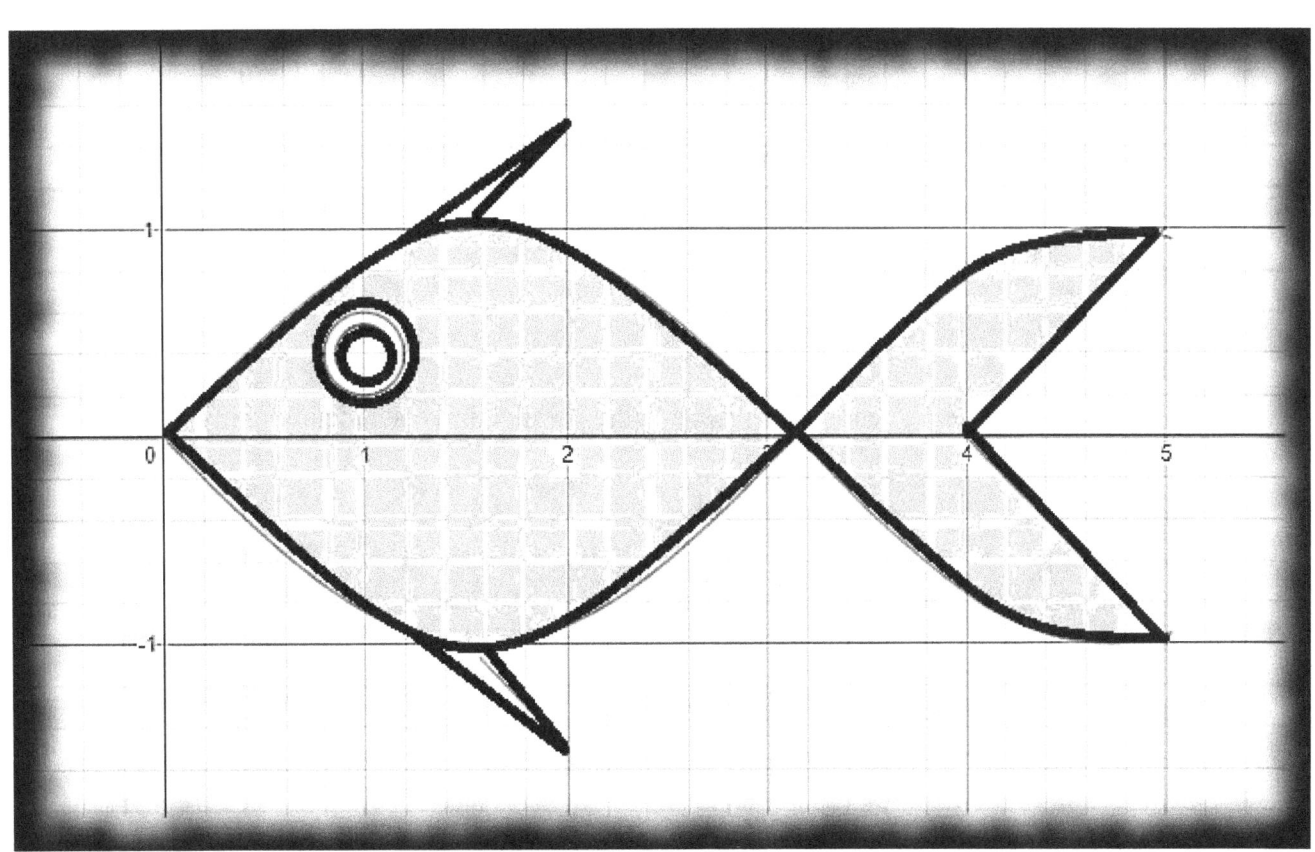

Author/Publisher: Georgia Gouros
©2025

Math Methods for ☺ sake!
Copyright © Georgia Gouros 2025

First Edition: 2019, This Edition: 2025
Author/Publisher: Georgia Gouros.
https://www.linkedin.com/today/author/georgia-gouros-590b25104

This publication can be purchased online at the url *http://lulu.com/spotlight/algorithmics*

Dedicated to every teenager in the entire universe, and the parallel universe, including my son Steven.

About the author and publisher: Georgia Gouros is a teacher of Mathematics and Algorithmics in Victoria, Australia.

A teacher of Mathematics since 2007, a fan of Mathematics for a lot longer.

Other books by this author:
An Introduction to Algorithmic Thinking: A student guide for Algorithmics (HESS)
Can be purchased online from *http://lulu.com/spotlight/algorithmics*

Math Methods for ☺ sake!

A guide for VCE Mathematical Methods 3&4

Contents

1 Algebra and Algorithms	5
Check your understanding	25
Solutions	27
2 Functions y=f(x) for real	33
Check your understanding	47
Solutions	53
3 Super Power Functions	61
Check your understanding	75
Solutions	79
4 Fabulous Functions	89
Check your understanding	109
Solutions	112
5 Deriving Derivatives	121
Check your understanding	140
Solutions	143
6 Integrating Integrals	149
Check your understanding	164
Solutions	169
7 Discrete Probability	175
Check your understanding	193
Solutions	197
8 Continuous Probability	203
Check your understanding	219
Solutions	223
9 Sampling Statistics	229
Check your understanding	243
Solutions	245
INDEX A-Z	249
TOPIC INDEX by Chapter 1-9	256

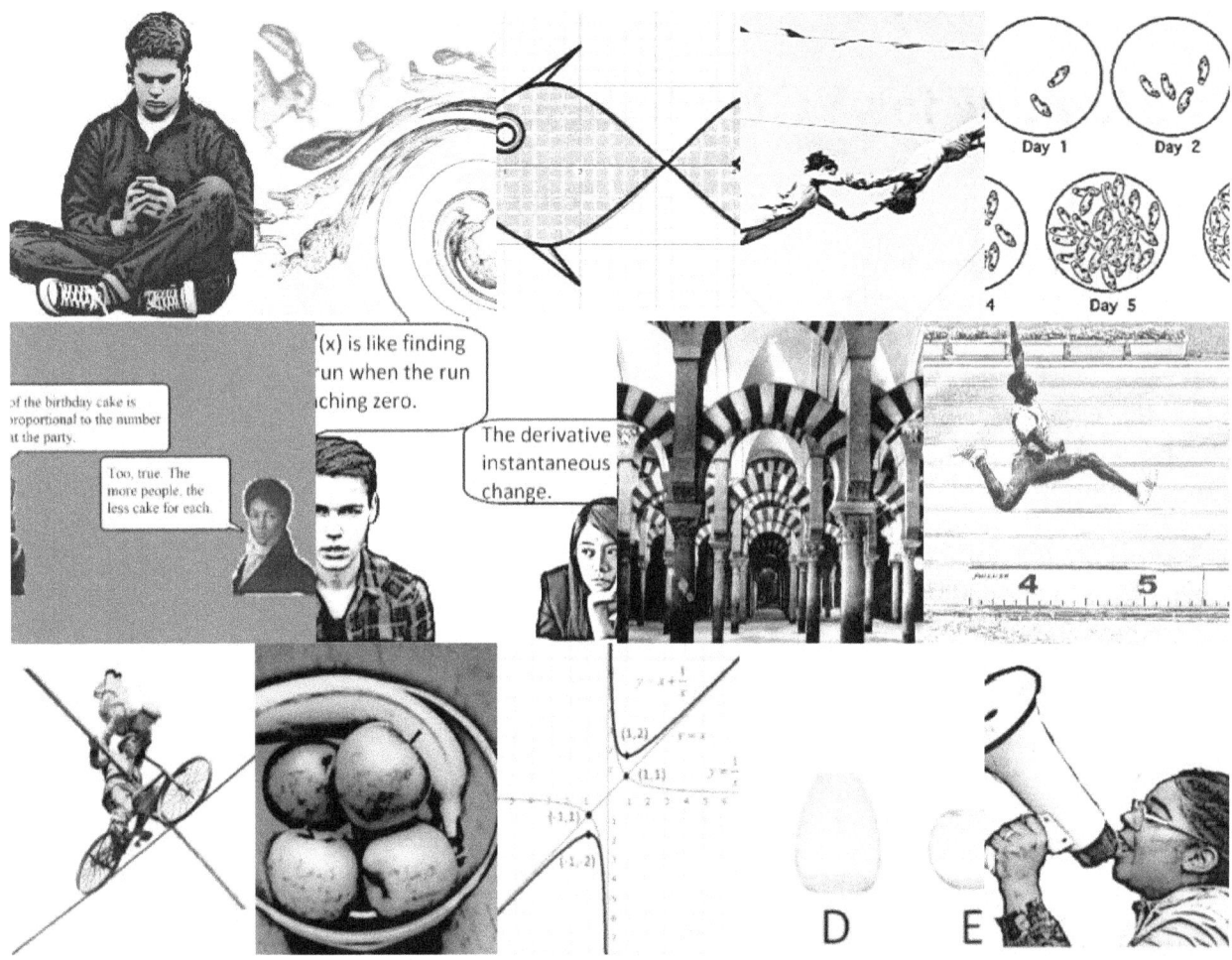

Internet resources used for creating graphs in this text.
MathGV http://www.mathgv.com/
Desmos https://www.desmos.com/calculator
Geogebra 3D https://www.geogebra.org/3d?lang=en

Internet resources used for creating images in this text: https://commons.wikimedia.org/wiki/Category:Images
Gimp https://www.gimp.org/

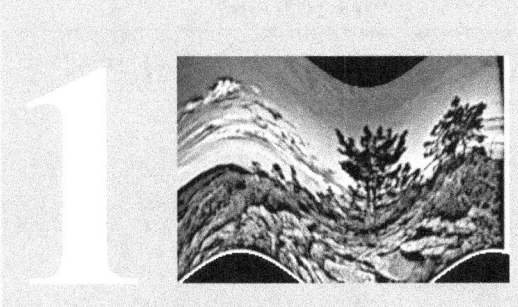

1 Algebra and Algorithms

Abstraction in Art, Science and Mathematics is used to simplify real life, so we can try to understand it.

 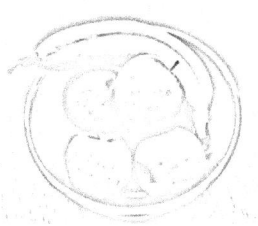 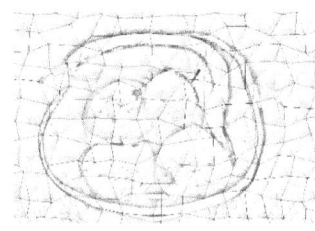

Image distortions using Gimp.
https://www.gimp.org/

Algebra is the foundation of modern mathematics, it allows the abstractions of real world problems that can then be solved using methods of linear algebra, calculus and trigonometry as well as other mathematical methods.

Algebra can also be used to detect patterns and relationships between variables which can then be used to predict outcomes using statistical methods.

There are many famous algebraic equations that model real world phenomena created by many mathematicians, logicians and scientists from all over the world since ancient times.

Examples of algebra used to define formulas.	
$a^2+b^2=c^2$ **Pythagoras** *c. 570 – c. 495 BC* 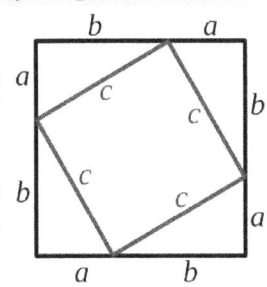	$1+2+3+.....+n=\dfrac{n(n+1)}{2}$ **Gauss** *1777-1855* 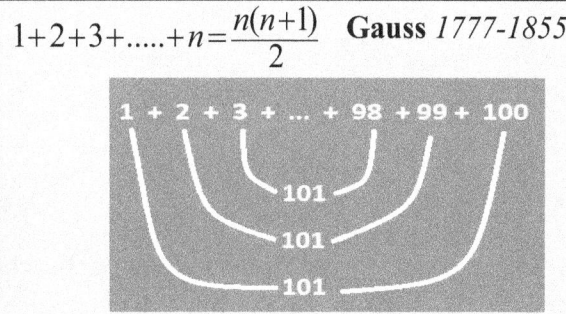

Algebra is used to represent mathematical problems with symbols that form equations and expressions. Solving problems with algebra involves manipulating equations to isolate and evaluate variables representing real world quantities.

Both algebra and algorithms involve step by step approaches to solving problems, they both require abstract thinking. Algorithms focus on solving computational problems by following instructions to achieve a desired outcome. Together algebra and algorithms are a powerful toolkit for solving complex problems in many fields.

Act 1, Scene 1. Simultaneous Linear Equations. The drama of setting up equations.

When you buy 3 apples and 2 bananas it costs $1.60. If you buy 7 apples and 1 banana in this supermarket the cost is $1.90.

Let's abstract the problem using algebra, and then we can solve it with many different techniques.

a=cost of apple

b=cost of banana

Let's ignore the weights of the fruits, let's ignore the type of the apples and the bananas.

$$3a + 2b = 160$$
$$7a + b = 190$$

Amazing

Algebrains.

I'm going to use Elimination to solve this one.

Double the second purchase, 14 apples and 2 bananas are $3.80.

Remove the first purchase from the doubled second purchase.

$$3a + 2b = 160$$
$$14a + 2b = 380$$

$$(14a - 3a) + (2b - 2b) = (380 - 160)$$
$$11a = 220$$
$$a = \frac{220}{11} = 20$$
$$3(20) + 2b = 160$$
$$2b = 100$$
$$b = 50$$

Act 1, Scene 2. Simultaneous Linear Equations. The drama continues….

I'm trying Substitution.

Isolate one of the variables and use it to replace the other variable.

Express the cost of bananas in terms of the cost of the apples.

Solve for one variable.

$$3a + 2b = 160$$
$$7a + b = 190 \Rightarrow b = 190 - 7a$$

$$3a + 2(190 - 7a) = 160$$
$$3a + 380 - 14a = 160$$
$$-11a = -220$$
$$a = 20$$
$$b = 190 - 7(20)$$
$$b = 50$$

Amazing

Algebrains.

With a bit of technology I'm going to use Graphing.

Map cost of apples to the x-axis

Map cost of bananas to the y-axis.

The intersection of the lines is the solution.

$(20, 50)$

$3a + 2b = 160$	$a \Rightarrow x$	$3x + 2y = 160$
$7a + b = 190$	$b \Rightarrow y$	$7x + y = 190$

$$3x + 2y = 160 \Rightarrow y = \frac{1}{2}(160 - 3x)$$

$$7x + y = 190 \Rightarrow y = 190 - 7x$$

$$Solution(20, 50)$$
$$x = 20, y = 50 \Rightarrow a = 20, b = 50$$

a=cost of apple, b=cost of banana

$$apples = 20^c, bananas = 50^c$$

7

🌼 + 🌼 + 🌼 = 60 🌼 + 🌸 + 🌸 = 30 🌸 − ❋ = 3 ❋ + 🌼 + 🌸 = ?	Set up the numbered equations using pronumerals. $3a = 60$ (1) $a + 2b = 30$ (2) $b - 2c = 3$ (3) $c + a + b = ?$	$a = 20$ from (1) sub into (2) $20 + 2b = 30 \Rightarrow b = 5$ Sub b into (3) $5 - 2c = 3 \Rightarrow 2c = 2 \Rightarrow c = 1$ Solution $c + a + b = 26$

🕐 + 🖊 + ❤ = 216

🕐 + ❤ = 🍪

🖊 + ❤❤ = 🕐

🕐 = ? 🖊 = ? ❤ = ?

Set up the numbered equations using pronumerals.

$a + b + c = 216$ (1)
$a + c = 3b$ (2)
$2b + 2c = a$ (3)

Sub (3) into (1)
$(2b + 2c) + b + c = 216$

Which simplifies to
$3b + 3c = 216 \Rightarrow b + c = 72$

Equation (1) becomes
$a + (b + c) = 216$

$a + (72) = 216$

$\Rightarrow a = 144$

Sub (3) into (2)
$(2b + 2c) + c = 3b \Rightarrow 3c = b$

Equation (3) becomes
$6c + 2c = 144$

$8c = 144$

$\Rightarrow c = 18$

Sub a and c into (1)
$a + b + c = 216$

$144 + b + 18 = 216$

$\Rightarrow b = 54$

Solution
a=144, b=54, c=18

Five numbers add to zero.
The first is equal to the sum of the second and the fourth.
The third is equal to the sum of the fourth and the fifth.
The sum of the first two numbers is 2 more than the fifth number.
The fifth number is three times the sum of the third and fourth numbers.
Find the five numbers.

Let a, b, c, d, e represent the five numbers and set up the numbered equations.

$a + b + c + d + e = 0$ (1)
$a = b + d$ (2)
$c = d + e$ (3)
$a + b = e + 2$ (4)
$e = 3(c + d)$ (5)

Sub (4) and (3) into (1)
$(e + 2) + (d + e) + d + e = 0$
Simplify
$3e + 2d + 2 = 0$ (6)
Sub (5) into (6)
$3(3(c + d)) + 2d + 2 = 0$

$9c + 9d + 2d + 2 = 0$

$9c + 11d = -2$ (7)
Sub (5) into (3)
$c = d + 3(c + d)$
Simplify
$c - d = 3c + 3d$

$-2c = 4d$

$c = -2d$

Sub $c = -2d$ into (7)
$-18d + 11d = -2$

$-7d = -2$

$d = \dfrac{2}{7}, hence \ c = \dfrac{-4}{7}$

Sub c and d into (3)
$e = c - d \Rightarrow e = \dfrac{-4}{7} - \dfrac{2}{7}$

$e = \dfrac{-6}{7}$

Sub d into (2)
$a = b + \dfrac{2}{7}$

$a - b = \dfrac{2}{7}$ (8)

Sub e into (4)
$a + b = e + 2$

$a + b = \dfrac{-6}{7} + \dfrac{14}{7}$

$a + b = \dfrac{8}{7}$ (9)

Add (8) to (9) to eliminate b

$a = \dfrac{5}{7}, hence \ b = \dfrac{3}{7}$

Solution
$\dfrac{5}{7}, \dfrac{3}{7}, \dfrac{-4}{7}, \dfrac{2}{7}, \dfrac{-6}{7}$

Example 1.1: A parabola with the equation $y = ax^2 + bx + c$ **goes through the 3 points** $(-1, -10), (2, -1), (3, -6)$. **Find the values of** a, b **and** c.

$y = ax^2 + bx + c$

Set up the simultaneous equations with the given (x, y) values.

$(-1, -10) \Rightarrow -10 = a - b + c$

$(2, -1) \Rightarrow -1 = 4a + 2b + c$

$(3, -6) \Rightarrow -6 = 9a + 3b + c$

$a - b + c = -10 \quad (1)$

$4a + 2b + c = -1 \quad (2)$

$9a + 3b + c = -6 \quad (3)$

A bit of elimination.

$(2) - (1) \Rightarrow 3a + 3b = 9$

$\Rightarrow a + b = 3$

$(3) - (1) \Rightarrow 8a + 4b = 4$

$\Rightarrow 2a + b = 1$

Then a bit of substitution.

$a + b = 3 \Rightarrow b = 3 - a$

$2a + b = 1 \Rightarrow 2a + (3 - a) = 1$

$\Rightarrow a + 3 = 1$

$\therefore a = -2, b = 5$

$(1) \quad a - b + c = -10$

$-2 - 5 + c = -10$

$\therefore c = -3$

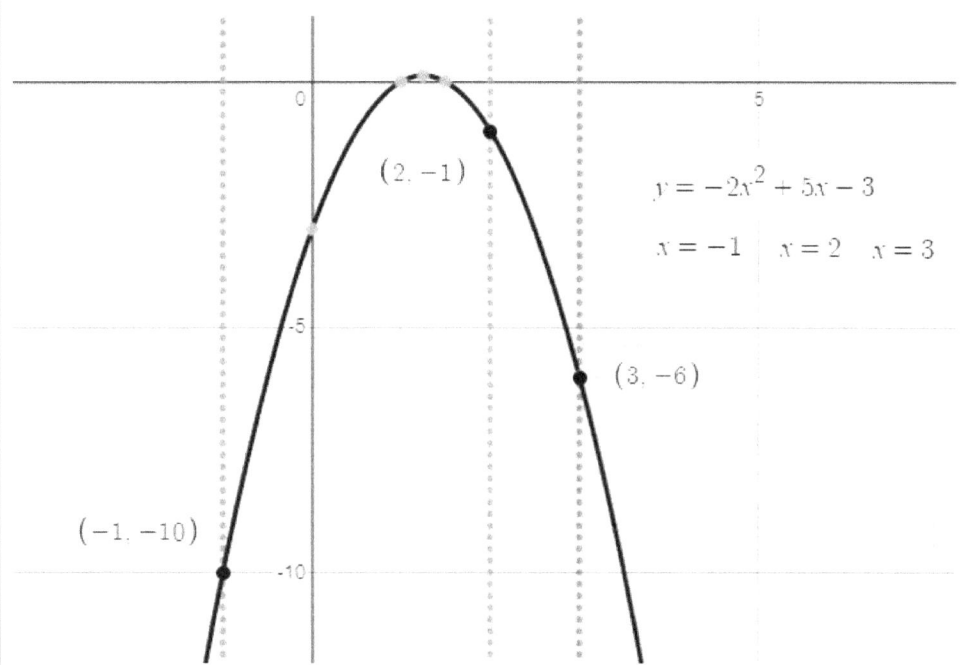

$y = -2x^2 + 5x - 3$

$x = -1 \quad x = 2 \quad x = 3$

$(2, -1)$

$(3, -6)$

$(-1, -10)$

And finally check the solution $y = -2x^2 + 5x - 3$ is correct against the given information.

Fibonacci Blue/Flickr

Architecture and Parabolas in postcards.

Moorish Empire – Spain

Roman Empire – location France   Persian Empire – location Iraq

Revision of some Linear Coordinate Geometry

Gradient of a Straight Line connecting two points (x_1, y_1) and (x_2, y_2) is $m = \frac{y_2 - y_1}{x_2 - x_1}$	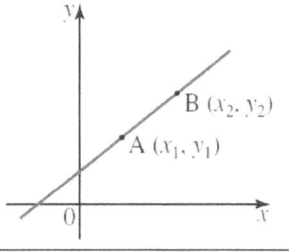
Finding the equation of a line given a point (x_1, y_1) and the gradient $m = \frac{y_2 - y_1}{x_2 - x_1}$ $$(y - y_1) = m(x - x_1)$$ Rearrange to get the gradient intercept form of the line $$y = m(x - x_1) + y_1$$	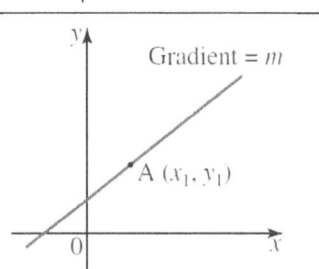

- Gradient Intercept Form of linear equation $y = mx + c$ where m is the gradient and c is the y-intercept
- Intercept form of linear equation $Ax + By = C$, intercepts can be found easily:
  - y intercept ($x = 0 \rightarrow By = C \therefore y = \frac{C}{B}$) x intercept ($y = 0 \rightarrow Ax = C \therefore x = \frac{C}{A}$)
- Parallel lines have the same gradient, Perpendicular lines meet at right angles, the product of their gradients is -1 $m_1 \times m_2 = -1$
- Midpoint of a line connecting two points (x_1, y_1) and (x_2, y_2) is $(\frac{x_1 + x_2}{2}, \frac{y_1 + y_2}{2})$
- Distance between two points (x_1, y_1) and (x_2, y_2)– using Pythagoras $d = \sqrt{(x_2 - x_1)^2 + (y_2 - y_1)^2}$
- Angle of slope $\tan \theta = m = \frac{y_2 - y_1}{x_2 - x_1}$
- Angle between intersecting lines $\alpha = \theta_2 - \theta_1$
- Simultaneous equations

Solving simultaneous equations with the substitution method - involves removing one of the unknown variables by substituting a transformed expression in place of that variable.	Solving simultaneous equations with the elimination method - combines two equations to make a third one that eliminates one of the variables either the x or the y. Look for an addition or subtraction that will eliminate one of the variables after applying a constant multiplication factor.
Example of substitution x + y = 6 (1) 2x + 4y = 20 (2) Equation (1) can be transformed to y = -x + 6 then substitute y into Equation 2. - Substitution 2x + 4(**-x+6**) = 20 - Simplify 2x – 4x + 24 = 20 - Answer x=2, y=4	Example of elimination -2x -3y = -9 (1) 4x +y = 8 (2) If equation (1) is multiplied by constant factor of 2 then the x coefficients can be eliminated when (1') added to (2) -4x -6y = -18 (1') Add (1') to (2) to get -5y = -10, which is y=2 Answer x=1.5, y=2

The Mathematical Methods 2023-2027 study in the VCE Curriculum allows the use of CAS technology for solving algebraic equations and other mathematical problems in some situations where it is permitted in examinations and assessments.

Example 1.2: Using CAS technology for solving algebraic problems.

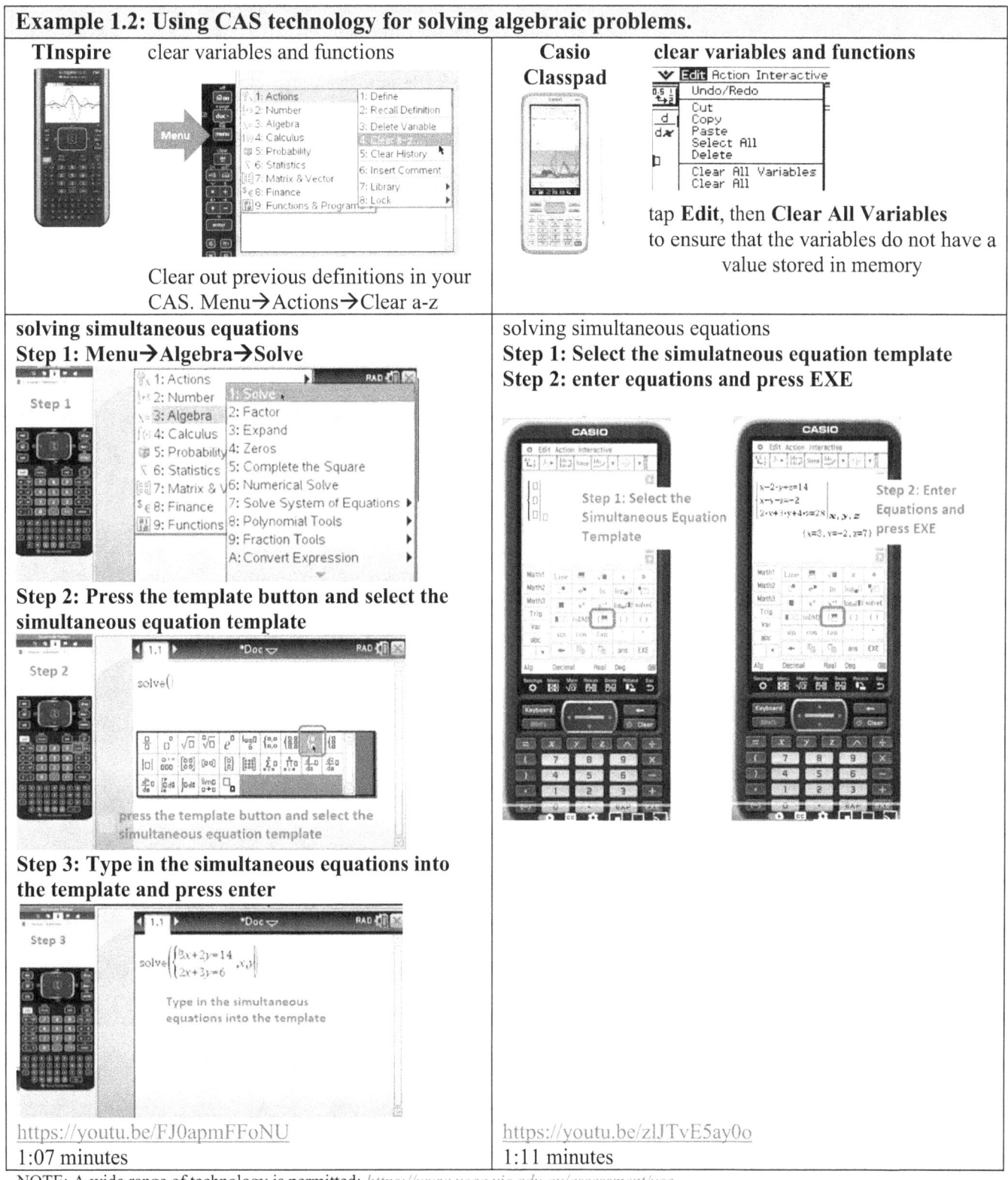

TInspire clear variables and functions

Clear out previous definitions in your CAS. Menu→Actions→Clear a-z

Casio Classpad **clear variables and functions**

tap **Edit**, then **Clear All Variables** to ensure that the variables do not have a value stored in memory

solving simultaneous equations
Step 1: Menu→Algebra→Solve

Step 2: Press the template button and select the simultaneous equation template

Step 3: Type in the simultaneous equations into the template and press enter

$$\text{solve}\left\{\begin{matrix}3x+2y=14\\2x+3y=6\end{matrix}\right., x,y\}$$

Type in the simultaneous equations into the template

https://youtu.be/FJ0apmFFoNU
1:07 minutes

solving simultaneous equations
Step 1: Select the simulatneous equation template
Step 2: enter equations and press EXE

Step 1: Select the Simultaneous Equation Template

$x-2 \cdot y+z=14$
$x-y=-2$
$2 \cdot x+(-y+4 \cdot y=28$ x,y,z
$\{x=3, y=-2, z=7\}$
Step 2: Enter Equations and press EXE

https://youtu.be/zlJTvE5ay0o
1:11 minutes

NOTE: A wide range of technology is permitted: *https://www.vcaa.vic.edu.au/assessment/vce-assessment/materials/Pages/calculators.aspx#:~:text=For%20approved%20schools%20only%2C%20students,a%20USB%20for%20examinations%20in* This text will show examples using the most popular technology at the present time of writing which is the **TInspire CAS** and the **Casio Classpad** technologies and sometimes will also show examples using **Mathematica.**

Act 2, Scene 1. Simultaneous Linear Equations. The drama where there is no unique solution.

I'm wondering what will happen if the equations that are set up are parallel or the same line.

How is the solution interpreted?

I agree, let's explore that.
Case 1: Take the lines 3x-2y=3 and 6x-4y=0.
Case 2: Take the lines 3x-2y=3 and 6x-4y=6.

Amazing

Algebrains

Case 1: Parallel lines: No solution.

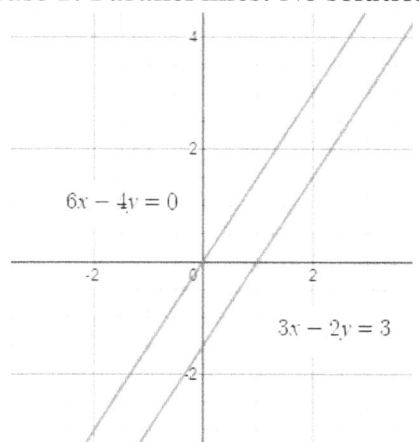
$6x - 4y = 0$
$3x - 2y = 3$

Case 2: Identical lines: Infinite solutions.

Case 1: Parallel lines
$$3x - 2y = 3 \qquad (1)$$
$$6x - 4y = 0 \qquad (2)$$

Use elimination multiply equation (1) by -2
$$-6x + 4y = -6 \qquad (1')$$
Add (1') to equation (2)
$$0 + 0 = -6$$

Which doesn't make sense and indicates that there are no solutions.

Case 2: Identical lines
$$3x - 2y = 3 \qquad (1)$$
$$6x - 4y = 6 \qquad (2)$$

Use elimination multiply equation (1) by -2
$$-6x + 4y = -6 \qquad (1')$$
Add (1') to equation (2)
$$0 + 0 = 0$$

Which is always true and indicates that there are infinite solutions.

Act 2, Scene 2. Simultaneous Linear Equations. The drama of 3 linear equations with 3 unknowns.

What about simultaneous equations with more than 2 unknown variables?

Elimination could be done, which would take some time.
Or use technology.

?????

Ok, let's try some elimination on 3 linear equation with 3 unknowns.

$3x+4y-z=8$ (1)
$5x-2y+z=4$ (2)
$2x-2y+z=1$ (3)

(1)+(2) will eliminate z
$8x+2y=12$ (4)
(1)+(3) will eliminate z
$5x+2y=9$ (5)

(4)-(5) will eliminate y
$3x=3 \Rightarrow x=1$
sub x into (5)
$5+2y=9 \Rightarrow y=2$
sub x,y into (3)
$2-4+z=1 \Rightarrow z=3$

Solution is (1,2,3)

Using technology set up this 3×3 matrix system and solve it using the inverse matrix.

$$\begin{bmatrix} 3 & 4 & -1 \\ 5 & -2 & 1 \\ 2 & -2 & 1 \end{bmatrix}\begin{bmatrix} x \\ y \\ z \end{bmatrix} = \begin{bmatrix} 8 \\ 4 \\ 1 \end{bmatrix}$$

$$\begin{bmatrix} x \\ y \\ z \end{bmatrix} = \begin{bmatrix} 3 & 4 & -1 \\ 5 & -2 & 1 \\ 2 & -2 & 1 \end{bmatrix}^{-1}\begin{bmatrix} 8 \\ 4 \\ 1 \end{bmatrix}$$

$$\begin{bmatrix} x \\ y \\ z \end{bmatrix} = \begin{bmatrix} 1 \\ 2 \\ 3 \end{bmatrix}$$

Solution found.

Act 3, Scene 3. Simultaneous Linear Equations. The drama when there are infinite solutions.

So if you try elimination on a 3x3 linear system and you get 0=N, where N is some non-zero number this is an inconsistent system and no solution can be found.

And if you try elimination on a 3x3 linear system for every variable and you keep getting 0=0, there are infinite solutions.

Amazing

Algebrains

Taking this linear system.	$x - y + z = 6$ (1) $x - 4y + 5z = 7$ (2) $x + 2y - 3z = 5$ (3)	
Matrix representation.	$\begin{bmatrix} 1 & -1 & 1 \\ 1 & -4 & 5 \\ 1 & 2 & -3 \end{bmatrix} \begin{bmatrix} x \\ y \\ z \end{bmatrix} = \begin{bmatrix} 6 \\ 7 \\ 5 \end{bmatrix}$	$\det\left(\begin{bmatrix} 1 & -1 & 1 \\ 1 & -4 & 5 \\ 1 & 2 & -3 \end{bmatrix} \right) = 0$

I wonder if this system is consistent?

Eliminate x	**Eliminate y**	**Eliminate z**
$(1)-(2) \Rightarrow \; 3y - 4z = -1$	$(2)+2(3) \Rightarrow \; 3x - z = 17$	$3(1)+(3) \Rightarrow \; 4x - y = 23$
$(1)-(3) \Rightarrow \underline{-3y + 4z = 1}$	$(2)-4(1) \Rightarrow \underline{-3x + z = -17}$	$5(1)-(2) \Rightarrow \underline{4x - y = 23}$
add $0 = 0$	*add* $0 = 0$	*subtract* $0 = 0$
$(2)-(3) \;\; \Rightarrow \; -6y + 8z = 2$	$(3)+2(1) \Rightarrow \; 3x - z = 17$	$3(2)+5(3) \;\; \Rightarrow \; 8x - 2y = 46$
$2[(2)-(1)] \Rightarrow \underline{\; -6y + 8z = 2}$	$2(3)+(2) \Rightarrow \underline{3x - z = 17}$	$2[5(1)-(2)] \Rightarrow \underline{8x - 2y = 46}$
subtract $0 = 0$	*subtract* $0 = 0$	*subtract* $0 = 0$

In every case and combination we get the true equation 0=0, this means there are infinite solutions. It's a consistent system.

Let's explore that by trying some constant values of z in this linear system.

Act 3, Scene 4. Simultaneous Linear Equations. The drama when there are infinite solutions.

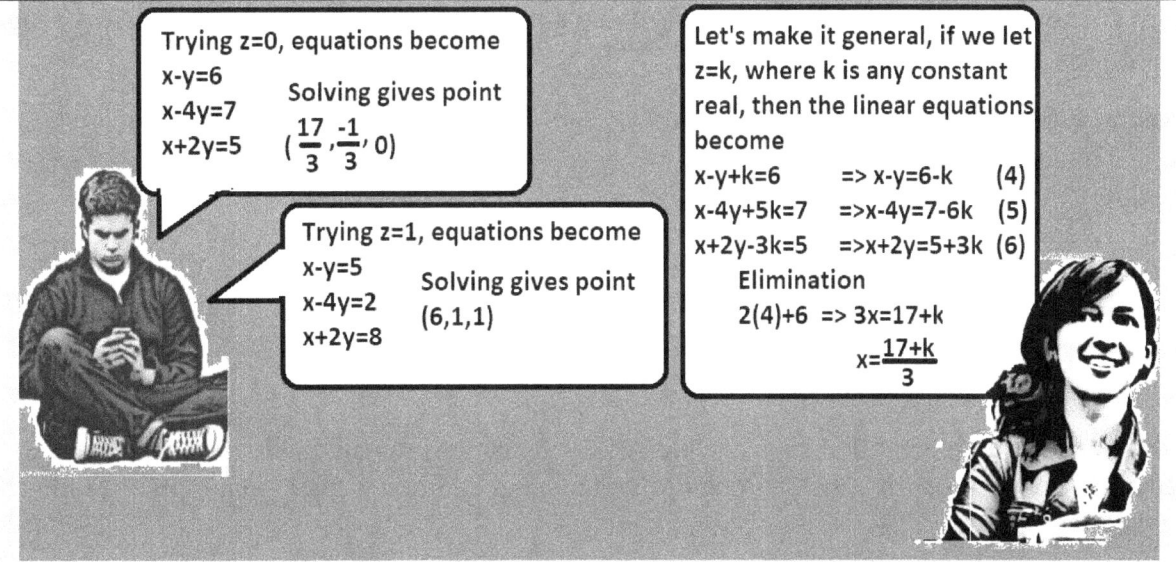

Trying z=0, equations become
x-y=6
x-4y=7
x+2y=5
Solving gives point
$(\frac{17}{3}, \frac{-1}{3}, 0)$

Trying z=1, equations become
x-y=5
x-4y=2
x+2y=8
Solving gives point
(6,1,1)

Let's make it general, if we let z=k, where k is any constant real, then the linear equations become
x-y+k=6   => x-y=6-k   (4)
x-4y+5k=7   =>x-4y=7-6k   (5)
x+2y-3k=5   =>x+2y=5+3k   (6)
Elimination
2(4)+6 => 3x=17+k
$x=\frac{17+k}{3}$

So far we've found that

$z=k, x=\frac{17+k}{3}$

Substituting this back into the equation (1)

$\frac{17+k}{3} - y+k=6$

$\frac{17+k}{3} +k-6=y$

$\Rightarrow$

$y= \frac{4k-1}{3}$

The infinite solution can be expressed as: $x=\frac{17+k}{3}$ $y=\frac{4k-1}{3}$ z=k

Amazing

Algebrains

By introducing another variable k, which is any real number, the solution (x,y,z) can be written as

$\left(\frac{17+k}{3}, \frac{4k-1}{3}, k\right) k \in R$

This is called the parametric form.

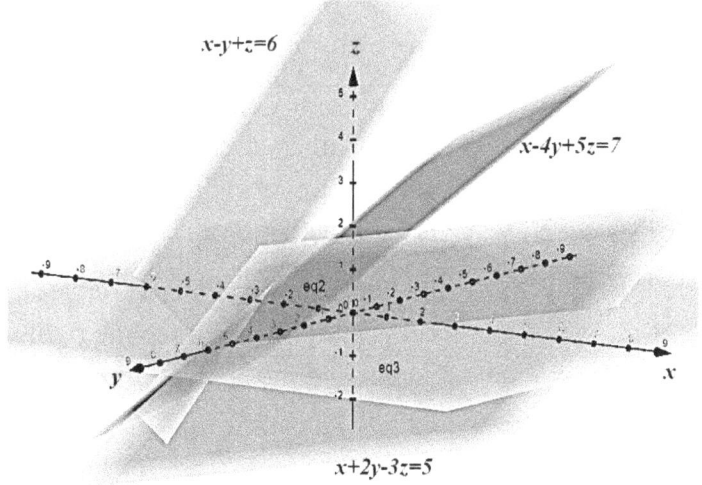

x-y+z=6

x-4y+5z=7

x+2y-3z=5

We can represent the infinite solutions of this linear system by introducing another parameter **k**, the infinite parametric solutions for this linear system can expressed in terms of **k**.

Planar images generated in Geogebra https://www.geogebra.org/3d?lang=en and enhanced with Gimp https://www.gimp.org/

Algorithms are used everywhere in every field of human endeavour as they define processes and methods for solving many kinds of problems using computation including mathematical problems and equations.

Algorithms define precise actions to create products, such as origami or to solve problems, such as factorizing quadratic functions.

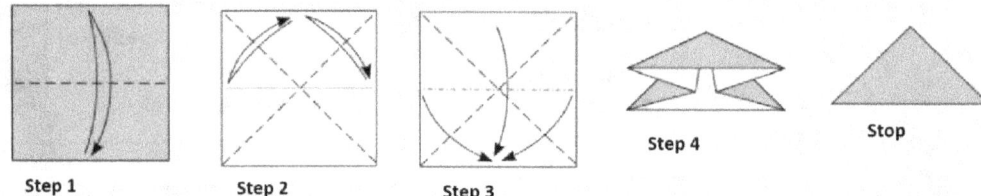

Over the course of their education students learn many algorithms for solving many arithmetic and mathematical problems, such as performing decimal number addition, subtraction, multiplication and other operations.

Addition			Subtraction			Multiplication		
1		1			1	5	3	
9	6	4	9	6⁵	4	9	6	4
+	4	2	8	-	4	2	8	×			8
=	1	3	9	2	=	5	3	6	=	7	7	1	2

Algorithms are often defined in a language called **pseudocode** which is a hybrid language combining key English words with coding language conventions and mathematical operations to form a precise and unambiguous computational solution for many real world problems.

This text defines algorithms that use the style of pseudocode following the *style guide (accessed January 2025) for pseudocode in VCE Mathematics:* [*https://www.vcaa.vic.edu.au/curriculum/vce/vce-study-designs/Pages/PseudoCode.aspx*](https://www.vcaa.vic.edu.au/curriculum/vce/vce-study-designs/Pages/PseudoCode.aspx)

Basic elements of pseudocode for defining algorithms - check your understanding with your class.		
Algorithm element	**Pseudocode definition**	**Example**
Variables	As in mathematics algorithms have variables hold values for performing computations. The ← arrow is used to assign values to variables. Example shown shows variables x and y on the left being assigned values to the right of the arrow ←	x ← 7 y ← x + 10
Sequential execution of actions	Actions defined in algorithms are run from top to bottom in sequence. Example shown runs actions sequentially <ul><li>assigns 7 to the variable x</li><li>assigns x + 10 to the variable y</li><li>print the content of the variable y</li></ul>	x ← 7 y ← x + 10 print y

Basic elements of pseudocode for defining algorithms – check your understanding with your class.

Pseudocode style guide (accessed January 2025) for VCE Mathematics:
https://www.vcaa.vic.edu.au/curriculum/vce/vce-study-designs/Pages/PseudoCode.aspx

Algorithm element	Pseudocode definition	Example

Conditional execution of actions	In many algorithms the actions are controlled by the values of the variables and whether certain conditions are true or false.
	
	if (condition is true) then
	    perform these indented actions
	end if
	
	The indentation of actions within the **if – then- end** is a very important part of defining an algorithm in pseudocode as it tells the reader of the algorithm that these indented actions belong to the if condition and are only actioned if the condition is true.
		if $(x > 0)$ then
		    $y \leftarrow \sqrt{x}$
		end if

Conditional execution of actions	Actions are controlled by **if-then-else-end if** and the use of indentation.
	
	if (condition is true) then
	    perform these indented actions
	else
	    perform these indented actions
	end if
	
	Indentation is very important for algorithms defined in pseudocode as it informs the reader of the algorithm which actions belong to which part of the **if-then-else-endif** condition and which of these should be done based on whether the condition is true or false.
		if $(x > 0)$ then
		    $y \leftarrow \sqrt{x}$
		else
		    print x is $\leq$ zero
		end if

Repeating actions for a fixed count of repetitions	**For** loops repeat actions based on a predefined number of repetitions referred to as iterations in algorithms.
	
	for i from 1 to n
	    perform these indented actions
	end for
		sum $\leftarrow$ 0
		for i from 1 to 10
		    $x \leftarrow i^3$
		    print x
		    sum $\leftarrow$ sum + x
		end for
		print sum

Repeating actions conditionally	**While** is used with a **condition** to repeat actions based on whether a condition is and remains true in an algorithm.
	    while (condition is true)
	        perform these indented actions
	    end while
		sum $\leftarrow$ 0
		k $\leftarrow$ 1
		while (k < 100)
		    sum $\leftarrow$ sum + i
		    k $\leftarrow$ k + 1
		end for
		print sum

Algorithms are not new, around 250 BC, the mathematician, Archimedes devised an algorithm of exhaustive calculations for approximating the value of **π**. Many translations of Archimedes' methods exist, some involve calculating the circumference using polygons and triangles, while others use polygons and triangles to calculate area.

The area of a unit circle is approximated by summation of the area of n isosceles triangles, which have their apex at the centre of the circle, with two of the sides equal to the radius. *Area=0.5×a×b×sin(θ),* where *a=b=1 units* for a unit circle*,* and the apex angle *θ* is *360/n* . The area of *n* triangles is used to approximate the area of this circle. As the number of triangles *n* approaches infinity, using mathematical notation, *n→∞*, the approximation becomes more accurate.

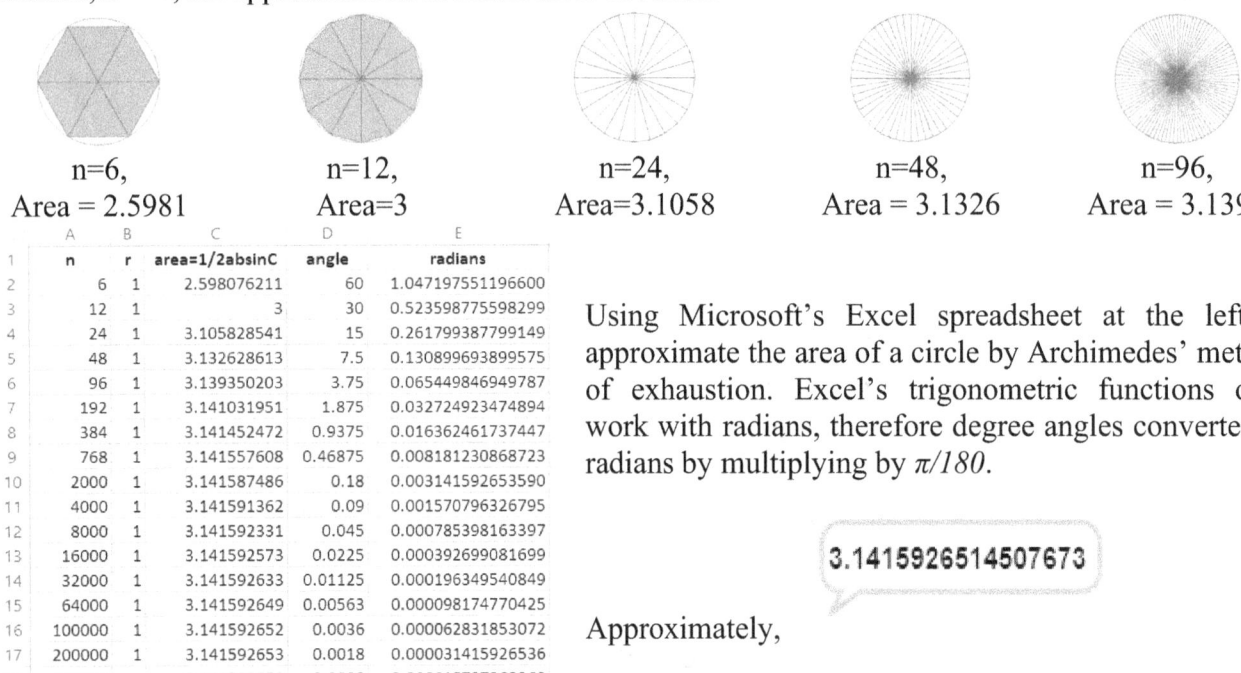

| | n=6, | n=12, | n=24, | n=48, | n=96, |
| Area = 2.5981 | Area=3 | Area=3.1058 | Area = 3.1326 | Area = 3.1395 |

	A	B	C	D	E
1	n	r	area=1/2absinC	angle	radians
2	6	1	2.598076211	60	1.047197551196600
3	12	1	3	30	0.523598775598299
4	24	1	3.105828541	15	0.261799387799149
5	48	1	3.132628613	7.5	0.130899693899575
6	96	1	3.139350203	3.75	0.065449846949787
7	192	1	3.141031951	1.875	0.032724923474894
8	384	1	3.141452472	0.9375	0.016362461737447
9	768	1	3.141557608	0.46875	0.008181230868723
10	2000	1	3.141587486	0.18	0.003141592653590
11	4000	1	3.141591362	0.09	0.001570796326795
12	8000	1	3.141592331	0.045	0.000785398163397
13	16000	1	3.141592573	0.0225	0.000392699081699
14	32000	1	3.141592633	0.01125	0.000196349540849
15	64000	1	3.141592649	0.00563	0.000098174770425
16	100000	1	3.141592652	0.0036	0.000062831853072
17	200000	1	3.141592653	0.0018	0.000031415926536
18	400000		3.141592653	0.0009	0.000015707963268

Using Microsoft's Excel spreadsheet at the left to approximate the area of a circle by Archimedes' method of exhaustion. Excel's trigonometric functions only work with radians, therefore degree angles converted to radians by multiplying by *π/180.*

3.1415926514507673

Approximately,

Archimedes would have computed this result by hand, it is no wonder that it was called an exhaustion method. We can create an algorithm to action Archimedes' method in pseudocode, and implement the algorithm in Python code environment, where a computer performs the exhaustive calculations.

Example 1.3 Area of a circle by exhaustive calculation.	
Archimedes' algorithm in pseudocode	Encoded algorithm in the Python coding language
| radius ← 1<br>tolerance ← 0.0001      *# acceptable error*<br>area0 ← 4<br>area1 ← 3<br>n ← 6   *# number n of triangles to begin with*<br>**while** (\|area1-area0\| > tolerance)<br>    area0 ← area1<br>    angleDegrees ← 360/n      *# apex angle*<br>    *# calculate the area of n triangles*<br>    area1 ← n×0.5×(radius)²sin(angleDegrees)<br>    n ← 2×n      *# double n, triangles for estimation*<br>**end while**<br>print area1      *# as the approximation* | ```
import math

radius = 1
tolerance = 0.0001
area0 = 4
area1 = 3
n = 6

while abs(area1 - area0) > tolerance:
    area0 = area1
    angle_degrees = 360 / n
    angle_radians = math.radians(angle_degrees)
    area1 = n * 0.5 * (radius**2) * math.sin(angle_radians)
    n = 2*n

print(area1)
``` |
```

Example 1.4: Algorithms and algebra combined for solving mathematical problems.

Numerical methods are algorithms that solve mathematical equations. The algorithm below finds $\sqrt{N}$ with a numerical method of approximation.

Step by step algorithm using basic algebra	The same algorithm for finding $\sqrt{N}$
1. Begin with an arbitrary positive number as the starting value x_0, that is somewhere between 1 and N, the closer to the actual square root of N, the better.	numerically is defined in pseudocode with variables and algebraic components with a margin of error.
2. Find the next approximation $x_{(n+1)}$ the arithmetic mean $x_{(n+1)} = \frac{1}{2}\left(x_n + \frac{N}{x_n}\right)$, and use it to approximate the geometric mean	x ← N/2 error ← 1 **while** x^2 is not in interval [N − error, N + error] x ← 0.5 × (x + N/x) **end while** print x
3. Check if $\left(x_{(n+1)}\right)^2 \approx N$ to the required accuracy, if it is then you are done, otherwise repeat from step 2	

The algorithm is used here to find the square root of N=844 to an accuracy dependent on an error of ± 1. The table shows the trace of the variable contents as the algorithm executes the actions.

While repetition	N=844, error = 1 Calculate variable x in while loop	x	x^2	N ± error interval
1		422	178084	[843,845]
2	x ← 0.5 × (422 + 844/422)	212	44944	
3	x ← 0.5 × (212 + 844/212)	107.99	11661.96	
4	x ← 0.5 × (107.99 + 844/107.99)	57.90	3352.73	
5	x ← 0.5 × (57.9 + 844/57.9)	36.24	1313.22	
6	x ← 0.5 × (36.24 + 844/36.24)	29.76	885.93	
7	x ← 0.5 × (29.76 + 844/29.76)	29.06	844.49	

The while loop in the algorithm finishes here as $29.06^2 = 844.49$ which is within the interval [843,845] which is **844 ± 1** making the condition on the while loop false.

Hence *since* $\sqrt{844} \approx 29.06$, the value 29.06 will be printed by the algorithm.

Algorithmic Functions

Functions are used in mathematics and in algorithms, algorithmic functions like mathematical functions accept input data, apply their computational logic, and then return computed values to the calling algorithm, which then uses these values to continue on with its own further computations. Algorithm functions are portable, reusable and shareable algorithms that once invented, tested and fully known to be correct can be used by lots of other algorithms, or lots of times by the one algorithm.

Mathematics uses functions to calculate values defined with one or more input variables using a mathematical rule. For example define a mathematical function f(x) that has one input variable x, f(x)=2x+1, then use it f(β)=2(β)+1=β+β+1. Mathematical functions defined with multiple input values, for example f(x,y) with two input variables x and y, f(x,y)=x+3y, f(♣,▲)=♣+3(▲)=♣+▲+▲+▲.

Defining algorithmic functions in pseudocode

Using the VCAA pseudocode style guide algorithmic functions in all examples in this text are defined in the following format:

define function_name(input for function):
.......perform these indented actions
return output

Example 1.5A: n! the factorial of n function in pseudocode.	Example 1.5B: minimum of two numbers a, b function in pseudocode.

define factorial(n):
    product ← 1
    for i from 1 to n
        product ← product × i
    end for
    return product
	define minimum (a, b):
	    if $(a \leq b)$ **then**
	        return a
	    else
	        return b
	    end if

Example 1.5C: function for finding the factors of an integer c in pseudocode.	Example 1.5D: function for finding the area of a triangle of base b, and height h in pseudocode

define factors(c):
    for a from 1 to c
        for b from 1 to c
            if $(a \times b$ equals c) **then**
                return a,b
            end if
        end for
    end for
	define areaTriangle(b, h):
	    return $0.5 \times b \times h$

Example 1.5E: function in pseudocode for finding the factors of a quadratic $f(x) = ax^2 + bx + c$	Example 1.5F: Euclids algorithm in pseudocode for finding the greatest common divisor (gcd) of two positive integers a, b

define factorise(a, b, c):
    discriminant ← $b^2 - 4ac$
    if (discriminant ≥ 0) **then**
        x1 ← $-b - \sqrt{discriminant}$
        x2 ← $-b + \sqrt{discriminant}$
        return x1, x2
    else
        return no solution
    end if
	Define gcd(a, b):
	    while $a \neq b$
	        if $(a > b)$ **then**
	            a ← $a - b$
	        else
	            b ← $b - a$
	        end if
	    end while
	    return a

Note: Indentation is very important for algorithms defined in pseudocode as it informs the reader of the algorithm which actions belong to which function and which actions belong in the **for loop** or the **if-then-else-endif** condition and which of these should be done based on whether the condition is true or false.	

Pseudocode style guide (accessed January 2025) for VCE Mathematics:

https://www.vcaa.vic.edu.au/curriculum/vce/vce-study-designs/Pages/PseudoCode.aspx

Encoding algorithms and algorithmic functions in Python code can easily be generated from pseudocode using chatbot technology such as Google Gemini or ChatGPT.

Pseudocode function syntax **	Python function syntax
define function_name(inputs for function) ....perform these indented actions **return output**	**def function_name(inputs for function):** ....perform these indented actions **return output**

Compare the previous algorithm examples created in pseudocode to the Python coded versions below that can be computed on a computer or CAS technology.

Example 1.6A: n! the factorial of n function in Python.	**Example 1.6B:** minimum of two numbers a, b function in Python.

``
def factorial(n):
  for i in range(1, n + 1):
    product *= i
  return product
``
	``
	def minimum(a, b):
	  if a <= b:
	    return a
	  else:
	    return b
	``

Example 1.6C: function for finding the factors of an integer c in Python.	**Example 1.6D:** function for finding the area of a triangle of base b, and height h in Python.

``
def factors(c):
  for a in range(1, c + 1):
    for b in range(1, c + 1):
      if a * b == c:
        return a, b
  return None
``
	``
	def areaTriangle(b, h):
	  return 0.5 * b * h
	``

Example 1.6E: function in Python for finding the factors of a quadratic $f(x)=ax^2 + bx + c$	**Example 1.5F:** Euclids algorithm in Python for finding the greatest common divisor (gcd) of two positive integers a, b

``
import math
 	
def factorise(a, b, c):
  discriminant = b**2 - 4*a*c
 	
  if discriminant >= 0:
    x1 = (-b + math.sqrt(discriminant)) / (2*a)
    x2 = (-b - math.sqrt(discriminant)) / (2*a)
    return x1, x2
  else:
    return "no solution"
``
	``
	def gcd(a, b):
	  while a != b:
	    if a > b:
	      a = a - b
	    else:
	      b = b - a
	  return a
	``

Note: There are many similarities that exist between the functions defined in pseudocode and in Python, however the Python syntax is very strict with its symbols and precise indentations and it requires the use of imported libraries of pre-defined functions, this is done by the import action. In these Python examples the `import math` action imports pre-defined math functions for use in the code.	

*** Pseudocode style guide (accessed January 2025) for VCE Mathematics:*

https://www.vcaa.vic.edu.au/curriculum/vce/vce-study-designs/Pages/PseudoCode.aspx

The CAS technology as used in Mathematical Methods has many built in functions that have been installed into the technology, so that when executed they perform mathematical operations to solve problems.

Example 1.7: Using CAS technology with built-in functions.

TInspire   Built in function **Define**

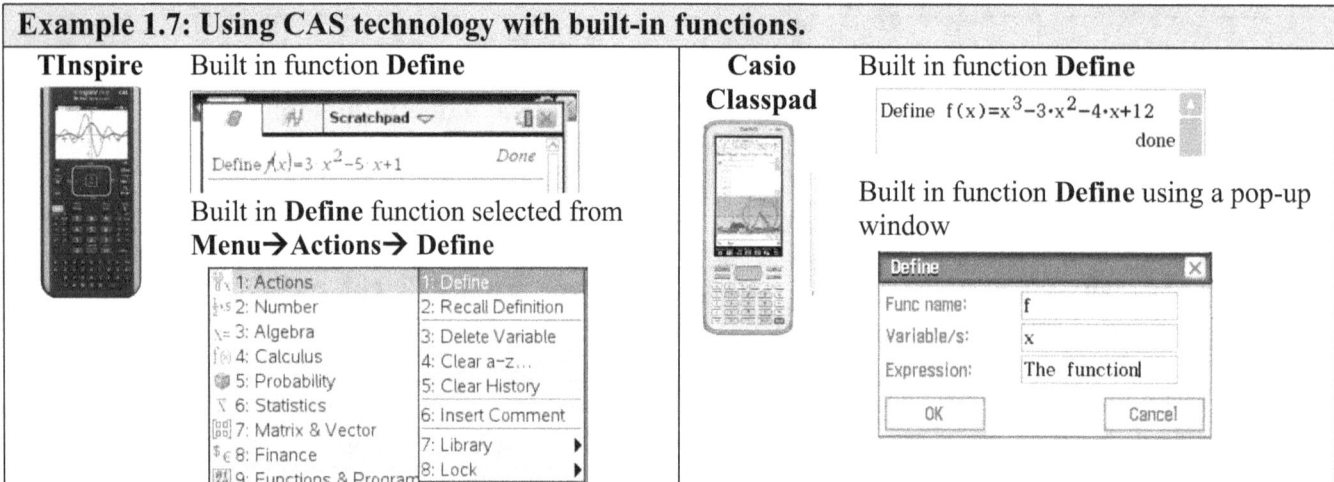

Built in **Define** function selected from
Menu→Actions→ Define

Casio Classpad   Built in function **Define**

Built in function **Define** using a pop-up window

The built in functions **factor** and **expand** for expressions from the command line if you know all the required parameters.

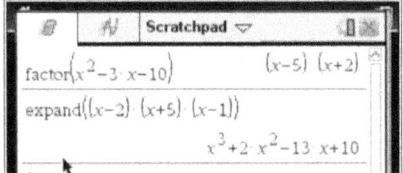

Alternatively the built in functions **factor** and **expand** can be selected from the menu.

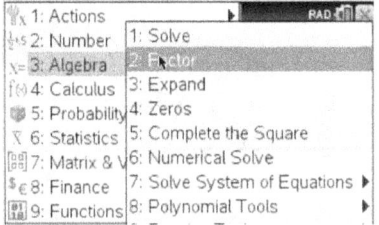

The built in function **solve** can be used to solve equations.

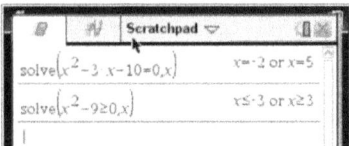

Alternatively the built in function **solve** can be selected from the menu: **Menu→Algebra→Solve**

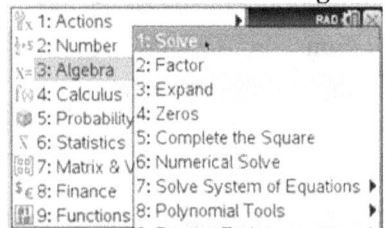

Many other built in functions have been implemented on the the TInspire.

The built in function **expand** can be used to expand brackets in expressions from the command line if you know all the required parameters.

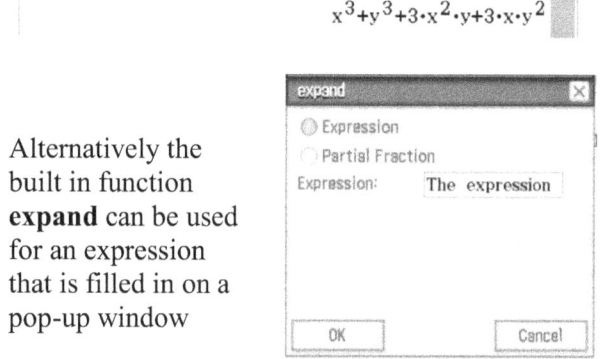

Alternatively the built in function **expand** can be used for an expression that is filled in on a pop-up window

The built in function **solve** can be used to solve equations.

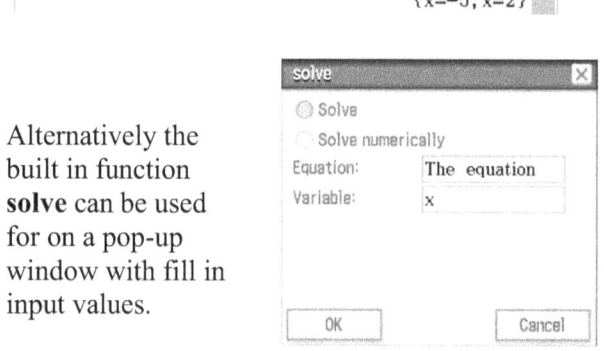

Alternatively the built in function **solve** can be used for on a pop-up window with fill in input values.

Many other built in functions have been implemented on the the Casio Classpad.

22

Example 1.8: An algorithm is defined in pseudocode with matrix methods to solve 2 simultaneous equations with two unknowns x, y.

$$2x + 4y = 5$$
$$-4x - 10y = 3$$

If we use variables as matrices where A has the coefficients of x,y and B has the constant values on the RHS of the equals sign.

$$A = \begin{bmatrix} 2 & 4 \\ -4 & -10 \end{bmatrix} \quad X = \begin{bmatrix} x \\ y \end{bmatrix} \quad B = \begin{bmatrix} 5 \\ 3 \end{bmatrix}$$

then $AX = B$ represents this linear system of simultaneous equations.

$$\begin{bmatrix} 2x+4y \\ -4x-10y \end{bmatrix} = \begin{bmatrix} 5 \\ 3 \end{bmatrix} \Rightarrow \begin{array}{l} 2x+4y = 5 \\ -4x-10y = 3 \end{array}$$

First let's find the inverse matrix

$$A^{-1} = \frac{1}{(2)(-10)-(-4)(4)} \begin{bmatrix} 10 & -4 \\ 4 & 2 \end{bmatrix}$$

$$= \frac{-1}{4} \begin{bmatrix} -10 & -4 \\ 4 & 2 \end{bmatrix}$$

$$= \begin{bmatrix} \frac{5}{2} & 1 \\ -1 & \frac{-1}{2} \end{bmatrix}$$

Check that $AA^{-1} = \begin{bmatrix} 1 & 0 \\ 0 & 1 \end{bmatrix}$

Then if we multiply both sides of $AX = B$ by the inverse matrix.

$$AA^{-1}X = A^{-1}B$$

$$\begin{bmatrix} 1 & 0 \\ 0 & 1 \end{bmatrix} \begin{bmatrix} x \\ y \end{bmatrix} = \begin{bmatrix} \frac{5}{2} & 1 \\ -1 & \frac{-1}{2} \end{bmatrix} \begin{bmatrix} 5 \\ 3 \end{bmatrix}$$

$$\begin{bmatrix} x \\ y \end{bmatrix} = \begin{bmatrix} \frac{25}{2} + 3 \\ -5 - \frac{3}{2} \end{bmatrix}$$

$$\begin{bmatrix} x \\ y \end{bmatrix} = \begin{bmatrix} \frac{31}{2} \\ \frac{-13}{2} \end{bmatrix}$$

You can use your CAS technology to do this too.

An algorithmic function can be defined to solve this problem.

define solver (A, B)
    if inverse(A) exists then
        AI = inverse(A)
        if (A × AI equals I) then
            return AI × B
        else
            return -1
        end if
    else
        return -1
    end if

We get the solution to the simultaneous linear equations with an algorithmic function using two nested **if-then-else-end if** conditions.

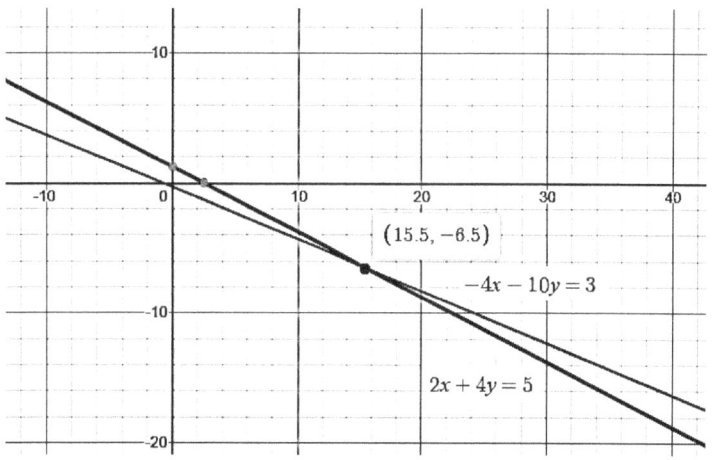

$(15.5, -6.5)$

$-4x - 10y = 3$

$2x + 4y = 5$

23

ALGEBRA and ALGORITHMS SUMMARY:

Algebraic equations are abstractions that model real world phenomena. Pronumerals representing entities should be defined and be able to be explained in the real world context.

Simultaneous equations are represented using algebra and are solved using many methods including by elimination and by substitution.

Mathematical Methods 2023-2027 allows the use of CAS technology for solving mathematical problems. This technology has many built-in functions to perform specific operations on equations and expressions.

Example: Using CAS technology with **solve** built-in function for simultaneous equations.

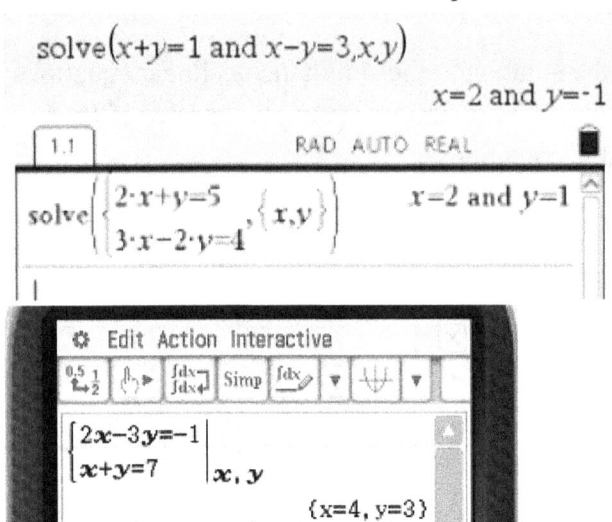

$$\text{solve}(x+y=1 \text{ and } x-y=3, x, y)$$
$$x=2 \text{ and } y=-1$$

Algorithms are defined in pseudocode to perform actions that solve problems using computational methods.

Variables in algorithms hold values and are defined using the arrow $\leftarrow$

$x \leftarrow 7$

$y \leftarrow x + 10$

Conditional actions in algorithms defined in pseudocode are done with **if-then-endif** or **if-then-else-endif** structures.

if (condition is true) **then**
     do indented actions
 end if
	if (condition is true) **then**
	     do indented actions
	 else
	     do indented actions
	 end if

Repetitive actions are done with **for** loop or **while** loop structures.

for i from 1 to n
     do indented actions
 end for
	while (condition is true)
	     do indented actions
	 end while

Functions can be defined in algorithms to execute specific tasks.

define function_name(inputs for function):
     do indented actions
     return output

Example: Solving simultaneous algebraic equations by substitution.

x + y = 6   (1)

2x + 4y = 20   (2)

Transform equation (1) to y = -x + 6 then substitute y into Equation 2. 2x + 4(**-x+6**) = 20

Simplify 2x − 4x + 24 = 20

Answer x=2, y=4

Example: Solving simultaneous algebraic equations by elimination.

-2x -3y = -9   (1)

4x +y = 8   (2)

Multiply equation (1) by constant 2

-4x -6y = -18   (1')

Add (1') to (2) eliminating the x variable to get -5y = -10, which is y=2

Answer x=1.5, y=2

ALGEBRA and ALGORITHMS: Check your Understanding.

1. Riddles.

a) Solve the puzzle

$$\bigcirc + \bigcirc = 10$$

$$\bigcirc \times \square + \square = 12$$

$$\bigcirc \times \square - \blacktriangle \times \bigcirc = \bigcirc$$

$$\blacktriangle = ?$$

b)

Old Granny Smith left half her money to her granddaughter and half that amount to her grandson.

She left a sixth to her brother, and the remainder, $1,000, to the lost dogs' home.

How much did she leave altogether?

c) Use logic and a matrix or table to help sort out these relationships between two sets of people.

{Alex, Boxuan and Con} are married to {Dora, Essie and Faye}.

1. Boxuan is Essie's brother and lives in Adelaide with his partner.
2. Con is shorter than Faye's partner.
3. Con works at a bank.
4. Dora and her partner live in Sydney.
5. Essie and her partner work in a restaurant.

Who is married to whom?

2. Fun with simultaneous equations.

a) Given two linear equations 2x−y=1 and 3x+2y=12 find the solution using algebraic methods without technology for x,y.

b) Solve and explain the solutions for:

    i. $2x + 3y = 7$ and $4x + 6y = 14$ using technology

    ii. $y = 2x + 3$ and $y = 2x - 3$ using technology.

c) Using algebraic methods without technology solve the system of three variable equations.

$$x - 3y + 3z = -4$$
$$2x + 3y - z = 15$$
$$4x - 3y - z = 19$$

d) Find the solution for the following systems of linear equations using technology.

$$x - y + z = 6 \qquad (1)$$
$$2x + z = 4 \qquad (2)$$
$$3x + 2y - z = 6 \qquad (3)$$

3. Algorithms are amazing

Consider the algorithm for the Greatest Common Divisor (GCD) – Euclid's Theorem circa 300 BC

Pseudocode definition	Python definition

Define gcd(a, b):
    while $a \neq b$
        if $(a > b)$ **then**
            $a \leftarrow a - b$
        else
            $b \leftarrow b - a$
        end if
    end while
    return a
	``
	def gcd(a, b):
	    while a != b:
	        if a > b:
	            a = a - b
	        else:
	            b = b - a
	    return a
	``

a) Trace the **gcd** algorithm with known multiples a=5, b=15 using a table to show the values of the variables changing and the actions during computation until it stops:

a	b	actions
5	15	a = b? no a > b? no

b) Trace the **gcd** algorithm with prime numbers a=7, b=11 using a table to show the values of the variables changing and the actions during computation until it stops:

a	b	actions
7	11	a = b? no a > b? no

4. Algorithm Challenge

Consider the step by step algorithm for Finding Primes using the sieve of Eratosthenes. This is an algorithm that dates from ancient times.

To find all the prime numbers from 1 to n by Eratosthenes algorithm:
1. Create a list of consecutive numbers from 2 to n.
2. Starting with c=2 the smallest prime number
3. Count in increments of c, that is *2c, 3c, 4c,...* and cross these numbers out in the list from 2 to n.
4. Find the next number greater than c that has not been crossed out, this is the next prime number and let that number be the new c
5. Continue from step 3 until c is greater or equal to n.

At the conclusion of the algorithm, the numbers that have not been crossed out in the integers from 1 to n are the primes.

- When c=2, the numbers crossed out will be multiples of 2 such as 4, 6, 8, 10,….
- When c=3, the numbers crossed out will be multiples of 3 that haven't already been crossed out, such as 9, 15, 21,….
- When c=5, the numbers crossed out will be multiples of 5 that haven't already been crossed out, such as 25, 35, 55, …..

Primes found by Eratosthenes sieve for integers 1 to 100 is shown below.

a) Create an algorithm in structured pseudocode using Eratosthenes sieve method which prints out all the non-prime integer number within the interval 1 to n.

b) Check and trace your algorithm for n=10

c) Use ChatGPT or Google Gemini to code your algorithm in Python.

ALGEBRA AND ALGORITHMS SOLUTIONS.

1. Riddles. (SOLUTIONS)

a)

$\bigcirc + \bigcirc = 10$

$\bigcirc \times \square + \square = 12$

$\bigcirc \times \square - \blacktriangle \times \bigcirc = \bigcirc$

$\blacktriangle = ?$

Let $a = \bigcirc$, $b = \square$, $c = \blacktriangle$

$2a = 10$ (1)

$ab + b = 12$ (2)

$ab - ca = a$ (3)

$a = 5$ $sub\,in\,(2)$

$6b = 12$

$\Rightarrow b = 2$ $sub\,in\,(3)$

$10 - 5c = 5$

$\Rightarrow 5 = 5c$

$\Rightarrow c = 1$

b)

Let m be the money Granny Smith left.

$$\frac{m}{2} + \frac{m}{4} + \frac{m}{6} + 1000 = m$$

$$m - \frac{m}{2} - \frac{m}{4} - \frac{m}{6} = 1000$$

$$\frac{12m}{12} - \frac{6m}{12} - \frac{3m}{12} - \frac{2m}{12} = 1000$$

$$\frac{m}{12} = 1000$$

$$m = 12000$$

Granny Smith left $12,000.

One half plus one quarter plus one-sixth equals eleven-twelfths. So, the remainder, $1,000, is one-twelfth of the whole, which must have been $12,000.

c) Use logic and a matrix or table to help sort out these relationships.

Use the numbered conditions to determine the couples

Assuming the couples are of the opposite gender.

1. Boxuan is Essie's brother and lives in Adelaide with his wife.
2. Con is shorter than Faye's husband.
3. Con works at a bank.
4. Dora and her husband live in Sydney.
5. Essie and her husband work in a restaurant.

	Alex	Boxuan	Con
Dora		No (4)	COUPLE
Essie	COUPLE	No (1)	No (5)(3)
Faye		COUPLE	No (2)

2. Fun with simultaneous equations. (SOLUTIONS)

a) Solve 2x−y=1, 3x+2y=12 for x,y

We can solve the system of equations by elimination.

$2x - y = 1$ (1)
$3x + 2y = 12$ (2)

Steps to solve:

Eliminate y: Multiply the first equation by 2 to get 4x -2y = 2 and then add this equation to the second equation.

$4x - 2y = 2$ (1')
$3x + 2y = 12$ (2)

Get 7x = 14, therefore x=2, subsitute x in the first equation: 2(2) – y = 1, which gives y=3.

b)

  i. Solve $2x + 3y = 7$ and $4x + 6y = 14$ using technology.

In this case infinite solutions exist as the equations are identical.

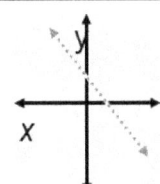

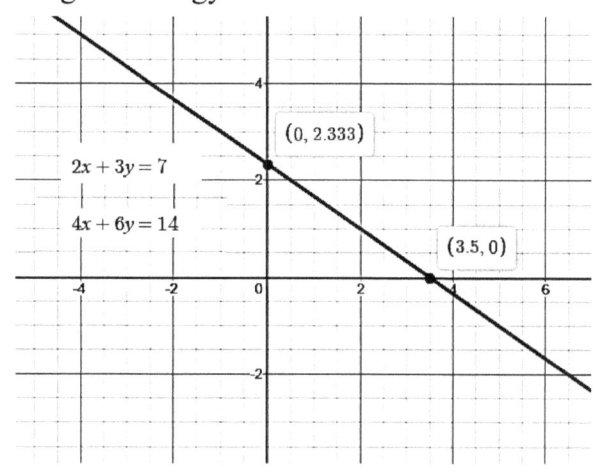

Graphing shows them to be an identical line.

  ii. Solve $y = 2x + 3$ and $y = 2x - 3$ using technology.

In this case no solutions exist as lines are parallel and do not intersect.

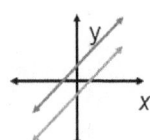

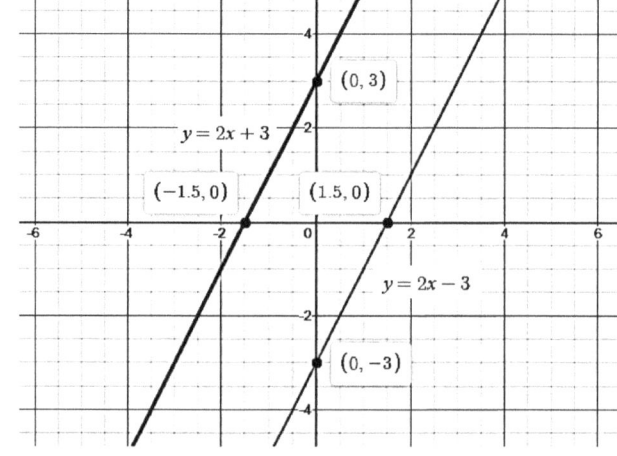

Graphing shows them to be parallel lines.

2. Fun with simultaneous equations. (SOLUTIONS)

c) Using elimination solve the following system of three variable equations.

$x - 3y + 3z = -4$

$2x + 3y - z = 15$

$4x - 3y - z = 19$

Confirm your answer using your calculator.

$x - 3y + 3z = -4$ (1)

$2x + 3y - z = 15$ (2)

$4x - 3y - z = 19$ (3)

Pair equations to eliminate the y variable. (1)+(2).

$x - 3y + 3z = -4$ (1)

$2x + 3y - z = 15$ (2)

- - - - - - - - -

$3x - 0 + 2z = 11$

$3x + 2z = 11$ (4)

Pair equations to eliminate the y variable. (2)+(3).

$2x + 3y - z = 15$ (2)

$4x - 3y - z = 19$ (3)

- - - - - - - - -

$6x - 0 - 2z = 34$

$6x - 2z = 34$ (5)

Solve (4) and (5)

$3x + 2z = 11$ (4)

$6x - 2z = 34$ (5)

- - - - - - - - -

$9x \qquad = 45$

$\Rightarrow x = 5$

Sub $x = 5$ into (4)

$15 + 2z = 11$

$\Rightarrow 2z = -4$

$\Rightarrow z = -2$

Sub $x = 5$ and $z = -2$ into (1)

$5 - 3y - 6 = -4$

$\Rightarrow -3y - 1 = -4$

$\Rightarrow -3y = -3$

$\Rightarrow y = 1$

Solution $(5, 1, -2)$

---

d)

Find the solution for the following system of linear equations using technology and matrix algebra.

$x - y + z = 6$ (1)

$2x + z = 4$ (2)

$3x + 2y - z = 6$ (3)

Set up the matrix system

$$\begin{bmatrix} 1 & -1 & 1 \\ 2 & 0 & 1 \\ 3 & 2 & -1 \end{bmatrix} \begin{bmatrix} x \\ y \\ z \end{bmatrix} = \begin{bmatrix} 6 \\ 4 \\ 6 \end{bmatrix}$$

Using technology with matrices

$$\begin{bmatrix} x \\ y \\ z \end{bmatrix} = \begin{bmatrix} 1 & -1 & 1 \\ 2 & 0 & 1 \\ 3 & 2 & -1 \end{bmatrix}^{-1} \begin{bmatrix} 6 \\ 4 \\ 6 \end{bmatrix}$$

$$\begin{bmatrix} x \\ y \\ z \end{bmatrix} = \frac{1}{3}\begin{bmatrix} 2 & -1 & 1 \\ -5 & 4 & -1 \\ -4 & 5 & -2 \end{bmatrix} \begin{bmatrix} 6 \\ 4 \\ 6 \end{bmatrix}$$

$$\begin{bmatrix} x \\ y \\ z \end{bmatrix} = \begin{bmatrix} \dfrac{14}{3} \\ \dfrac{-20}{3} \\ \dfrac{-16}{3} \end{bmatrix}$$

3. Algorithms are amazing. (SOLUTIONS)

a) **Trace the algorithm which known multiples a=5, b=15 using a table.**

a	b	actions
5	15	a = b? no a > b? no
	10	b <= b - a loop
5	10	a = b? no a > b? no
	5	b <= b – a loop
5	5	a = b? yes Print a stop

b) **Trace the algorithm with prime numbers a=7, b=11 using a table.**

a	b	actions
7	11	a = b? no a > b? no
	4	b <= b - a loop
7	4	a = b? no a > b? yes
3		a <= a - b loop
3	4	a = b? no a > b? no
	1	b <= b - a loop
3	1	a = b? no a > b? yes
2		a <= a - b loop
2	1	a = b? no a > b? yes
1		a <= a - b loop
1	1	a = b? yes print a stop

4. Algorithm Challenge. (SOLUTIONS)

a) Create an algorithm in structured pseudocode using Eratosthenes sieve method which prints out all the non-prime integer number within the interval 1 to n.

The algorithm needs to print all the non-prime composite numbers ie,1, 4, 6, 8, 9, … that are less than or equal to n.

Algorithm – First attempt

```
input n
i ← 1
print i
j ← 2
while (j < √n)
    while (i < n)
        print i × j
        i ← i + 1
    end while
    i ← 1
    j ← j + 1
end while
```

b) Trace your algorithm for n=10 and notice it has error as some values printed are greater than n

Iteration	n	i	j	sqrt(n)	j < sqrt(n)	i < n	Output	
1	10	1	2	3.162	TRUE	-	1	
2	10	1	2	-	TRUE	TRUE	2	
3	10	2	2	-	-	TRUE	4	
4	10	3	2	-	-	TRUE	6	
5	10	4	2	-	-	TRUE	8	
6	10	5	2	-	-	TRUE	10	
7	10	6	2	-	-	TRUE	12	error
8	10	7	2	-	-	TRUE	14	error
9	10	8	2	-	-	TRUE	16	error
10	10	9	2	-	-	TRUE	18	error
11	10	1	3	-	TRUE	TRUE	3	
12	10	2	3	-	-	TRUE	6	
13	10	3	3	-	-	TRUE	9	
14	10	4	3	-	-	TRUE	12	error
15	10	1	4	-	FALSE	-	4	

4. Algorithm Challenge. (SOLUTIONS)

a) Algorithm – second attempt with improved logic

```
input n
i ← 1
print i
j ← 2
while (j < √n
    while (i < n)
        if ((i × j) < n) then
            print i × j
        end if
        i ← i + 1
    end while
    i ← 1
    j ← j + 1
end while
```

b) Trace your algorithm for n=10 and notice it has error as some values printed are greater than n, further improvements are possible such as using a list to avoid repeating numbers.

Iteration	n	i	j	sqrt(n)	j < sqrt(n)	i < n	(i * j) <= n	Output
1	10	1	2	3.162	TRUE	-	-	1
2	10	1	2	-	TRUE	TRUE	TRUE	2
3	10	2	2	-	-	TRUE	TRUE	4
4	10	3	2	-	-	TRUE	TRUE	6
5	10	4	2	-	-	TRUE	TRUE	8
6	10	5	2	-	-	TRUE	TRUE	10
7	10	6	2	-	-	TRUE	FALSE	-
8	10	7	2	-	-	TRUE	FALSE	-
9	10	8	2	-	-	TRUE	FALSE	-
10	10	9	2	-	-	TRUE	FALSE	-
11	10	1	3	-	TRUE	TRUE	TRUE	3
12	10	2	3	-	-	TRUE	TRUE	6
13	10	3	3	-	-	TRUE	TRUE	9
14	10	4	3	-	-	TRUE	FALSE	-
15	10	1	4	-	FALSE	-	-	4

c) Use ChatGPT or Google Gemini to help code your algorithm in Python.

```python
import math
n = 10
i = 1
print(i)
j = 2
while j < math.sqrt(n):
    while i < n:
        if (i * j) <= n:
            print(i * j)
        i += 1
    i = 1
    j += 1
```

2 Functions y=f(x) for real!

We define *f(x)* as a function *f* that is dependent on the variable *x*. A function defines a mathematical computation that is actioned on the input independent value. Any letter can be used to define a function, commonly we use the letters *f*, *g* and *h*. Any letter can be used to define an independent variable, commonly we use the letters *x*, *θ* and *t*.

Example 2.1: A function accepting an input value and giving an output value.	
$f(x)=3x+1$ Function *f* accepts input *x* multiplies *x* by 3 and then adds 1 to it.	$f(2)=3(2)+1$ $f(2)=7$ $f(a+5)=3(a+5)+1$ $f(a+5)=3a+15+1$ $f(a+5)=3a+16$

Application 2.1: Aircraft Passenger Baggage Restrictions.

Aircraft Passenger Baggage Restrictions specify an allowance of up to and including *21kg* free per passenger. For baggage heavier than *21kg* a charge of *$50* per kilo weight over *21kg*, and applies up to a maximum of *100kg*.

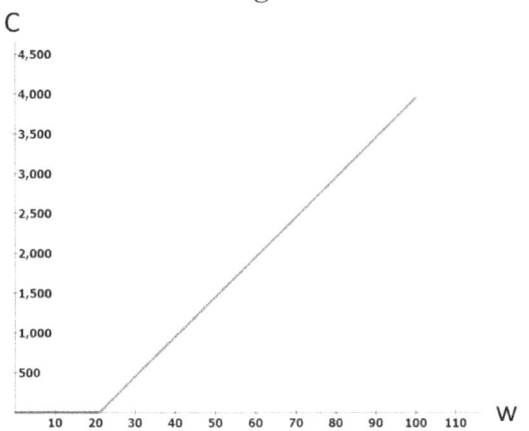

A graph of the cost *C* against the weight *w*.

Expressing the cost *C* as a function of weight *w* in kg has two definitions depending on the weight value interval range.

1. $C(w)=0$, when $0 \leq w \leq 21$
2. $C(w)=50(w-21)$, when $21 < w \leq 100$

Using the function *C* requires applying the weight restrictions on the variable *w*.

The cost of *20kg* of baggage is given by $C(20)=0$.
The cost of *63kg* of baggage is given by $C(63)=50(63-21)=2100$.

Predefined number sets are used in Mathematics for defining functions.

N Natural numbers are positive integers.
Z All Integers and Zero.
Q Quotients, numbers that are expressable as fractions.
Q' Non Quotients called irrationals such as π, e, $\sqrt{2}$, $\sqrt{5}$
R All real numbers.
R^+ All positive reals. $\{x:x > 0\}$
R^- All negative reals. $\{x:x<0\}$
$R/\{0\}$ All reals except zero.

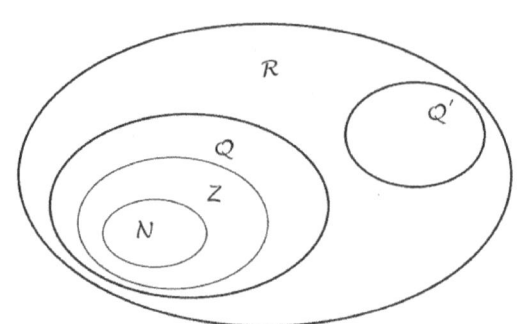

Number set relationships.

Example 2.2: Notation is used in Mathematics to define or restrict equations and functions.	
$f(x)= x^2+5$, where $x \in R^+ U\{0\}$ Function f accepts input x squares it, and then adds 5 to it. This function is restricted to only accept all positive reals and zero as inputs, and is not defined for other values.	$f(4)=(4)^2+5$ $f(4)=21$ $f(a+1)=(a+1)^2+5$ $f(a+1)=a^2+2a+1+5$ $f(a+1)=a^2+2a+6$ where $(a+1)\geq0$

Unlike optical illusions, functions have a unique interpretation.

Defining a function is an unambiguous description of the mathematical function and its operations, with the restriction defined on the input values it will accept.

$$\{(x,y):y=x+1, x \in R^+\}$$
$$f: (0, \infty) \rightarrow R, f(x)=x+1$$

The **domain** of function f is the set of x values for which the function is defined.

The **range** of function f are the y values where $y=f(x)$, which are derived from the x values for which the function is defined.

Functions must pass the ***vertical line test*** and which shows that the function will always return a unique y value for any x value that is input.

Example 2.3: The vertical line test for determining a function.	
Many x values map to one y value. **many:1**	One x value maps to one y value. **1:1**
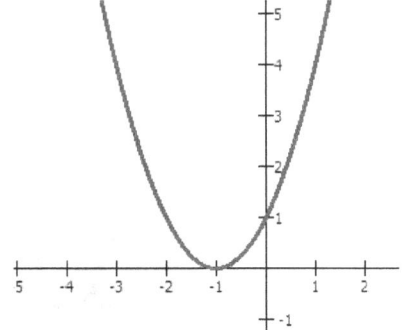	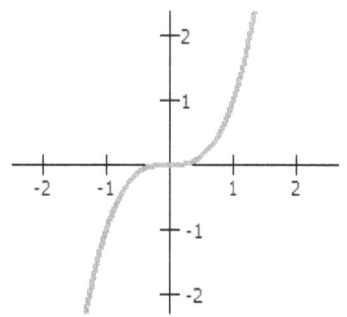
In the figure shown, for the parabola when $x=-2$ or when $x=0$, the y value is 1.	In the figure shown, for the cubic when $x=1$ the y value is 1.

Non-Functions fail the **_vertical line test_** and multiple values are returned for any x value that is input.

Example 2.4: Non-Functions fail the vertical line test for determining a function.	
One x value maps to many y values.**1:many**	Many x values map to many y values. **many:many**
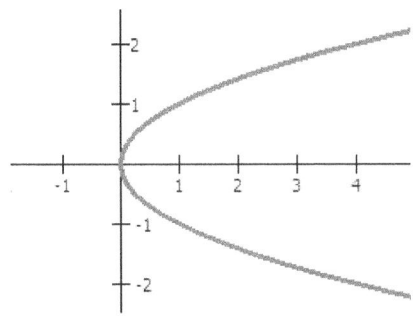	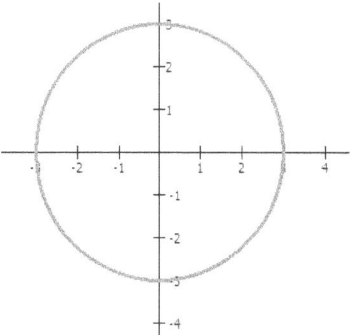
In the figure shown, for the parabola when $x=2$, there are two y values possible.	In the figure shown, for the circle when $x=2$, there are two y values possible. When $x=2$ or $x=-2$ the same y value of 2 is given.

Don't Forget: The functions that you need to know about in VCE Mathematical Methods 3&4.

Power functions in general form, $f(x) = A\big(n(x-b)\big)^P + c$

    Polynomial functions, the power is a positive integer, which may include zero.
    Negative power functions, the power is a negative integer.
    Quotient power functions, the power is a quotient or fraction.

Exponential functions with base of a and e in the general form, where $f(x) = Aa^{n(x-b)} + c$ or
$f(x) = Ae^{n(x-b)} + c$.

Logarithmic functions with base of a and e in the general form, where $f(x) = A\log_a\big(n(x-b)\big) + c$ or
$f(x) = A\log_e\big(n(x-b)\big) + c$.

Circular functions for anti-clockwise angle θ from the positive
x-axis where A, n, b, c are constant reals.

- Sine function $f(\theta) = A\sin(n(\theta - b)) + c$.
- Cosine function $f(\theta) = A\cos(n(\theta - b)) + c$.
- Tangent function $f(\theta) = A\tan(n(\theta - b)) + c$.

35

Example 2.5: Gallery of graphs y=f(x).

Linear $y = mx + c$	**Quadratic** $y = a(x-h)^2 + k$	**Cubic** $y = a(x-h)^3 + k$	**Quartic** $y = a(x-h)^4 + k$

Square Root $y = A\left(n(x-b)\right)^{1/2} + c$	**Hyperbola** $y = A\left(n(x-b)\right)^{-1} + c$	**Truncus** $y = A\left(n(x-b)\right)^{-2} + c$	**Power function** $y = x^{2/3}$

Exponential $y = Aa^{n(x-b)} + c$	**Logarithmic** $f(x) = A\log_a\left(n(x-b)\right) + c$	**Adding functions**	**Hybrid function**

Sine $y = A\sin(n(x-b)) + c$	**Cosine** $y = A\sin(n(x-b)) + c$	**Tangent** $y = A\sin(n(x-b)) + c$	**Harp function.**

36

Hybrid functions.

Hybrid functions are made up of pieces of functions that are specified in sub-domains, which are then placed together to form one overall function. The hybrid function formed meets the definition for a function where every y value has a unique x value. Sometimes the sub-function pieces will join together smoothly, and sometimes you have to jump between the pieces.

$$f(x) = \begin{cases} g(x), & x < 0 \\ h(x), & 0 \le x \le 10 \\ i(x), & x > 10 \end{cases}$$

Example 2.6: A graph of $y=f(x)$ where $f(x)$ is a hybrid function.

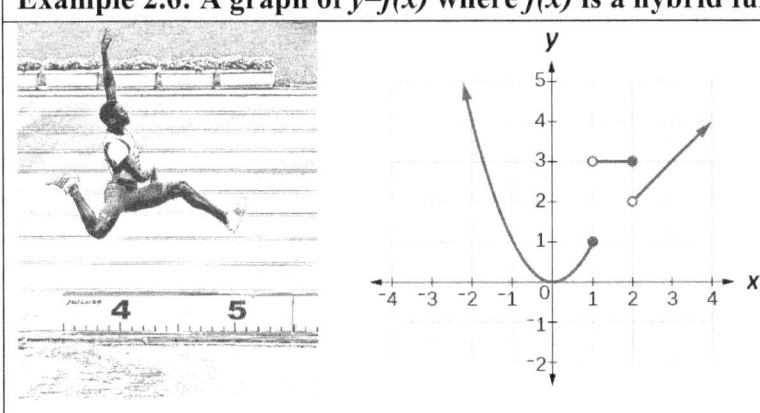

Piece by piece the function is defined within sub-domains.

$$f(x) = \begin{cases} x^2, & x \le 1 \\ 3, & 1 < x \le 2 \\ x, & x > 2 \end{cases}$$

The resulting Hybrid function may or may not be continuous.

Application 2.2: Creating Art with transformed and restricted Functions.

The symmetry of many functions can be combined by translation, dilation and reflection to form art compositions when the functions have artfully restricted domains.

$y = \sin(x) \left\{ 0 < x < \frac{32\pi}{20} \right\}$

$y = -\sin(x) \left\{ 0 < x < \frac{32\pi}{20} \right\}$

$y = x - 4 \left\{ 4 < x < 5 \right\}$

$y = -x + 4 \left\{ 4 < x < 5 \right\}$

$y = x - 0.5 \left\{ \frac{\pi}{2} < x < 2 \right\}$

$y = -x + 0.5 \left\{ \frac{\pi}{2} < x < 2 \right\}$

$(x-1)^2 + (y-0.4)^2 = 0.04$

$(x-1)^2 + (y-0.4)^2 = 0.02$

$y = 0.7x + 0.1 \left\{ \frac{2\pi}{5} < x < 2 \right\}$

$y = -0.7x - 0.1 \left\{ \frac{2\pi}{5} < x < 2 \right\}$

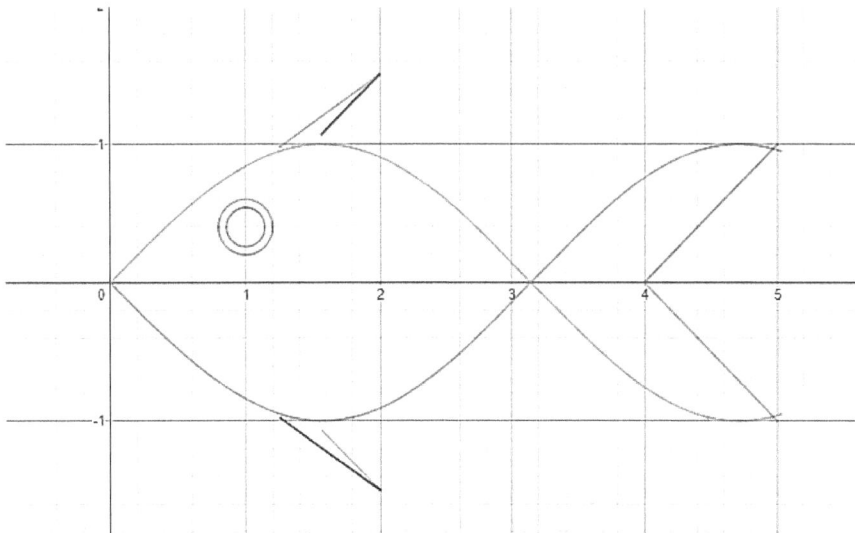

Fish created using restricted sine and linear functions and two concentric circle relations used to form the eye.
Technology used is desmos. https://www.desmos.com/calculator

Note: The final fishy result doesn't pass the vertical line test, and is not a function.

Composite functions.

Composite functions are nested functions, where the output of the inner function becomes the input of the outer function.

For the composed function $y = f(g(x))$ the output of $g(x)$ which is in the range of $g(x)$ becomes the input of $f(x)$ **only if** the range of $g(x)$ is in the domain of $f(x)$. So you always need to check the range of the inner function against the domain of the outer function before trying to compose them.

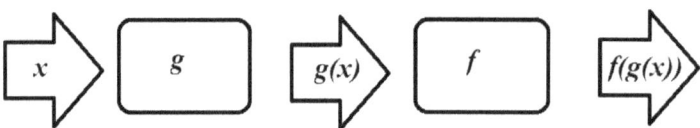

Composite functions can be thought of as chained functions or computation machines, where the output of one function becomes the input of another functions.

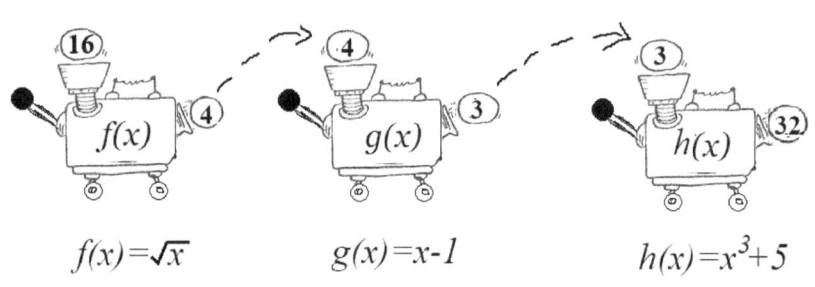

$$h(g(f(x))) = \left(\sqrt{x}-1\right)^3 + 5$$

The domain of the composed function h(x) is $x \geq 0$ because of the restriction on the innermost square root function f(x).

Example 2.7: Composite functions being combined.

Let $y = g(x)$, where $g(x) = \sqrt{x}$
  - The domain of $g(x)$ is $x \geq 0$
  - The range of $g(x)$ is $x \geq 0$

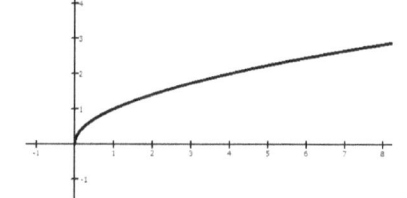

Let $y = f(x)$, where $f(x) = \sin(x)$
  - The domain of $f(x)$ is $(-\infty, \infty)$
  - The range of $f(x)$ is $[-1, 1]$

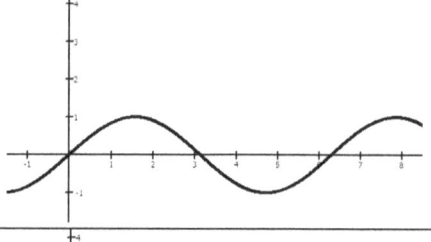

Let $y = f(g(x))$, where $f(g(x)) = \sin(\sqrt{x})$
  - The domain of $f(g(x))$ is $x \geq 0$
  - The range of $f(g(x))$ is $[-1, 1]$

This composition of functions worked because the range of $g(x)$ which is $x \geq 0$, was in the domain of $f(x)$ which is $(-\infty, \infty)$.

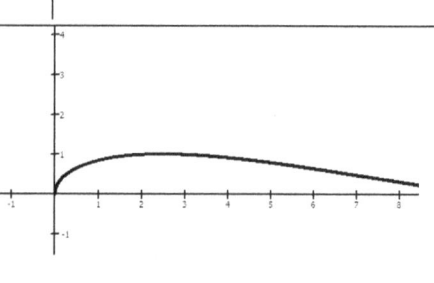

And of course, you can even compose a function within itself.

Example 2.8: Composing functions within themselves.

$f(x) = \left(\dfrac{x-1}{5}\right)$	$f(x) = 1 + x^2$	$f(x) = \left(\dfrac{x-2}{x}\right)$	$f(x) = \tan(x)$
$f(f(x)) = \dfrac{\dfrac{x-1}{5} - 1}{5}$	$f(f(x)) = 1 + \left(1 + x^2\right)^2$	$f(f(x)) = \dfrac{\dfrac{x-2}{x} - 2}{\dfrac{x-2}{x}}$	$f(f(x)) = \tan\left(\tan(x)\right)$
rewritten as		rewritten as	
$f(f(x)) = \dfrac{\dfrac{x-1-5}{5}}{5}$		$f(f(x)) = \dfrac{\dfrac{x-2-2x}{x}}{\dfrac{x-2}{x}}$	
that is not pretty, if we simplify we get		that is not pretty, if we simplify we get	
$f(f(x)) = \dfrac{x-6}{25}$		$f(f(x)) = \dfrac{-x-2}{x-2}$	

Decomposing functions.

Plastics can take up to 400 years to decompose.

Hopefully you'll be able to decompose functions in a shorter time.

Decomposing functions can make them simpler to work with.

The decomposed functions retain the domain of the original composed function.

Example 2.9: Decomposing functions.

$y = \left(\dfrac{x-2}{x}\right)^3$	$y = \sqrt{\sin(x)}$	$y = \dfrac{1}{1 - \sqrt{x}}$	$y = \log_e\left(\dfrac{x-2}{x}\right)^3$
Let $f(x) = \dfrac{x-2}{x}$ and $g(x) = x^3$ then $y = g(f(x))$	Let $f(x) = \sin(x)$ and $g(x) = \sqrt{x}$ then $y = g(f(x))$	Let $f(x) = \dfrac{1}{1-x}$ and $g(x) = \sqrt{x}$ then $y = g(f(x))$	Let $f(x) = \left(\dfrac{x-2}{x}\right)^3$ and $g(x) = \log_e x$ then $y = g(f(x))$

Inverse functions.

Inverse functions $f^{-1}(x)$ can reverse the computation of the original function $f(x)$. If the function $f(x)$ is one:one, that is one *x-value* corresponds to one *y-value* and vice versa, then the inverse function $f^{-1}(x)$ exists.

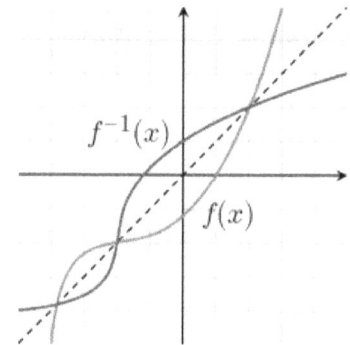

To find the inverse you take the original function and let $y=f(x)$ then interchange x and y and make y the subject, check the inverse function is a reflection of the original function in the line $y=x$ by graphing.

Domain $f^{-1}(x)$ = Range $f(x)$ and the Range $f^{-1}(x)$ = Domain $f(x)$.
It has the composition properties $f(f^{-1}(x))=x$ and $f^{-1}(f(x))=x$

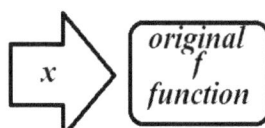 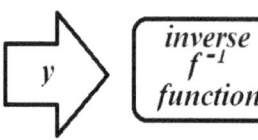 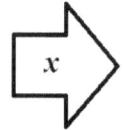

Functions which are not 1:1 can have their domains restricted so that they become 1:1 and then we can find the inverse function $f^{-1}(x)$. If $f(x)$ and $f^{-1}(x)$ intersect it will usually but not always occur on the line $y=x$.

Inverse functions are really useful for going forward and backwards with values.

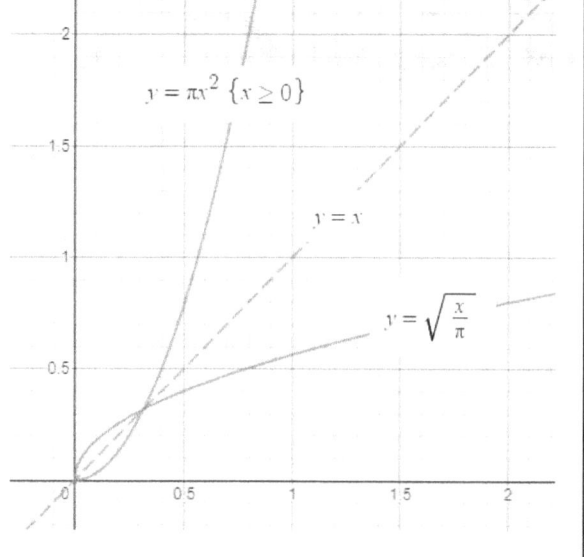

Application 2.3: Inverse functions used everyday.

Area of a circle $\quad A(r)=\pi r^2$, where A is the area in square units and r is the radius of a circle.

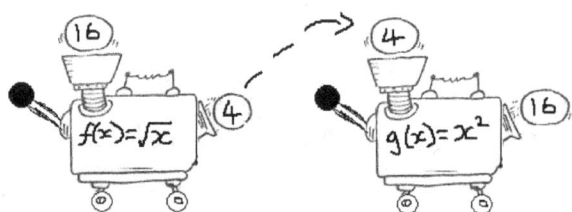

Let $y=A$, and $x=r$ where r is the radius and this is always greater than or equal to zero, then the equation in terms of x and y is
$y=\pi x^2 \quad y \leftrightarrow x \quad x=\pi y^2$.

Resolving for y gives $y=\pm\sqrt{\dfrac{x}{\pi}}$ since the radius can

only be positive for a real circle. $y=\sqrt{\dfrac{x}{\pi}}$ is

equivalent to the equation $r=\sqrt{\dfrac{A}{\pi}}$, which finds the

radius r given the Area A of a circle.

Graph $y=\pi x^2$ and $y=\sqrt{\dfrac{x}{\pi}}$ and $y=x$ and notice

the inverse properties.

Example 2.10: Inverse functions may require the original function to have a restricted domain.

Let $y = f(x)$, where $f(x) = x^2$

- The domain of $f(x)$ is $(-\infty, \infty)$.
- The range of $f(x)$ is $x \geq 0$
- The function $f(x)$ is many to one, to find the inverse the function must be restricted so it is one:one.
- Let $f(x) = x^2$ where $x \geq 0$, graph is in the first quadrant.

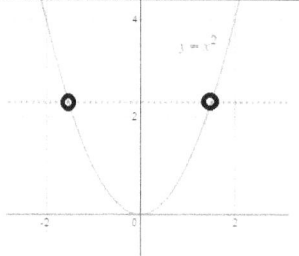

$f(x) = x^2$ has many x values map to one y value.

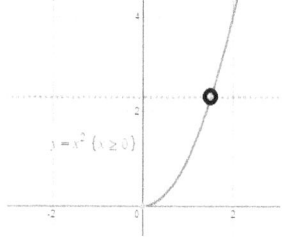

$f(x) = x^2$ is restricted so one x value maps to one y value.

Let $y = f(x)$, which is restricted then $y = x^2$.

- Interchange x and y ➜ $x = y^2$
- Resolve for y ➜ $y = \pm\sqrt{x}$ which is not one:one.
- A quick glance at the graph informs us of the answer which is a reflection of $y = f(x)$ in the x-axis which is $y = \sqrt{x}$ therefore

$$f^{-1}(x) = \sqrt{x}$$

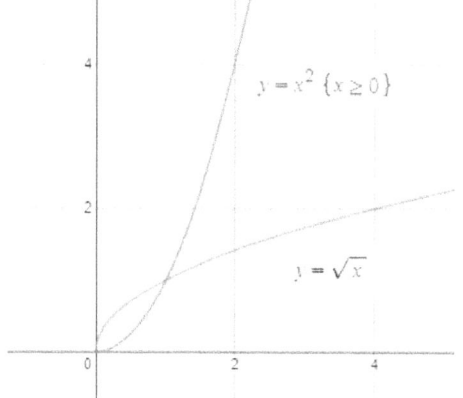

Check $f^{-1}(f(x)) = x$

This composition of functions worked because the range of $f(x)$ which is restricted to $x \geq 0$, was in the domain of $f^{-1}(x)$ which is $x \geq 0$.

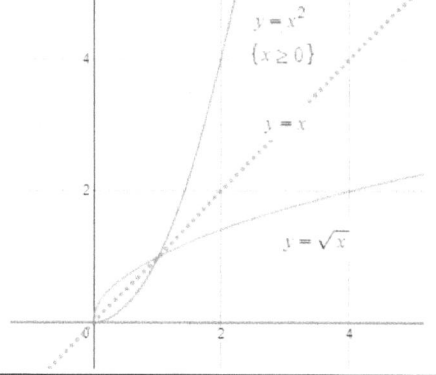

The function story so far, **Hybrid** involves cutting and piecing together, **Composite** involves nesting functions within each other, **Inverse** functions are used to reverse the original function.

But wait there is more, bend me shape me, move me.
Getting functions to change shape and move

Transformation of functions.

From $y=f(x)$ to $y=Af(n(x-b))+c$ where A, n, b and c are real numbers and n is not equal to zero.

- Dilation from x-axis by a factor of A
- Dilation from y-axis by a factor of $1/n$
- Translation horizontally by b units
- Translation vertically by c units

The following table shows a summary of function transformation and the effect on the coordinates (x,y).

Dilations	Reflections	Translations
Dilation from the *x-axis* • $y=Af(x)$ $(x,y) \rightarrow (x,Ay)$ $f(x) = x^3 - 2x^2 + 1$	Reflection in the *x-axis* • $y=-f(x)$ $(x,y) \rightarrow (x,-y)$ 	Translating $+b$ units in the direction of the *x-axis*. • $y=f(x-b)$ $(x,y) \rightarrow (x+b,y)$ $f(x+3)$ (-3,-4) (0,-4)
Dilation from the *y-axis* • $y=f(nx)$ $(x,y) \rightarrow (x/n,y)$ $f(x) = x^3 - 2x^2 + 1$	Reflection in the *y-axis* • $y=f(-x)$ $(x,y) \rightarrow (-x,y)$ Reflection in the both axes • $y=-f(-x)$ $(x,y) \rightarrow (-x,-y)$	Translating c units in the direction of the *y-axis*. • $y=f(x)+c$ $(x,y) \rightarrow (x,y+c)$ $f(x)+5$ (4,5) (4,0)

Transformations: Warning Matrix algebra ahead!

Translation

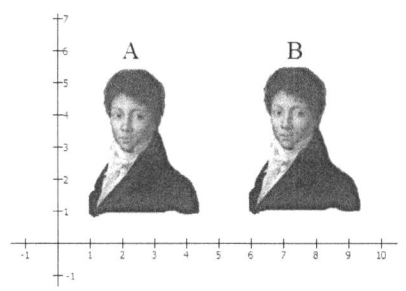

Image A becomes Image B when moved 5 units to the right

$$\begin{bmatrix} x \\ y \end{bmatrix} + \begin{bmatrix} 5 \\ 0 \end{bmatrix} = \begin{bmatrix} x+5 \\ y \end{bmatrix}$$

$$x^{new} = x+5, y^{new} = y$$

$$(x, y) \rightarrow (x+5, y)$$

Translation

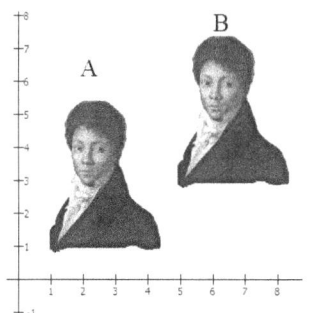

Image A becomes Image B when moved 4 units to the right, and 2 units up.

$$\begin{bmatrix} x \\ y \end{bmatrix} + \begin{bmatrix} 4 \\ 2 \end{bmatrix} = \begin{bmatrix} x+4 \\ y+2 \end{bmatrix}$$

$$x^{new} = x+4, y^{new} = y+2$$

$$(x, y) \rightarrow (x+4, y+2)$$

Reflection

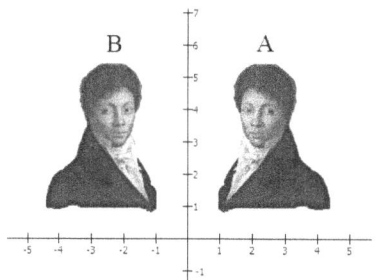

Image A becomes Image B when reflected in the y-axis.

$$\begin{bmatrix} -1 & 0 \\ 0 & 1 \end{bmatrix} \begin{bmatrix} x \\ y \end{bmatrix} = \begin{bmatrix} -x \\ y \end{bmatrix}$$

$$x^{new} = -x, y^{new} = y$$

$$(x, y) \rightarrow (-x, y)$$

Reflection

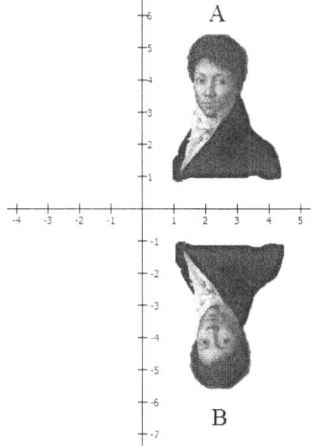

Image A becomes Image B when reflected in the x-axis.

$$\begin{bmatrix} 1 & 0 \\ 0 & -1 \end{bmatrix} \begin{bmatrix} x \\ y \end{bmatrix} = \begin{bmatrix} x \\ -y \end{bmatrix}$$

$$x^{new} = x, y^{new} = -y$$

$$(x, y) \rightarrow (x, -y)$$

Reflections

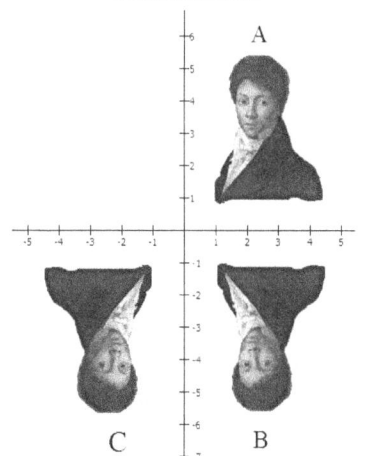

Image A becomes Image B when reflected in the x-axis, Image B becomes Image C when reflected in the y-axis.

$$\begin{bmatrix} -1 & 0 \\ 0 & 1 \end{bmatrix} \begin{bmatrix} 1 & 0 \\ 0 & -1 \end{bmatrix} \begin{bmatrix} x \\ y \end{bmatrix} = \begin{bmatrix} -x \\ -y \end{bmatrix}$$

$$x^{new} = -x, y^{new} = -y$$

$$(x, y) \rightarrow (-x, -y)$$

Reflection and Translation

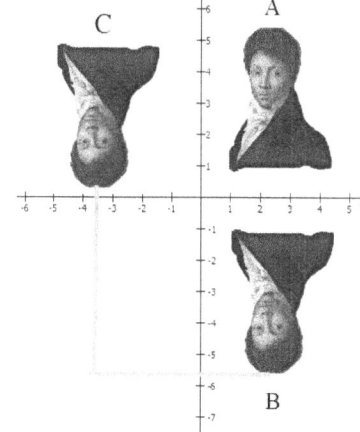

Image A becomes Image B when reflected in the x-axis, Image B becomes Image C when moved.

$$\begin{bmatrix} 1 & 0 \\ 0 & -1 \end{bmatrix} \begin{bmatrix} x \\ y \end{bmatrix} + \begin{bmatrix} -6 \\ 6 \end{bmatrix} = \begin{bmatrix} x-6 \\ -y+6 \end{bmatrix}$$

$$x^{new} = x-6, y^{new} = -y+6$$

$$(x, y) \rightarrow (x-6, -y+6)$$

Dilation & Reflection

Image A becomes Image B when dilated by a factor of 0.5 from the x-axis and a factor of 0.5 from the y-axis, and then reflected in the y-axis.

$$\begin{bmatrix} -1 & 0 \\ 0 & 1 \end{bmatrix}\begin{bmatrix} 0.5 & 0 \\ 0 & 0.5 \end{bmatrix}\begin{bmatrix} x \\ y \end{bmatrix} = \begin{bmatrix} 0.5x \\ -0.5y \end{bmatrix}$$

$$x^{new} = 0.5x, \ y^{new} = -0.5y$$

$$(x, y) \rightarrow (0.5x, -0.5y)$$

Dilation & Translation

Image A becomes Image B when dilated by a factor of 0.5 from the x-axis and a factor of 0.5 from the y-axis, and then moved.

$$\begin{bmatrix} 0.5 & 0 \\ 0 & 0.5 \end{bmatrix}\begin{bmatrix} x \\ y \end{bmatrix} + \begin{bmatrix} 0 \\ -4 \end{bmatrix} = \begin{bmatrix} 0.5x \\ 0.5y - 4 \end{bmatrix}$$

$$x^{new} = 0.5x, \ y^{new} = 0.5y - 4$$

$$(x, y) \rightarrow (0.5x, 0.5y - 4)$$

Reflection in line y=x

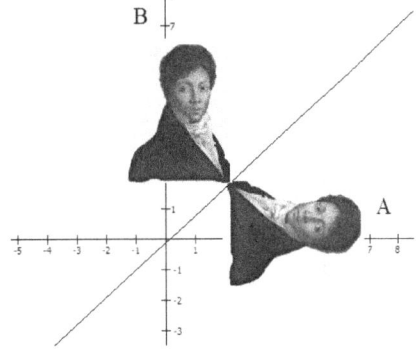

Image A becomes Image B when reflected in the line y=x. That is the x values become the y values and the y values become the x values. An inverse image.

$$\begin{bmatrix} 0 & 1 \\ 1 & 0 \end{bmatrix}\begin{bmatrix} x \\ y \end{bmatrix} = \begin{bmatrix} y \\ x \end{bmatrix}$$

$$x^{new} = y, \ y^{new} = x$$

$$(x, y) \rightarrow (y, x)$$

Rotation, just for fun.

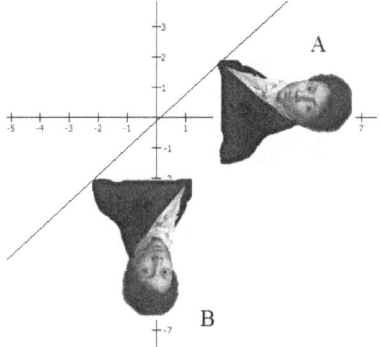

Image B becomes Image A when rotated 90° anti-clockwise. The general form of a rotation matrix.

$$\begin{bmatrix} \cos\theta & -\sin\theta \\ \sin\theta & \cos\theta \end{bmatrix}\begin{bmatrix} x \\ y \end{bmatrix} = \begin{bmatrix} x\cos\theta - y\sin\theta \\ x\sin\theta + y\cos\theta \end{bmatrix}$$

$$x^{new} = x\cos\theta - y\sin\theta, \ y^{new} = x\sin\theta + y\cos\theta$$

$$(x, y) \rightarrow (x\cos\theta - y\sin\theta, x\sin\theta + y\cos\theta)$$

When $\theta = 90^o$

$$(x, y) \rightarrow (x\cos 90^o - y\sin 90^o, x\sin 90^o + y\cos 90^o)$$

$$(x, y) \rightarrow (-y, x)$$

Operations with functions.

Just like numbers, functions can be combined using the operations of addition, subtraction, multiplication and division. A bit of thinking is required to determine the resulting domain of the combined function that results. The intersection of the sub-domains is usually the domain of the combined function, however you always need to check for division by zero when dividing functions.

- Addition of functions $y = f(x) + g(x)$
- Subtraction of functions $y = f(x) + g(x)$
- Multiplication of functions $y = f(x) \times g(x)$
- Division of functions $y = f(x) \div g(x)$

Example 2.11: Operations on functions, where subtraction is equivalent to adding the function that is reflected in the x-axis, and division is equivalent to multiplying by the reciprocal.

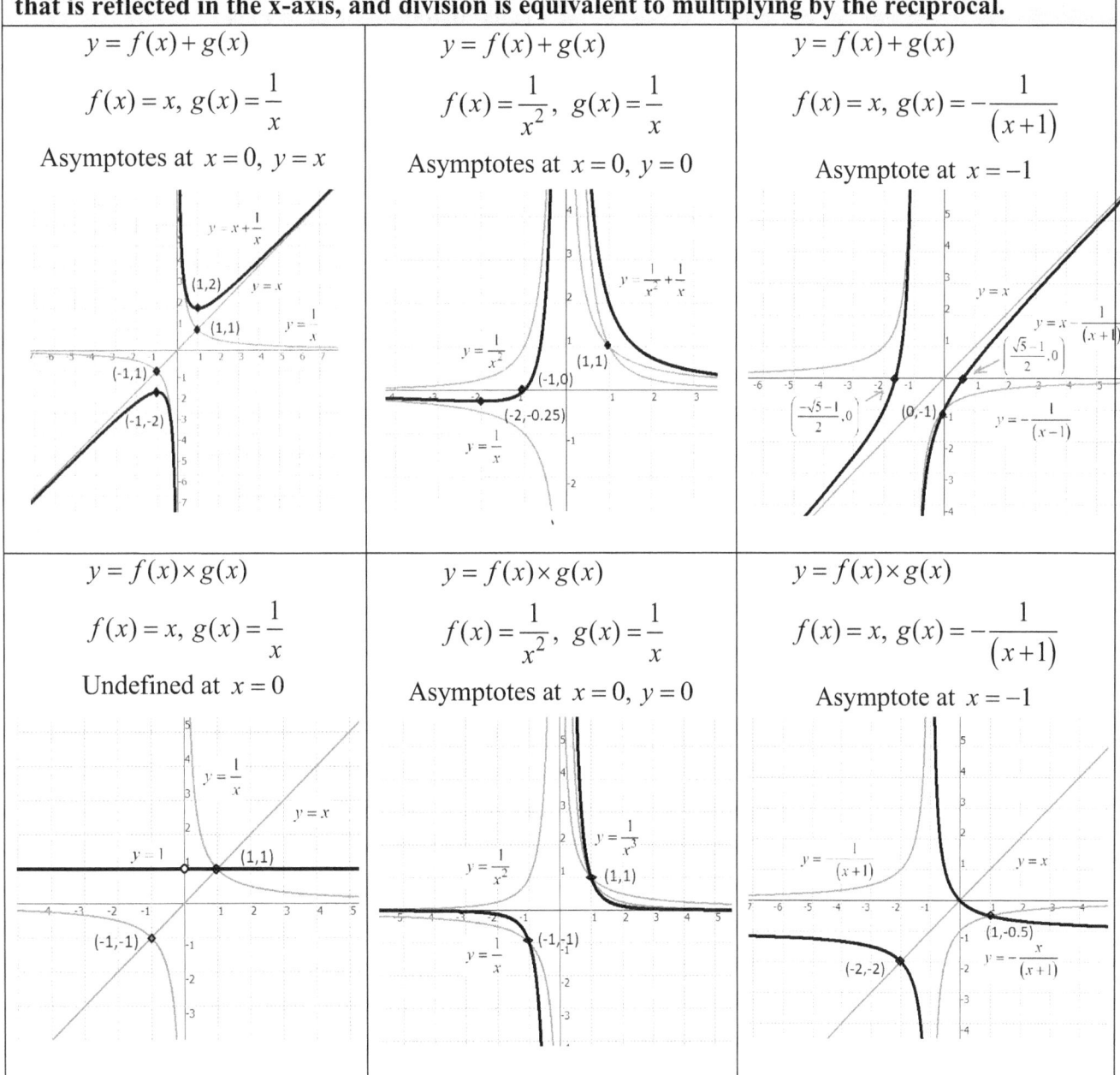

$y = f(x) + g(x)$

$f(x) = x,\ g(x) = \dfrac{1}{x}$

Asymptotes at $x = 0,\ y = x$

$y = f(x) + g(x)$

$f(x) = \dfrac{1}{x^2},\ g(x) = \dfrac{1}{x}$

Asymptotes at $x = 0,\ y = 0$

$y = f(x) + g(x)$

$f(x) = x,\ g(x) = -\dfrac{1}{(x+1)}$

Asymptote at $x = -1$

$y = f(x) \times g(x)$

$f(x) = x,\ g(x) = \dfrac{1}{x}$

Undefined at $x = 0$

$y = f(x) \times g(x)$

$f(x) = \dfrac{1}{x^2},\ g(x) = \dfrac{1}{x}$

Asymptotes at $x = 0,\ y = 0$

$y = f(x) \times g(x)$

$f(x) = x,\ g(x) = -\dfrac{1}{(x+1)}$

Asymptote at $x = -1$

FUNCTIONS y=f(x) FOR REAL! SUMMARY:

Functions by operations	Defining Functions $y=f(x)$
➢ Power $f(x)=x^n$ ➢ Trigonometric $f(x)=sin(x), cos(x), tan(x)$ ➢ Exponential $f(x)=a^n$ ➢ Logarithmic $f(x)=log_a x$	➢ Predefined Numbers sets **R, Z, R⁺** ➢ Domain and Range $x \, \varepsilon \, (0,\infty), y \, \varepsilon \, (0,\infty)$ ➢ One:One functions ➢ Many:One functions ➢ Vertical line test for functions

Functions types	Operations with functions
➢ Hybrid ➢ Composite $x \Rightarrow \boxed{g} \Rightarrow g(x) \Rightarrow \boxed{f} \Rightarrow f(g(x))$ ➢ Inverse $x \Rightarrow \boxed{\text{original } f \text{ function}} \Rightarrow y \Rightarrow \boxed{\text{inverse } f^{-1} \text{ function}} \Rightarrow x$	Take care the resulting domain is the common subdomains. ➢ Addition ➢ Subtraction ➢ Multiplication ➢ Division **Transformation of functions** ➢ Dilations ➢ Reflection ➢ Translation ➢ See the table below

They say Mathematics is a language of symbols.

For the graph of $y = f(x)$	Equivalent Mapping & Matrices	Graphically
$y = -f(x)$ is a reflection of $f(x)$ in the x-axis	$(x,y) \rightarrow (x,-y)$ $\begin{bmatrix} 1 & 0 \\ 0 & -1 \end{bmatrix} \begin{bmatrix} x \\ y \end{bmatrix} = \begin{bmatrix} x' \\ y' \end{bmatrix}$	
$y = f(-x)$ is a reflection of $f(x)$ in the y-axis	$(x,y) \rightarrow (-x,y)$ $\begin{bmatrix} -1 & 0 \\ 0 & 1 \end{bmatrix} \begin{bmatrix} x \\ y \end{bmatrix} = \begin{bmatrix} x' \\ y' \end{bmatrix}$	
$y = Af(x)$ is a dilation by factor A of $f(x)$ from the x-axis	$(x,y) \rightarrow (x,Ay)$ $\begin{bmatrix} 1 & 0 \\ 0 & A \end{bmatrix} \begin{bmatrix} x \\ y \end{bmatrix} = \begin{bmatrix} x' \\ y' \end{bmatrix}$	
$y = f\left(\frac{x}{n}\right)$ is a dilation by factor n of graph $f(x)$ from the y-axis	$(x,y) \rightarrow (nx,y)$ $\begin{bmatrix} n & 0 \\ 0 & 1 \end{bmatrix} \begin{bmatrix} x \\ y \end{bmatrix} = \begin{bmatrix} x' \\ y' \end{bmatrix}$	
$y = f(x - b)$ is a translation of b units of $f(x)$ parallel to the x-axis	$(x,y) \rightarrow (x+b,y)$ $\begin{bmatrix} x \\ y \end{bmatrix} + \begin{bmatrix} b \\ 0 \end{bmatrix} = \begin{bmatrix} x' \\ y' \end{bmatrix}$	
$y = f(x) + c$ is a translation of c units of $f(x)$ parallel to the y-axis	$(x,y) \rightarrow (x,y+c)$ $\begin{bmatrix} x \\ y \end{bmatrix} + \begin{bmatrix} 0 \\ c \end{bmatrix} = \begin{bmatrix} x' \\ y' \end{bmatrix}$	

FUNCTIONS y=f(x) FOR REAL! Check your Understanding.

2.1 Define the Hybrid function for the following scenario.

The cross-section of the mountain range meeting the sea, where sea level is *y=0*.

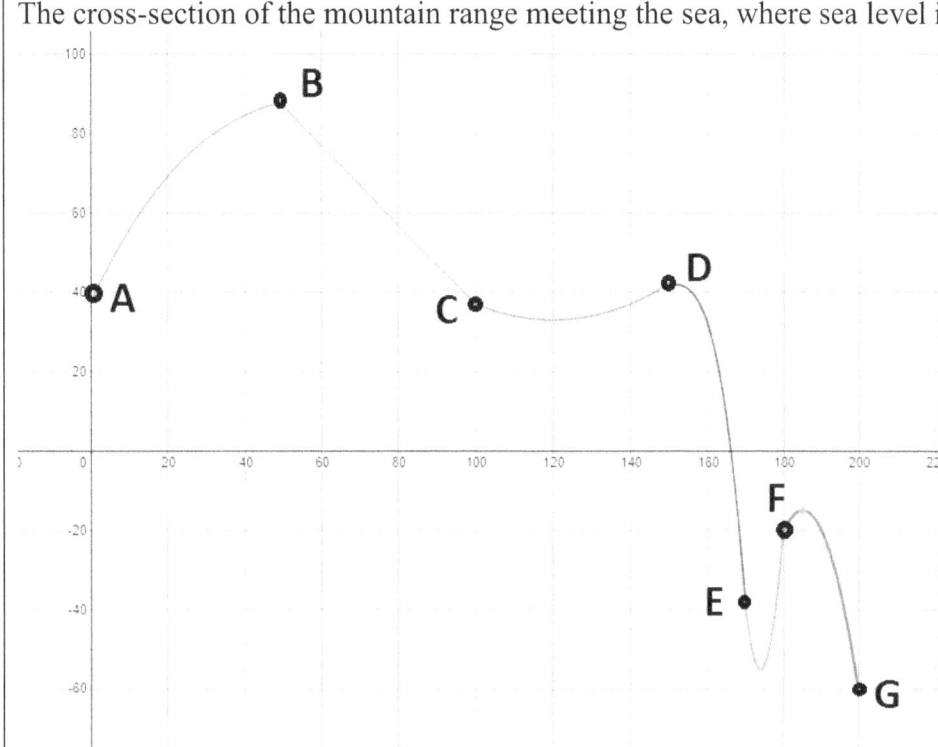

Coordinates
0 (0,0)
A (0,38)
B (49,88)
C (100,38)
D (150,42)
E (170,-36
F (180,-21)
G (200,-59)

A to B $y_1 = \left(0.05\left(x-75\right)\right)^3 + 90$

B to C $y_2 = x + 137$

C to D $y_3 = 0.01\left(x-120\right)^2 + 30$

D to E $y_4 = -0.01\left(x-150\right)^2 + 42$

E to F $y_5 = \left(x-174\right)^2 - 55$

F to G $y_6 = -0.2\left(x-185\right)^2 - 15$

a) Construct the hybrid function definition *f(x)* for the cross-section.

b) Will an inverse function exist? Justify your answer

2.2 Inverse functions.

a) Radians to degrees, define the function $D(r)$ where r is the radians, and $D(r)$ returns the degrees.

b) Degrees to radians, find the function $R(d)$ where d is the degrees and $R(d)$ returns the radians.

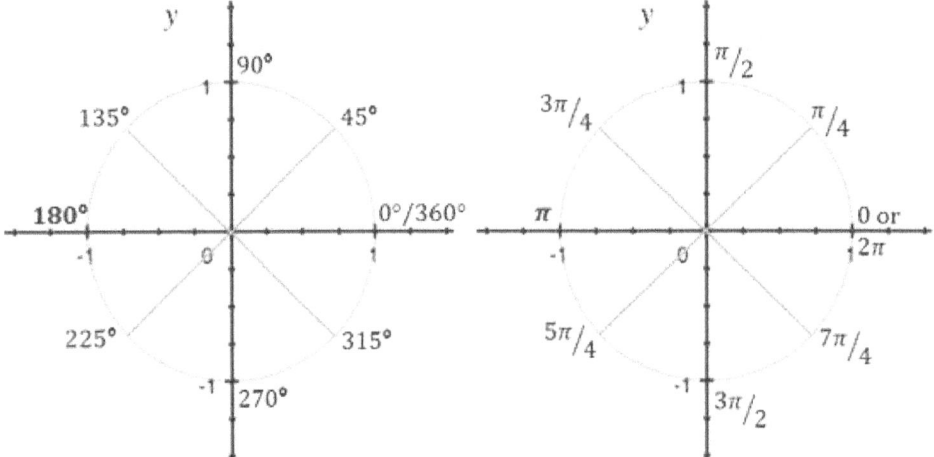

Figure 1: Unit circle measured in degrees.   Figure 2: Unit circle measured in radians.

c) Show the functions $D(r)$ and $R(d)$ are inverse functions of each other.

d) Celsius to Fahrenheit and Fahrenheit to Celsius, find the conversion functions.

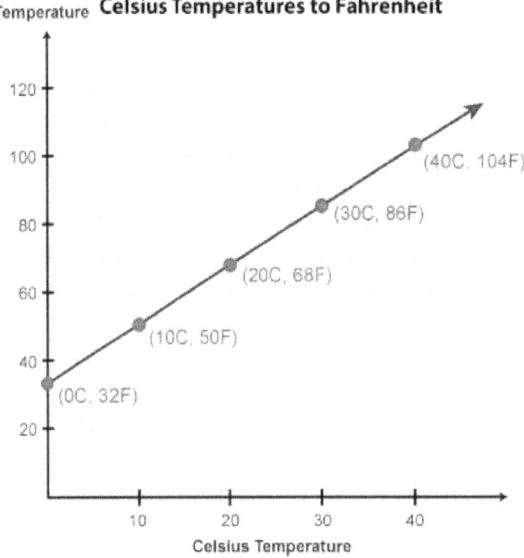

e) Show that the functions found in part d) are inverses of each other.

Intersections of functions and their inverses commonly occur on the y=x line. But not always.

f) Find the inverse function $f^{-1}(x)$ of $f(x)=e^{\left(-x+\frac{1}{2}\right)}$ and graph both functions.

g) Find the inverse function $f^{-1}(x)$ of $f(x)=1-x^3$ and graph both functions.

2.3 Composing functions for solving larger problems.

a) What do onion layers and composite functions have in common?

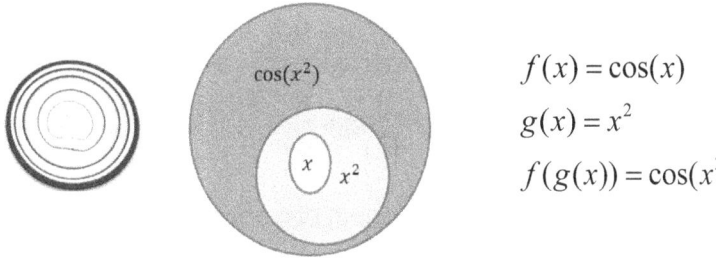

$$f(x) = \cos(x)$$
$$g(x) = x^2$$
$$f(g(x)) = \cos(x^2)$$

b) State the volume V of a sphere as a function of the radius r, that is $V(r)$, you may need to look it up.

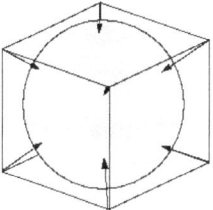

A sphere is inscribed inside a cube of side length l. The sides of the cube are just touching the sides of the sphere.

c) State the radius r of the sphere as a function of the side length of the cube l. That is find $r(l)$.

d) By composing the functions, $V(r)$ and $r(l)$, state the volume V of the sphere as a function of the side length of the cube l. that is find $V(l)$.

If $T(h)$ is the temperature T in the atmosphere as a function of height h,
$T(h) = 100 - 10\sqrt{h}$ $h \in [0,100]$ and $h(t)$ is the height of a weather balloon as a function of time t
$$h(t) = \frac{100}{1+100e^{-0.1t}} t \in [0,100].$$

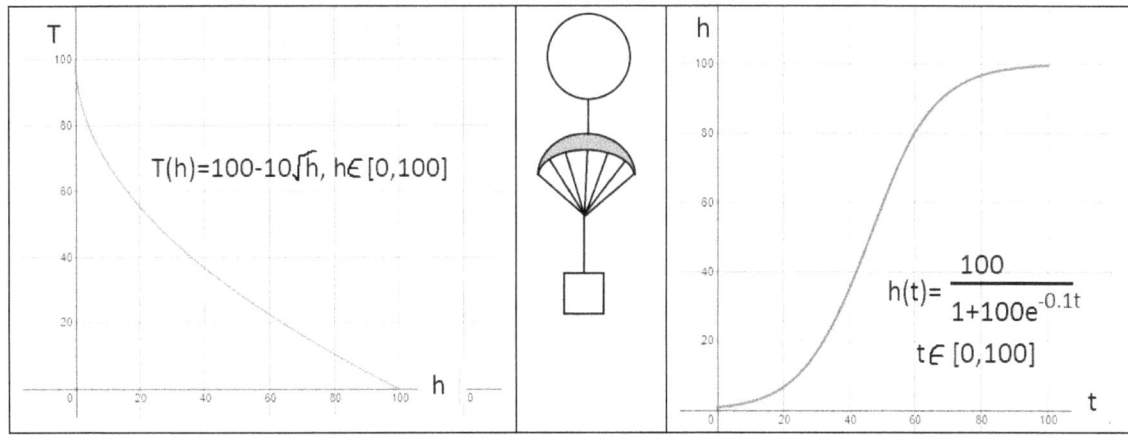

e) What is the temperature at the location of the weather balloon as a function of time?

f) Graph $T(t)$. Consider if the graph makes sense.

2.4 Moving transformations.

A stream of water comes out of a fountain in the shape of a quadratic function *f(x)* where *x* is the horizontal distance from the tap and is measured in centimetres.

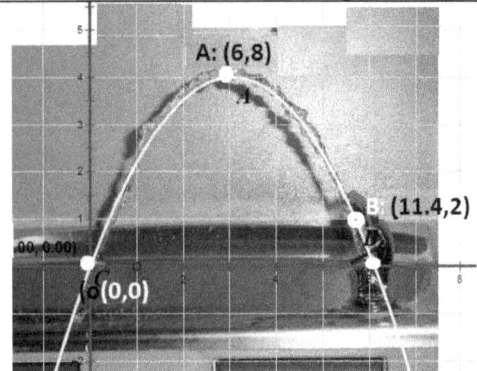

The function is approximately equal to

$$f(x) = -\frac{2}{9}x(x-12)$$

where *x* is measured in *cm*, and the height *f(x)* is in *cm* from the origin.

Overlay original image with a grid. Each 1 grid cell is 2cm by 2cm in real life.

a) If the water fountain is moved down the end of the hall which is *1000cm* from the original position what is the effect on the original quadratic function?

b) If the water fountain is then moved downstairs by *600cm* what is the effect on the original quadratic function?

The distance in *cm* of the minute hand of an analog round wall clock from the ceiling C over 12 hours where time is measured in hours is represented by the function *C(t)=Asin(n(t-b)+c*.

The distance in *cm* of the minute hand from the floor *F* over 12 hours where time is measured in hours is represented by the function *F(t)=Dsin(n(t-g)+h*

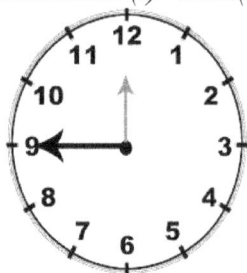

Warning: Don't try this problem with a digital clock.

Time hours t	0	1	2	3	4	5	6	7	8	9	10	11
Distance ceiling (cm) C	30	31.3	35	40	45	48.7	50	48.7	45	40	35	31.3
Distance floor (cm) F	170	168.7	165	160	155	151.3	150	151.3	155	160	165	168.7

c) Can you work out the two functions *C(t)* and *F(t)*?

d) What is the effect on the functions if the clock is moved lower down the wall by 21 centimetres?

2.5 Honey, I shrunk the functions.

A very popular logo is drawn on a Cartesian plane, and it is approximately 13 units high.

a) Define the hybrid function for the top of the M logo $T(x)$ made up of two parabolas with turning points of *(8,19)* and *(18,19)* respectively.

b) Define the hybrid function for the bottom of the M logo $B(x)$ made up of two parabolas with turning points of *(8,15)* and *(18,15)* respectively.

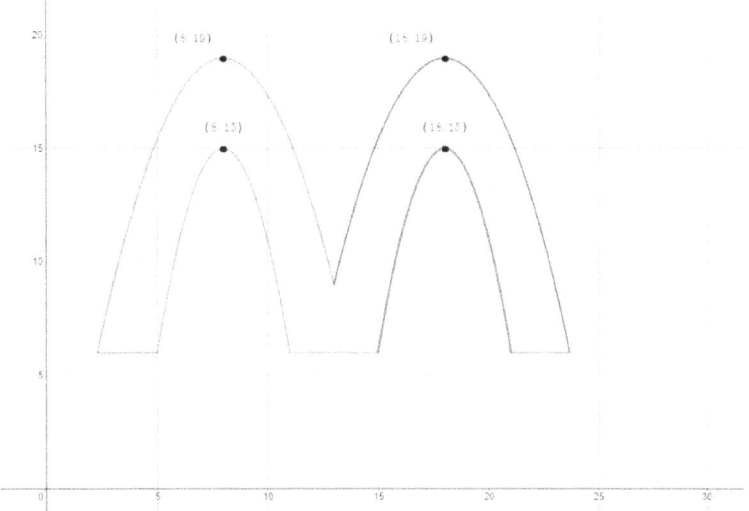

c) Describe the transformation for shrinking the M logo to a quarter of the original size. Use as many notations as you can to describe the transformations, including matrix notation.

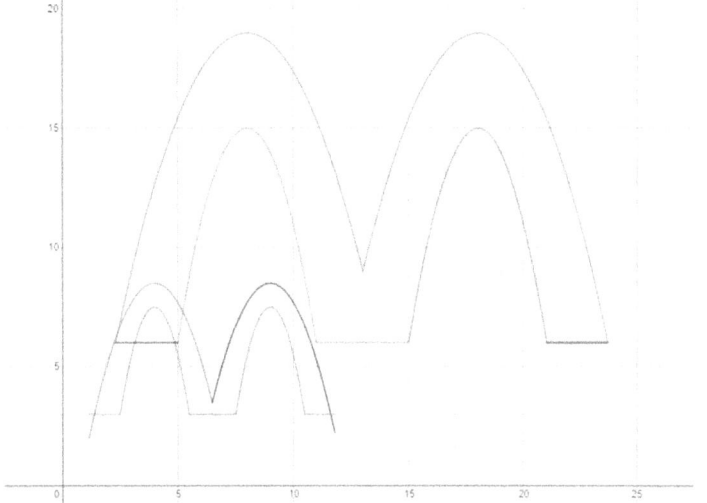

d) Describe the transformations for shrinking the M logo to half of the original size using function notation and matrix notation.

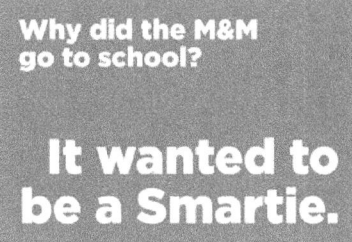

Why did the M&M go to school?

It wanted to be a Smartie.

51

2.6 More Transformations

Given below is a table that gives the populations of foxes and rabbits in a national park over a 12 month period. Note that each value of t is in months and corresponds to the beginning of the month and t=0 corresponds to the beginning of January.

(source: https://tasks.illustrativemathematics.org/content-standards/HSF/TF/B/5/tasks/816)

t, month	0	1	2	3	4	5	6	7	8	9	10	11
R, rabbits	1000	750	567	500	567	750	1000	1250	1433	1500	1433	1250
F, Foxes	150	143	125	100	75	57	50	57	75	100	125	143

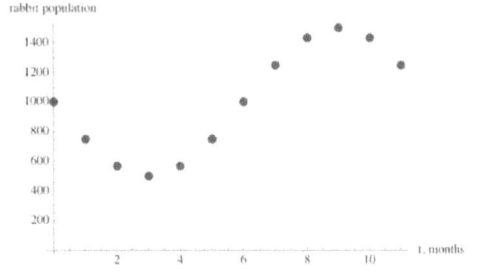  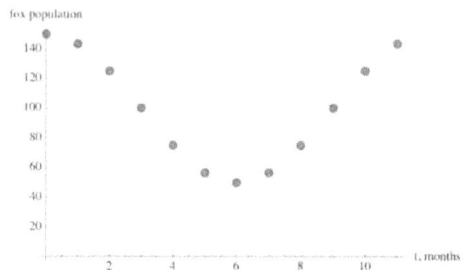

a) Explain why it is appropriate to model the number of rabbits and foxes as trigonometric functions of time.

b) Find an appropriate trigonometric function that models the number of rabbits, R(t), as a function of time, t, in months.

c) Find an appropriate trigonometric function that models the number of foxes, F(t), as a function of time, t, in months.

d) Graph both functions and give one possible explanation why one function seems to "chase" the other function.

e) Transform the rabbits R(t) function to the foxes function F(t) using function notation and matrix notation.

52

FUNCTIONS y=f(x) FOR REAL! SOLUTIONS.

2.1 Define the Hybrid function for the following scenario. (SOLUTIONS)

a) The cross-section of the mountain range meeting the sea, where sea level is *y=0*.

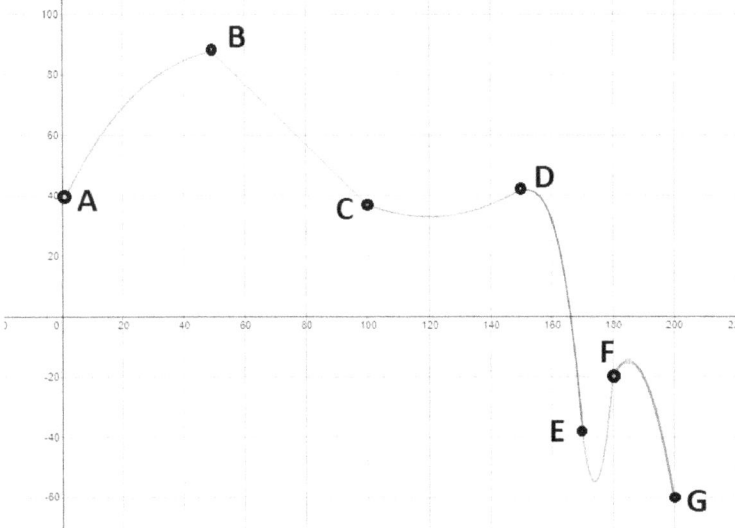

$y = \left(0.05(x-75)\right)^3 - 90 \; \{0 < x < 30\}$

$y = -x - 137 \; \{49 < x < 100\}$

$y = 0.01(x-120)^2 - 33 \; \{100 < x < 150\}$

$y = -0.01(x-150)^3 - 42 \; \{150 < x < 170\}$

$y = (x-174)^2 - 55 \; \{170 < x < 180\}$

$y = -0.2(x-185)^2 - 15 \; \{180 < x < 200\}$

Graphed with desmos https://www.desmos.com/calculator

A to B $y_1 = \left(0.05(x-75)\right)^3 + 90$

B to C $y_2 = x + 137$

C to D $y_3 = 0.01(x-120)^2 + 30$

D to E $y_4 = -0.01(x-150)^2 + 42$

E to F $y_5 = (x-174)^2 - 55$

F to G $y_6 = -0.2(x-185)^2 - 15$

$$f(x) = \begin{cases} \left(0.05(x-75)\right)^3 + 90 & 0 \le x \le 49 \\ x + 137 & 49 < x \le 100 \\ 0.01(x-120)^2 + 30 & 100 < x \le 150 \\ -0.01(x-150)^3 + 42 & 150 < x \le 170 \\ (x-174)^2 - 55 & 170 < x \le 180 \\ -0.2(x-185)^2 - 15 & 180 < x \le 200 \end{cases}$$

b) The inverse function cannot be found, since the function has intervals where it is many *x* values to one *y* value. To find the inverse function, the original function must be one to one.

2.2 Inverse functions. (SOLUTIONS)

a) Radians to degrees, $D(r)$ where r is the radians. $D(r) = \dfrac{180}{\pi} r$

b) Degrees to radians, $R(d)$ where d is the degrees. $R(d) = \dfrac{\pi}{180} d$

c) Show the functions $D(r)$ and $R(d)$ are inverse functions of each other.

$$D(R(d)) = \frac{180}{\pi} \times \frac{\pi}{180} d = d$$

$$R(D(r)) = \frac{\pi}{180} \times \frac{180}{\pi} r = r$$

d) Celsius to Fahrenheit and Fahrenheit to Celsius, find the functions.

$$F(c) = \frac{9}{5}c + 32$$

$$C(f) = \frac{5}{9}(f - 32)$$

e) Show that the functions found in part d) are inverses of each other.

$$F(C(f)) = \frac{9}{5}\left(\frac{5}{9}(f - 32)\right) + 32 \qquad\qquad C(F(c)) = \frac{5}{9}\left(\left(\frac{9}{5}c + 32\right) - 32\right)$$

$$= (f - 32) + 32 \qquad\qquad\qquad\qquad = \frac{5}{9}\left(\left(\frac{9}{5}c\right)\right)$$

$$= f \qquad\qquad\qquad\qquad\qquad\qquad = c$$

f) $f^{-1}(x) = -\ln(x) - \dfrac{1}{2}$ of $f(x) = e^{\left(-x + \frac{1}{2}\right)}$

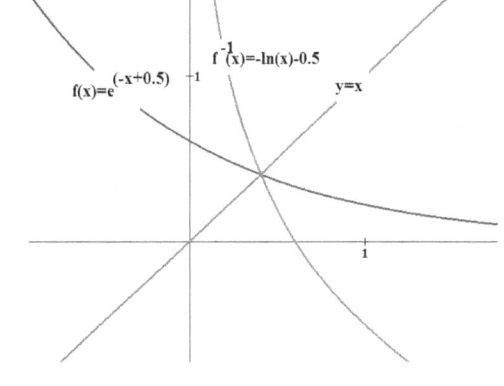

g) $f^{-1}(x) = (1 - x)^{1/3}$ of $f(x) = 1 - x^3$

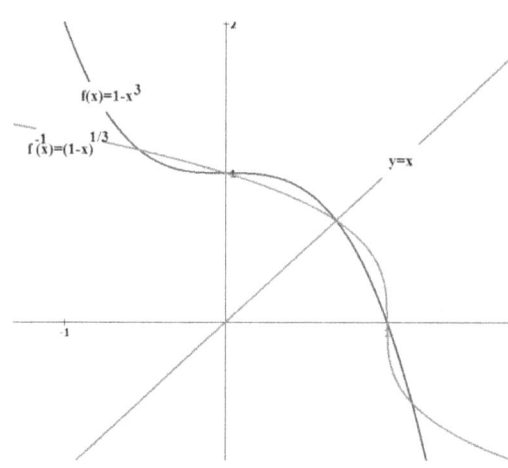

Hey! Not all the intersections are on y=x.

2.3 Composing functions for solving larger problems. (SOLUTIONS)

a) What do onion layers and composite functions have in common?

Onions are made up of layers. The inside layers support the outer layers.

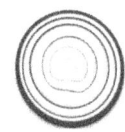

Composite functions are also layered, the inside function(s) support the outer function(s), by passing the output value to the outer function.

b) Volume V of a sphere as a function of the radius r. $V(r) = \dfrac{4}{3}\pi r^3$

composing the functions

$$V(r(l)) = \frac{4}{3}\pi \left(r(l)\right)^3$$

c) Radius r of the sphere as a function of the side length of the cube l.

$$r(l) = \frac{l}{2}$$

$$V(r(l)) = \frac{4}{3}\pi \left(\frac{l}{2}\right)^3$$

$$V(l) = \frac{4}{3}\pi \left(\frac{l^3}{8}\right)$$

d) By composing the functions, $V(r)$ and $r(l)$, find $V(l)$.

$$V(r) = \frac{4}{3}\pi r^3 \text{ and } r(l) = \frac{l}{2}$$

$$V(l) = \frac{\pi l^3}{6}$$

e) The temperature at the location of the weather balloon as a function of time.

$$T(h) = 100 - 10\sqrt{h} \quad h \in [0,100]$$

$$h(t) = \frac{100}{1 + 100e^{-0.1t}} \quad t \in [0,100]$$

the range of $h(t)$ is approximately $(1,100)$

Composing the functions

$$T(h(t)) = 100 - 10\sqrt{h(t)} \quad h(t) \in (1,100)$$

$$T(h(t)) = 100 - 10\sqrt{h(t)} \quad h(t) \in (1,100)$$

$$T(t) = 100 - 10\sqrt{\frac{100}{1 + 100e^{-0.1t}}}$$

$$T(t) = 100 - 100\sqrt{\frac{1}{1 + 100e^{-0.1t}}}$$

$$T(t) = 100 - \frac{100}{\sqrt{1 + 100e^{-0.1t}}}$$

f) Graph $T(t)$ and compare it to $T(h)$ and $h(t)$. Does it make sense?

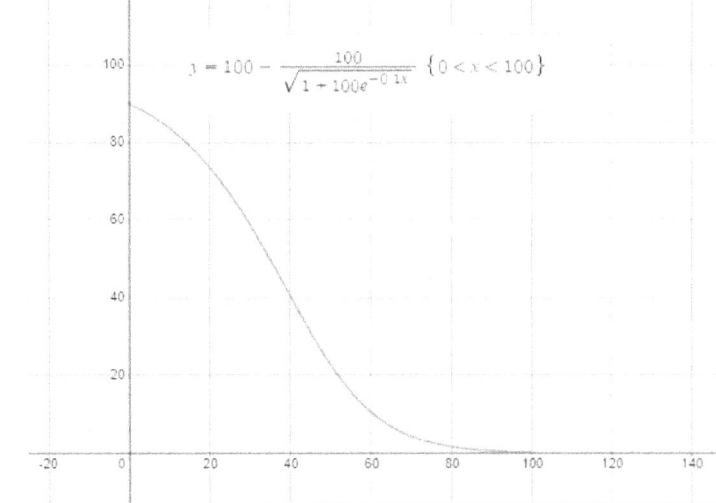

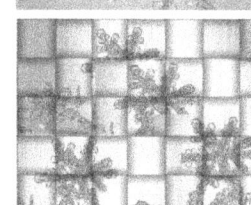

2.4 Moving transformations. (SOLUTIONS)

A stream of water comes out of a fountain in the shape of a quadratic function *f(x)*.

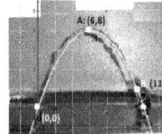

$$f(x) = -\frac{2}{9}x(x-12)$$

where *x* is measured in *cm*, and the height *f(x)* is in *cm* from the origin.

a) Water fountain is moved horizontally *1000cm* from the original position

$$f(x \pm 1000) = -\frac{2}{9}(x \pm 1000)\big((x \pm 1000) - 12\big)$$

b) Water fountain is then moved downstairs by *600cm* what is the effect on the original quadratic function? $f(x \pm 1000) - 600 = -\frac{2}{9}(x \pm 1000)\big((x \pm 1000) - 12\big) - 600$

c) *C(t)* ceiling and *F(t)* floor

$$C(t) = 10\sin\left(\frac{\pi}{6}(t-3)\right) + 40, \ 0 \le t \le 12 \qquad F(t) = -10\sin\left(\frac{\pi}{6}(t-3)\right) + 160, \ 0 \le t \le 12$$

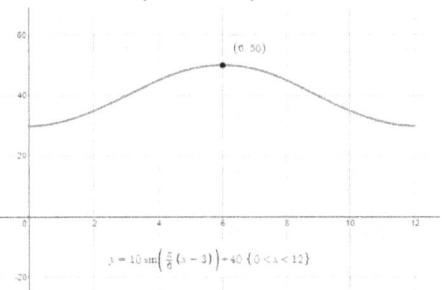

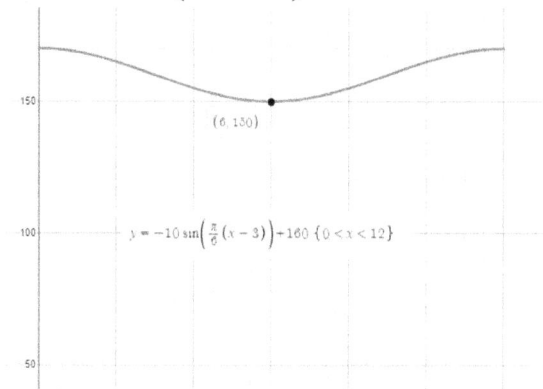

d) What is the effect on the functions if the clock is moved lower down the wall by 21 centimetres?

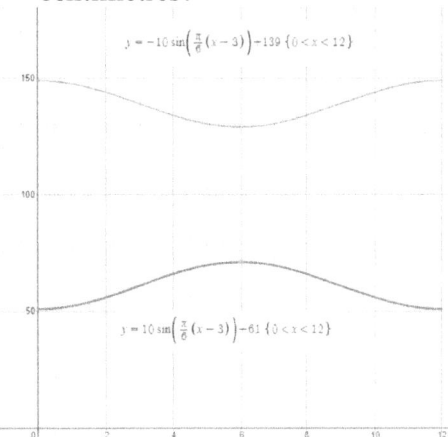

$$C(t) = 10\sin\left(\frac{\pi}{6}(t-3)\right) + 61, \ 0 \le t \le 12$$

$$F(t) = -10\sin\left(\frac{\pi}{6}(t-3)\right) + 139, \ 0 \le t \le 12$$

Only one analog clock was used in this problem.

2.5 Honey, I shrunk the functions. (SOLUTIONS)

A very popular logo is drawn on a Cartesian plane, and it is approximately 13 units high.

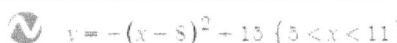

$y=-(x-8)^2-13\ \{5<x<11\}$

$y=-(x-18)^2-13\ \{15<x<21\}$

$y=6\ \{11<x<15\}$

$y=-0.4(x-8)^2-19\ \{2.3<x<13\}$

$y=-0.4(x-18)^2-19\ \{13<x<23.7\}$

$y=6\ \{2.3<x<5\}$

$y=6\ \{21<x<23.7\}$

Check your answers on desmos
https://www.desmos.com/calculator
or your graphing calculator.

a) Top of the M logo *T(x)* made up of two parabolas with turning points of *(8,19)* and *(18,19)* respectively.

$$T(x)=\begin{cases} -0.4(x-8)^2+19, & 2.3<x<13 \\ -0.4(x-18)^2+19, & 13<x<23.7 \end{cases}$$

b) Bottom of the M logo *B(x)* made up of two parabolas with turning points of *(8,15)* and *(18,15)* respectively.

$$B(x)=\begin{cases} -(x-8)^2+15, & 5<x<11 \\ -(x-18)^2+15, & 15<x<21 \\ 6, & x\in\{(2.3,5)\cup(11,15)\cup(21,23.7)\} \end{cases}$$

c) Describe the transformation for shrinking the logo to a quarter of the original size.
- Dilation, by 0.5 from y-axis,
- Dilation by 0.5 from x-axis
- Remember to restrict the intervals: eg. $2.3<2x<13$

$$0.5T(2x)=\begin{cases} -0.2(2x-8)^2+\dfrac{19}{2}, & \dfrac{2.3}{2}<x<\dfrac{13}{2} \\ -0.2(2x-18)^2+\dfrac{19}{2}, & \dfrac{13}{2}<x<\dfrac{23.7}{2} \end{cases}$$

$$0.5B(2x)=\begin{cases} -0.5(2x-8)^2+\dfrac{15}{2}, & \dfrac{5}{2}<x<\dfrac{11}{2} \\ -0.5(2x-18)^2+\dfrac{15}{2}, & \dfrac{15}{2}<x<\dfrac{21}{2} \\ 3, & x\in\left\{\left(\dfrac{2.3}{2},\dfrac{5}{2}\right)\cup\left(\dfrac{11}{2},\dfrac{15}{2}\right)\cup\left(\dfrac{21}{2},\dfrac{23.7}{2}\right)\right\} \end{cases}$$

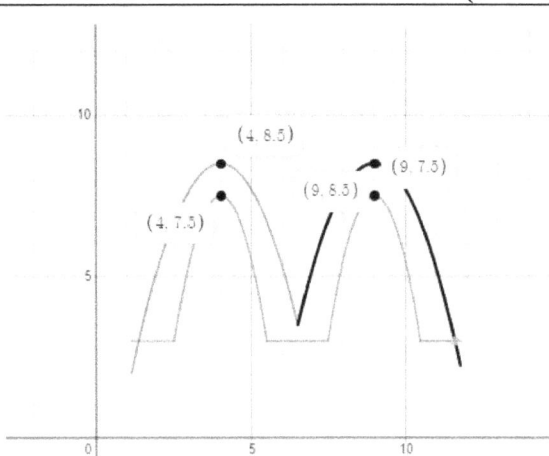

Graph equations:
$$y = -0.5(2x-8)^2 + 7.5 \ \{2.5 < x < 5.5\}$$
$$y = -0.5(2x-18)^2 + 7.5 \ \{7.5 < x < 10.5\}$$
$$y = 3 \ \{1.15 < x < 2.5\}$$
$$y = -0.2(2x-8)^2 - 8.5 \ \{1.15 < x < 6.5\}$$
$$y = -0.2(2x-18)^2 - 8.5 \ \{6.5 < x < 11.8\}$$
$$y = 3 \ \{5.5 < x < 7.5\}$$
$$y = 3 \ \{10.5 < x < 11.8\}$$

$$\begin{bmatrix} 0.5 & 0 \\ 0 & 0.5 \end{bmatrix}\begin{bmatrix} x \\ y \end{bmatrix} = \begin{bmatrix} 0.5x \\ 0.5y \end{bmatrix}$$

$$x^{new} = 0.5x, \ y^{new} = 0.5y$$

$$(x,y) \to (0.5x, 0.5y)$$

Matrix notation for transformations.

d) Describe the transformation for shrinking the logo to a half of the original size.

- Dilation, by $\dfrac{1}{\sqrt{2}}$ from y-axis,

- Dilation by $\dfrac{1}{\sqrt{2}}$ from x-axis

$$\begin{bmatrix} \dfrac{1}{\sqrt{2}} & 0 \\ 0 & \dfrac{1}{\sqrt{2}} \end{bmatrix}\begin{bmatrix} x \\ y \end{bmatrix} = \begin{bmatrix} \dfrac{1}{\sqrt{2}}x \\ \dfrac{1}{\sqrt{2}}y \end{bmatrix}$$

- Remember to restrict the intervals:
  eg. $2.3 < \sqrt{2}x < 13$

$$x^{new} = \frac{1}{\sqrt{2}}x, \ y^{new} = \frac{1}{\sqrt{2}}y \Rightarrow (x,y) \to (\frac{1}{\sqrt{2}}x, \frac{1}{\sqrt{2}}y)$$

$$\frac{1}{\sqrt{2}}T(\sqrt{2}x) = \begin{cases} -\dfrac{0.4}{\sqrt{2}}(\sqrt{2}x-8)^2 + \dfrac{19}{\sqrt{2}}, & \dfrac{2.3}{\sqrt{2}} < x < \dfrac{13}{\sqrt{2}} \\ -\dfrac{0.4}{\sqrt{2}}(\sqrt{2}x-18)^2 + \dfrac{19}{\sqrt{2}}, & \dfrac{13}{\sqrt{2}} < x < \dfrac{23.7}{\sqrt{2}} \end{cases}$$

$$\frac{1}{\sqrt{2}}B(\sqrt{2}x) = \begin{cases} -\dfrac{1}{\sqrt{2}}(\sqrt{2}x-8)^2 + \dfrac{15}{\sqrt{2}}, & \dfrac{5}{\sqrt{2}} < x < \dfrac{11}{\sqrt{2}} \\ -\dfrac{1}{\sqrt{2}}(\sqrt{2}x-18)^2 + \dfrac{15}{\sqrt{2}}, & \dfrac{15}{\sqrt{2}} < x < \dfrac{21}{\sqrt{2}} \\ \dfrac{6}{\sqrt{2}}, & x \in \left\{ \left(\dfrac{2.3}{\sqrt{2}}, \dfrac{5}{\sqrt{2}}\right) \cup \left(\dfrac{11}{\sqrt{2}}, \dfrac{15}{\sqrt{2}}\right) \cup \left(\dfrac{21}{\sqrt{2}}, \dfrac{23.7}{\sqrt{2}}\right) \right\} \end{cases}$$

2.6 More Transformations (SOLUTIONS)

There is Cyclic fluctuations in populations due to the Predator/Prey relationship between foxes and rabbits.

a) Number of rabbits, R(t), as a function of time, t, in months.

$$R(t) = A\sin\left(n(t-b)\right) + c$$

$$A = \frac{500 - 1500}{2} = -500$$

$$\frac{2\pi}{n} = 12 \therefore n = \frac{\pi}{6}, \; b = c = 0$$

$$R(t) = -500\sin\left(\frac{\pi}{6}t\right)$$

b) Number of foxes, F(t), as a function of time, t, in months.

$$F(t) = A\cos\left(n(t-b)\right) + c$$

$$A = \frac{150 - 50}{2} = 50$$

$$\frac{2\pi}{n} = 12 \therefore n = \frac{\pi}{6}, \; b = c = 0$$

$$F(t) = 50\cos\left(\frac{\pi}{6}t\right)$$

c) The functions seems to "chase" each other just like the Predator/Prey do.

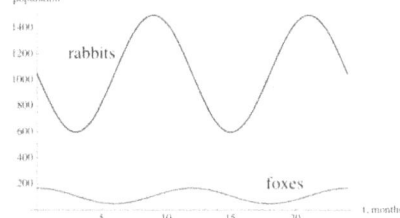

Functions model the Predator/Prey cyclic population control/growth phases.

d) To transform F(t) to R(t)
- Dilate by factor of 10 from x-axis
- Reflect in x-axis
- Translate by +6 months in x direction

$$-10F(t-6) = R(t)$$

$$LHS = -10F(t-6)$$

$$= -500\cos\left(\frac{\pi}{6}(t-6)\right)$$

$$= -500\sin\left(\frac{\pi}{6}t\right)$$

$$= R(t)$$

$$Note: \cos\left(\frac{\pi}{6}(t-6)\right) = \sin\left(\frac{\pi}{6}t\right)$$

Let $y = F(t), x = t$

$$\begin{bmatrix} 1 & 0 \\ 0 & -10 \end{bmatrix}\begin{bmatrix} x \\ y \end{bmatrix} + \begin{bmatrix} 6 \\ 0 \end{bmatrix} = \begin{bmatrix} x+6 \\ -10y \end{bmatrix}$$

$$x^{new} = x+6, \; y^{new} = -10y$$

$$(x, y) \to (x+6, -10y)$$

e) To transform R(t) to F(t)
- Dilate by factor of 0.1 from x-axis
- Reflect in x-axis
- Translate by -6 months in x direction

$$-\frac{1}{10}R(t+6) = F(t)$$

$$LHS = -\frac{1}{10}R(t+6)$$

$$= 50\sin\left(\frac{\pi}{6}(t+6)\right)$$

$$= 50\cos\left(\frac{\pi}{6}t\right)$$

$$= F(t)$$

$$Note: \sin\left(\frac{\pi}{6}(t+6)\right) = \cos\left(\frac{\pi}{6}t\right)$$

Let $y = R(t), x = t$

$$\begin{bmatrix} 1 & 0 \\ 0 & -0.1 \end{bmatrix}\begin{bmatrix} x \\ y \end{bmatrix} + \begin{bmatrix} 6 \\ 0 \end{bmatrix} = \begin{bmatrix} x+6 \\ -0.1y \end{bmatrix}$$

$$x^{new} = x+6, \; y^{new} = -0.1y$$

$$(x, y) \to (x+6, -0.1y)$$

GET IT?

"Just a darn minute! Yesterday you said X equals two!"

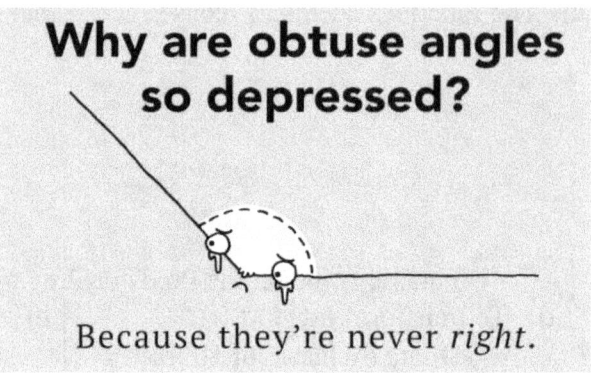

Super Power Functions

Don't Forget: The functions that you need to know about in VCE Mathematical Methods 3&4.

Power functions in general form, $f(x) = A(n(x-b))^p + c$

Polynomial functions, the power is a positive integer.

- Linear functions $f(x) = mx + c$
- Quadratic functions $f(x) = ax^2 + bx + c$ in turning point form $f(x) = a(x-h)^2 + k$.
- Cubic functions $f(x) = ax^3 + bx^2 + cx + d$ in turning point form $f(x) = a(x-h)^3 + k$.
- Quartic functions $f(x) = x^3 - ax^2 + 20x - 24$ in turning point form $f(x) = a(x-h)^4 + k$.

  where $a, b, c, d, e, h, k, m, n, p, q, A$ are constant reals.

Negative power functions, the power is a negative integer.

- Hyperbola functions $f(x) = A(n(x-b))^{-1} + c$

- Truncus functions $f(x) = A(n(x-b))^{-2} + c$

- General negative power functions $f(x) = A(n(x-b))^{-p} + c$

Quotient power functions, the power is a quotient or fraction.

- Square root functions $f(x) = A(n(x-b))^{1/2} + c$

- General quotient power functions $f(x) = A(n(x-b))^{p/q} + c$

Exponential functions with base of a and e in the general form, where $f(x) = Aa^{n(x-b)} + c$ or $f(x) = Ae^{n(x-b)} + c$.

Logarithmic functions with base of a and e in the general form, where $f(x) = A\log_a(n(x-b)) + c$ or $f(x) = A\log_e(n(x-b)) + c$.

Circular functions for anti-clockwise angle θ from the positive x-axis where A, n, b, c are constant reals.

- Sine function $f(\theta) = A\sin(n(\theta - b)) + c$.
- Cosine function $f(\theta) = A\cos(n(\theta - b)) + c$.
- Tangent function $f(\theta) = A\tan(n(\theta - b)) + c$.

Fibonacci Blue/Flickr

Power Functions.

$$f(x) = A\big(n(x-b)\big)^p + c$$

Polynomial, where p is a positive integer, with zero included, the graph may be straight or curvy.

Negative power functions, where p is a negative integer, the graph may have asymptotes.

Quotient power functions, where p is a quotient aka fraction the graph may have restricted domains.

Polynomial functions

A subset of the power functions, polynomial functions of x are functions of the form $f(x) = a_n x^n + a_{n-1} x^{n-1} + \cdots + a_1 x^1 + a_0 x^0$, where some of the coefficients are non-zero $a_n \neq 0$, and the powers of x are all positive integers.

For this polynomial function a_n is the leading coefficient, a_0 is the constant term, and the highest power of n is called the order or the degree of the polynomial.

Order	Type	Polynomials in function form.
0	Constant	$f(x) = a_0$
1	Linear	$f(x) = a_1 x + a_0$
2	Quadratic	$f(x) = a_2 x^2 + a_1 x + a_0$
3	Cubic	$f(x) = a_3 x^3 + a_2 x^2 + a_1 x + a_0$
4	Quartic	$f(x) = a_4 x^4 + a_3 x^3 + a_2 x^2 + a_1 x + a_0$

Recall that *x-intercepts* are found when $f(x) = 0$ and *y-intercepts* are found when $x = 0$ is substituted into the function, giving the value $f(0)$.

Example 3.1: Polynomial functions accepting an input value and giving a zero value, indicating that an x-intercept has been found. This is very handy to help with factorizing.	
$f(x) = x^2 + 3x - 10$	$g(x) = x^2 + 5x + 6$
$f(2) = (2)^2 + 3(2) - 10 = 0$	$g(-2) = (-2)^2 + 5(-2) + 6 = 0$
$f(-5) = (-5)^2 + 3(-5) - 10 = 0$	$g(-3) = (-3)^2 + 5(-3) + 6 = 0$
Factors have been found for $f(x)$	Factors have been found for $g(x)$
$f(x) = (x-2)(x+5)$	$g(x) = (x-2)(x-3)$

Polynomials, Quadratics or Parabolas have order of 2.

Standard Form, where a, b, c are real constants.	Turning Point form, where a, h, k are real constants.	
$f(x) = ax^2 + bx + c$	$f(x) = a(x-h)^2 + k$	
The y intercept is c and the x-intercept(s) $ax^2 + bx + c = 0$ can be found by the quadratic formula $x = \dfrac{-b \pm \sqrt{b^2 - 4ac}}{2a}$ where the axis of symmetry occurs at $x = \dfrac{-b}{2a}$	Turning point (h,k) is a minimum or a maximum The axis of symmetry goes through point (h,k). 	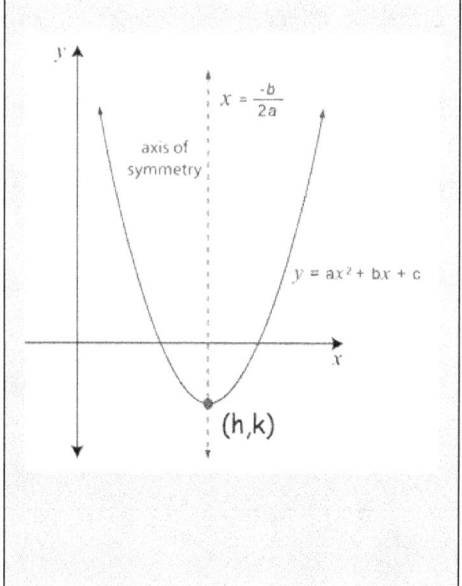

Some things you can do with a quadratic polynomial, include solving a quadratic intersecting with a line $ax^2 + bx + c = mx + k$, where a, b, c, m, k are real constants.

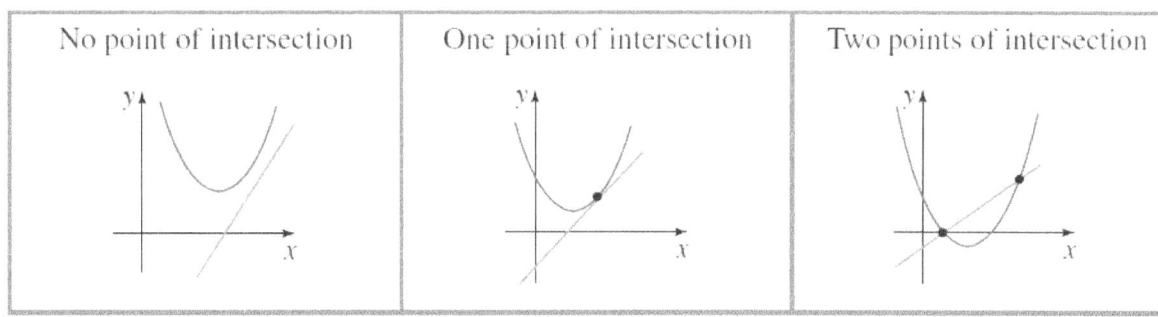

No point of intersection	One point of intersection	Two points of intersection

Solving a quadratic intersecting with a quadratic $ax^2 + bx + c = dx^2 + ex + f$, where a, b, c, d, e, f are real constants.

No points of intersection	One point of intersection	Two points of intersection

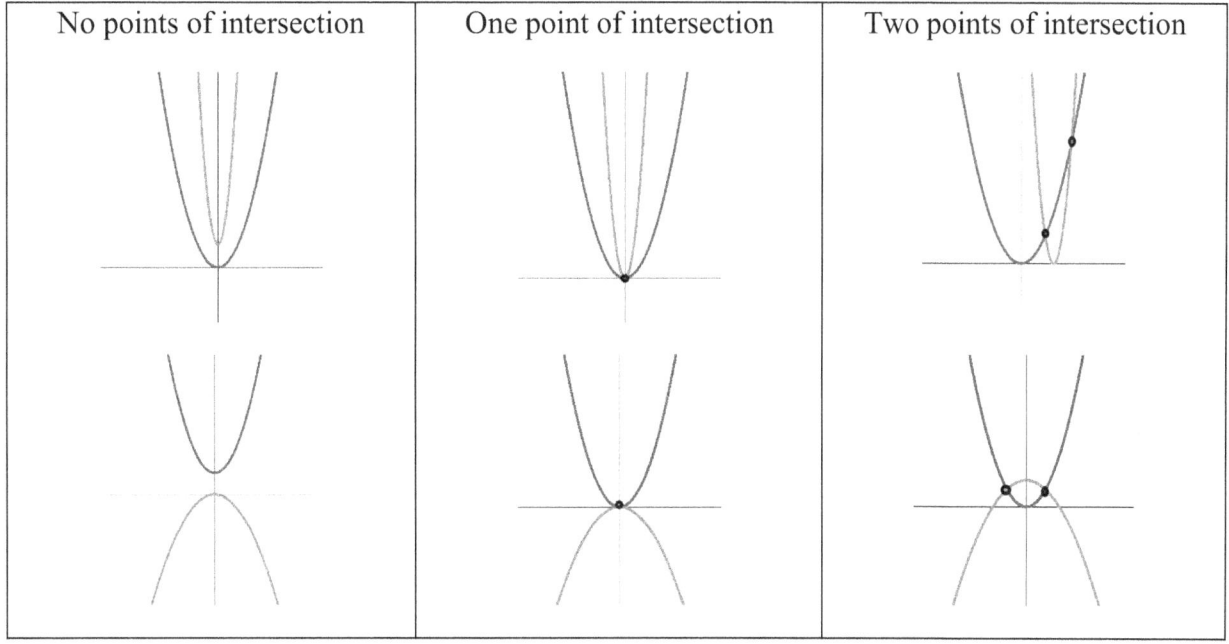

Cubic polynomials have an order of 3. The general cubic equation is $f(x) = ax^3 + bx^2 + cx + d$ and are very flexible.

Cubic graphs can have between 1 and 3 *x-intercepts* and between 0 and 2 stationary points.

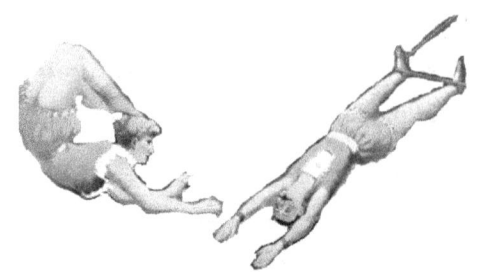

Cubics when the leading co-efficient a is positive. Extreme values as $x \to \infty, f(x) \to \infty$ and $x \to -\infty, f(x) \to -\infty$			
One *x-intercept*, zero stationary points. $f(x) = x^3 + x + 1$	One *x-intercept*, one stationary point. In turning point form. $f(x) = (x-2)^3 + 1$	Two *x-intercepts*, two stationary points. In repeated factor form. $f(x) = (x-2)^2(x+1)$	Three *x-intercepts*, two stationary points. Distinct factor form. $f(x) = (x-1)(x+1)(x+2)$
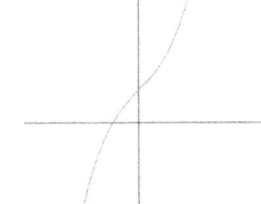		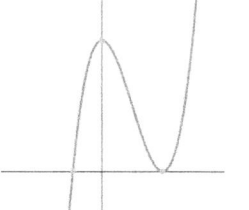	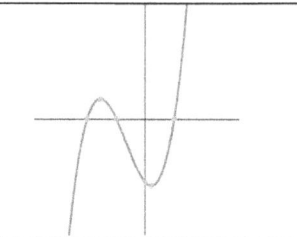

Cubics when the leading co-efficient a is negative. Extreme values as $x \to \infty, f(x) \to -\infty$ and $x \to -\infty, f(x) \to \infty$			
One *x-intercept*, zero stationary points. $f(x) = -x^3 - x - 1$	One *x-intercept*, one stationary point. In turning point form. $f(x) = -(x-2)^3 + 2$	Two *x-intercepts*, two stationary points. In repeated factor form. $f(x) = -(x-2)(x-1)^2$	Three *x-intercepts*, two stationary points. Distinct factor form. $f(x) = -\left(x - \dfrac{1}{2}\right)(x+1)(x-2)$
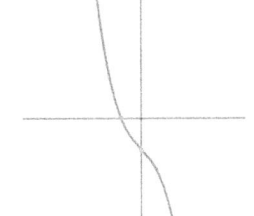	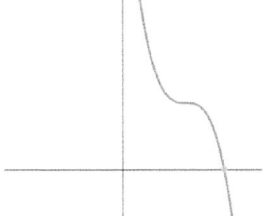	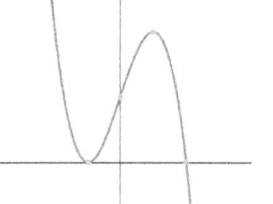	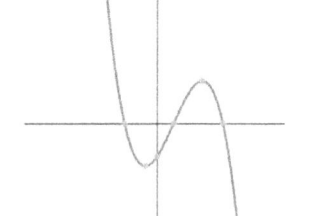

Example 3.2: Generate your own cubic. A cubic with 3 *x-intercepts*. There are an *infinite number of polynomials of degree 3 whose x-intercepts are -4, -2, and 3. They can be expressed in the form:*

$f(x) = (x+4)(x+2)(x-3)$

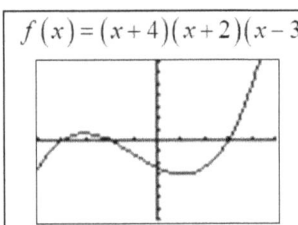

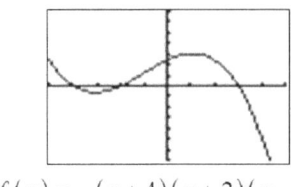

$f(x) = -(x+4)(x+2)(x-3)$

$f(x) = 2(x+4)(x+2)(x-3)$

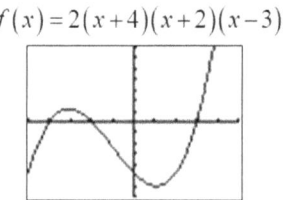

64

Even functions have one line of symmetry, are mirror images, are many to one functions.

$$f(x) + f(-x) = 2f(x)$$
$$f(x) - f(-x) = 0$$

Even polynomials, reflection in the y-axis, many x values map to one y value.

$$f(x) = x^2$$

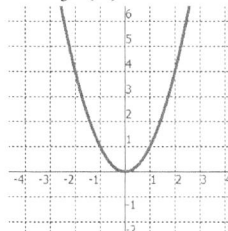

$$f(x) = -x^2$$

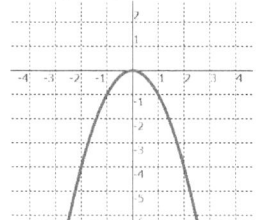

$$f(x) = x^4$$

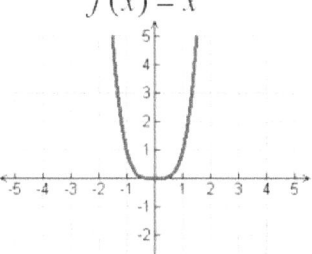

$$f(x) = -x^4$$

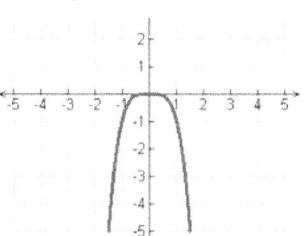

Odd functions have rotational symmetry, are one to one functions.

$$f(x) + f(-x) = 0$$
$$f(x) - f(-x) = 2f(x)$$

Odd polynomials, rotate the graph 180º, and it looks the same, one x value maps to one y-value.

$$f(x) = x^3$$

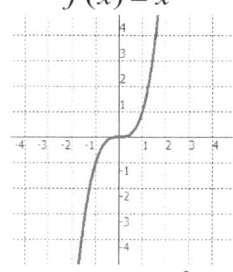

$$f(x) = -x^3$$

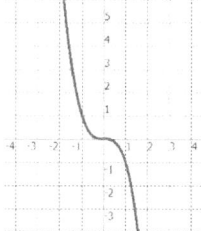

$$f(x) = x^5$$

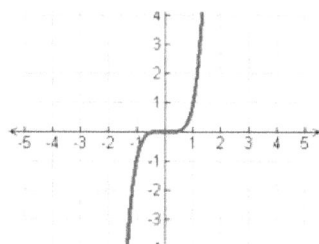

$$f(x) = -x^5$$

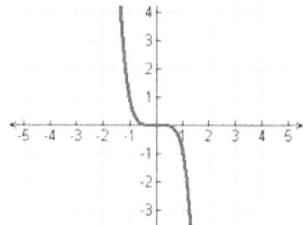

The quadratic formula is very handy for the factorization of quadratics, as is finding the x-intercepts by either guessing or factorizing as shown in Example 1.

For cubic and quartic factorization the **Remainder theorem** can be used to try and factorise if at least one *x-intercept* or factor can be determined.

Remainder theorem: Here's how it works. First get a polynomial function $P(x)$.

Determine an *x-intercept* $(a,0)$, where $P(a)=0$, expressed as a factor $(x-a)$ of the polynomial.

Divide $P(x)$ by $(x-a)$ you get a quotient function $Q(x)$ and a remainder function $R(x)$.

$$\frac{P(x)}{(x-a)} = Q(x) + \frac{R(x)}{(x-a)}$$

Rearranging the algebra gives,

$$P(x) = (x-a)Q(x) + R(x)$$

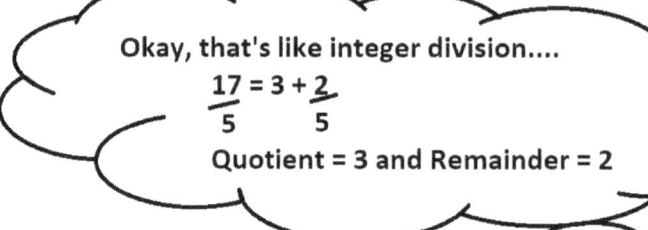

Okay, that's like integer division....

$$\frac{17}{5} = 3 + \frac{2}{5}$$

Quotient = 3 and Remainder = 2

Substitute $x=a$ into $P(x)$ gives
$$P(a) = (a-a)Q(x) + R(a)$$
$$P(a) = R(a)$$

Recall that $R(a)$ is the remainder, so if $P(a) = R(a) = 0$ then $(x-a)$ is a factor of $P(x)$, because the remainder is equal to zero.

What do you do when you find a factor of a cubic or quartic using the remainder theorem? If you have to continue using algebra you can use polynomial long division to find the quotient and repeat the remainder theorem again or factorise by recognition, or by using the quadratic formula if the quotient is a quadratic.

Polynomial long division is a similar technique to numeric long division, where remainders are carried over for further division.

$$
\begin{array}{r}
32 \\
12\overline{)384} \\
-36 \\
\hline
24 \\
-24 \\
\hline
0
\end{array}
\qquad
\begin{array}{r}
30+2 \\
10+2\overline{)300+80+4} \\
-(300+60) \\
\hline
20+4 \\
-(20+4) \\
\hline
0
\end{array}
\qquad
\begin{array}{r}
3x+2 \\
x+2\overline{)3x^2+8x+4} \\
-(3x^2+6x) \\
\hline
2x+4 \\
-(2x+4) \\
\hline
0
\end{array}
$$

Example 3.3: Polynomial long division with a factor results in a zero remainder.

$$\begin{array}{r} x+3 \\ x+7{\overline{\smash{\big)}\,x^2+10x+21}} \\ \underline{x^2+7x} \\ 3x+21 \\ \underline{3x+21} \\ 0 \end{array}$$

$$P(x)=x^2+10x+21$$
$$=(x+7)(x+3)$$

Notice that: $\begin{array}{l}P(-3)=0\\P(-7)=0\end{array}$

$$\begin{array}{r} x^2+6x+7 \\ x-2{\overline{\smash{\big)}\,x^3+4x^2-5x-14}} \\ \underline{-x^3+2x^2} \\ 6x^2-5x \\ \underline{-6x^2+12x} \\ 7x-14 \\ \underline{-7x+14} \\ 0 \end{array}$$

$$P(x)=x^3+4x^2-5x+14$$
$$=(x-2)(x^2+6x+7)$$

Notice that: $P(2)=0$

$$\begin{array}{r} 6x^2-7x+2 \\ x-2{\overline{\smash{\big)}\,6x^3-19x^2+16x-4}} \\ \underline{6x^3-12x^2} \\ -7x^2+16x \\ \underline{-7x^2+14x} \\ 2x-4 \\ \underline{2x-4} \\ 0 \end{array}$$

$$P(x)=6x^3-19x^2+16x-4$$
$$=(x-2)(6x^2-7x+2)$$

Notice that: $P(2)=0$

Polynomial long division with non-factors results in a non-zero remainder.

$$\begin{array}{r} x+3 \\ x-1{\overline{\smash{\big)}\,x^2+2x+6}} \\ \underline{-x^2+x} \\ 3x+6 \\ \underline{-3x+3} \\ 9 \end{array}$$

$$P(x)=x^2+2x+6$$
$$=(x-1)(x+3)+9$$

Notice that: $P(1)=9$

$$\begin{array}{r} X^2+2X+2 \\ X-1{\overline{\smash{\big)}\,X^3+X^2\quad-1}} \\ \underline{-X^3+X^2} \\ 2X^2 \\ \underline{-2X^2+2X} \\ 2X-1 \\ \underline{-2X+2} \\ 1 \end{array}$$

$$P(x)=x^3+x^2-1$$
$$=(x-1)(x^2+2x+2)+1$$

Notice that: $P(1)=1$

$$\begin{array}{r} x^2-6x+14 \\ x+2{\overline{\smash{\big)}\,x^3-4x^2+2x-3}} \\ \underline{x^3+2x^2} \\ -6x^2+2x \\ \underline{-6x^2-12x} \\ 14x-3 \\ \underline{14x+28} \\ -31 \end{array}$$

$$P(x)=x^3-4x^2+2x-3$$
$$=(x+2)(x^2-6x+14)-31$$

Notice that: $P(-2)=-31$

67

Act 1, Scene 1. The drama of functions with negative integer powers. That is $f(x) = A(x-b)^{-1} + c$

My share of the birthday cake is inversely proportional to the number of people at the party.

Too, true. The more people, the less cake for each.

Speed and travel time are also **Inversely Proportional**.

As our speed goes up, our travel time goes down and we get there earlier.
If our speed does down, our travel time goes up, and we get there later.

Renaissance

Figures.

When one quantity y is **Inversely Proportional** to another quantity x, this is a relationship between two variables in which the product is a constant, for example xy=k, where k is a constant.

When one variable increases the other decreases in proportion so that the product is always equal to the constant value.

In the equation $y = \dfrac{100}{x}$, y is inversely proportional to x. Doubling x causes y to halve.

The product of x and y is always 100. Some *(x,y)* coordinates for this equation are: {(1,100), (2,50), (4,25), (5,20), (10,10), (20,5), (25,4), (50,2), (100,1)}

Notice we cannot get a constant product if one of the variables is equal to zero.

The function $f(x) = A(x-b)^{-1} + c$ is called a hyperbola.

The wine tankard served in the tavern which is in the shape of a cylinder must always hold 250ml, according to the law. This constant means that the height of the tankard is **Inversely Proportional** to radius squared.

The volume of a cylinder is given by the formula: $V = \pi r^2 h$

If the volume V remains constant, since π is also a constant then we get the equation: $\dfrac{V}{\pi} = r^2 h$. Let $\dfrac{V}{\pi} = k$ a constant.

The height h is inversely proportional to the radius r squared. $h \propto \dfrac{1}{r^2} \Rightarrow hr^2 = k$.

Which leads to various shaped cylindrical tankards that hold the same volume. $h(r) = \dfrac{k}{r^2}$

Power functions with negative integer exponents.

The hyperbola function.
$f(x) = A(x-b)^{-1} + c$

The truncus function.
$f(x) = A(x-b)^{-2} + c$

Be careful not to try and divide by zero.

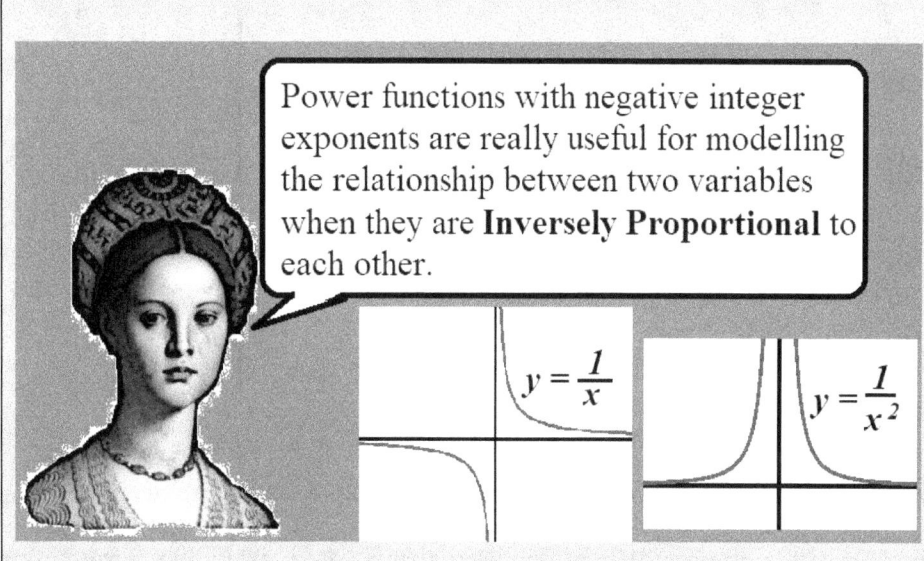

Power functions with negative integer exponents are really useful for modelling the relationship between two variables when they are **Inversely Proportional** to each other.

$y = \dfrac{1}{x}$

$y = \dfrac{1}{x^2}$

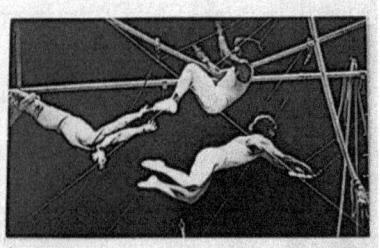

Negative power functions

The power or exponent is a negative integer.

- Hyperbola functions $f(x) = A\big(n(x-b)\big)^{-1} + c$

- Truncus functions $f(x) = A\big(n(x-b)\big)^{-2} + c$

- General negative power functions $f(x) = A\big(n(x-b)\big)^{-p} + c$

You can't divide by zero, the result is said to be undefined. When graphing a function there may be values for which the function is undefined.

An asymptote is an equation that represents a line on the Cartesian plane where a function is undefined and cannot exist.

The distance between the graph and the asymptote gets smaller and smaller, approaching zero for extreme values of x and or y.

Example 3.4: Graphing Negative power functions with asymptotes.

Graphing $f(x) = \dfrac{1}{x-2}$

When $x = 2$, this function will be undefined. The line $x = 2$ is an asymptote for this function.

There are no real values of x that will satisfy $f(x) = 0 \Rightarrow \dfrac{1}{x-2} = 0 \Rightarrow 1 = 0$. The line $y = 0$ is also an asymptote for this graph.

We need to explore the values near the asymptotes.
Exploring around the value $x=2$.

$$f(2.001) = \frac{1}{2.001-2} = 1000$$

$$f(1.999) = \frac{1}{1.999-2} = -1000$$

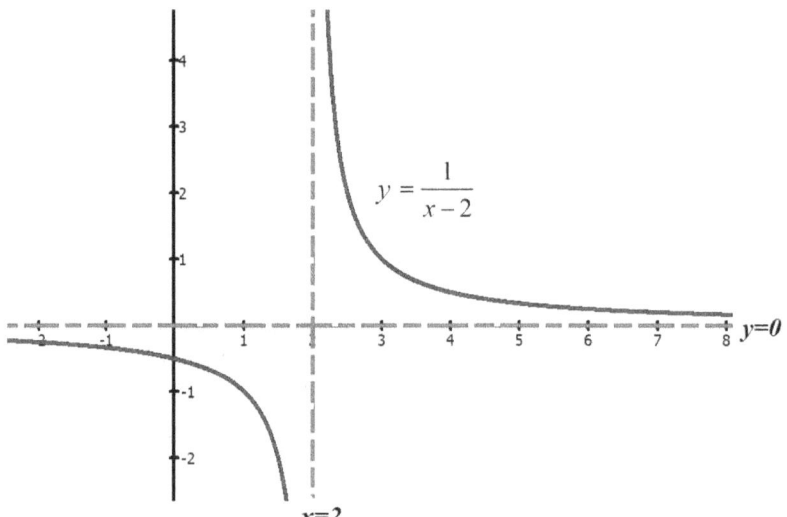

In general terms, a tiny bit more than 2 can be represented as 2^+ , and a tiny bit less than 2 can be represented as 2^- .
As $x \to 2^+, f(x) \to \infty$.
As $x \to 2^-, f(x) \to -\infty$

Extreme denominator exploration. As $x \to \infty, f(x) \to 0^+$. As $x \to -\infty, f(x) \to 0^-$.

So part of the graph will approach zero, that is $y = 0$ from above the x-axis, and part of the graph will approach zero from below the x-axis.

Quotient power functions,

The power is a quotient or fraction.

- Square root functions $f(x) = A\left(n(x-b)\right)^{\frac{1}{2}} + c$

- General quotient power functions $f(x) = A\left(n(x-b)\right)^{\frac{p}{q}} + c$

In the real number system you cannot find an "even" root of a negative number, the result is said to be undefined. When graphing a function there may be values for which the function is undefined.

Are fractions and decimals odd or even? An interesting question for those Mathematical philosophers amongst us. The division of two whole numbers does not necessarily result in a whole number. For example, 1 divided by 4 equals 1/4, which is neither even nor odd, since the concepts even and odd apply only to integers. Or do they?

When working with quotient power functions $f(x) = A\left(n(x-b)\right)^{\frac{p}{q}} + c$ we need to note details about the quotient $\frac{p}{q}$.

- If the denominator q is even, we cannot find real roots for negative values, and if q is odd then roots exist across the whole real number system.
- If the numerator p is even, negative values to even powers become positive, and if p is odd then values to odd powers retain their original sign.

Example 3.5: Graphing Quotient power functions with domain and range restrictions.	
$f(x) = (5-x)^{\frac{1}{2}}$ and $g(x) = (5-x)^{\frac{1}{3}}$ $f(x) = \sqrt{5-x}$ $\quad g(x) = \sqrt[3]{5-x}$ note the quotient powers When $x > 5$, the function $f(x)$ will be undefined, as even roots of negative numbers are undefined in the real number system. The point $x = 5$ is an endpoint for the function $f(x)$, and needs to be acknowledged and labelled with its full (x, y) coordinate.	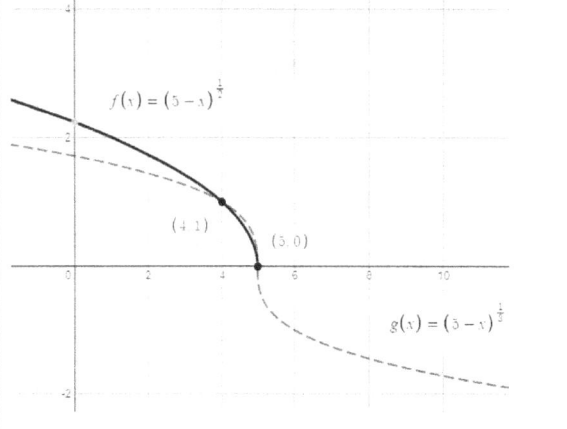
$f(x) = (5-x)^{\frac{3}{2}}$ and $g(x) = (5-x)^{\frac{4}{5}}$ note the quotient powers. When $x > 5$, the function $f(x)$ will be undefined, the point $x = 5$ is an endpoint for the function $f(x)$, and is labelled with its full (x, y) coordinate. Since the numerator of the quotient power of $g(x)$ is even, all negative values are raised to an even power resulting in positive results.	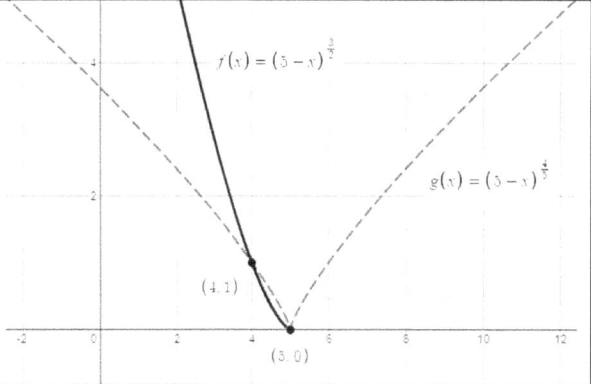

Act 2, Scene 1. The drama of finding solutions for $f(x) = 0$ **without technology.**

Mathematics has a *numerical* method called the **Bisection Method** for solving equations of the type *f(x)=0*.

We can use it to find *x-intercepts* for functions such as polynomials that we cannot easily factorise.

It's looking for the *x-intercept* by repeatedly bisecting (cutting in half) intervals intelligently until an *x-intercept* is found.

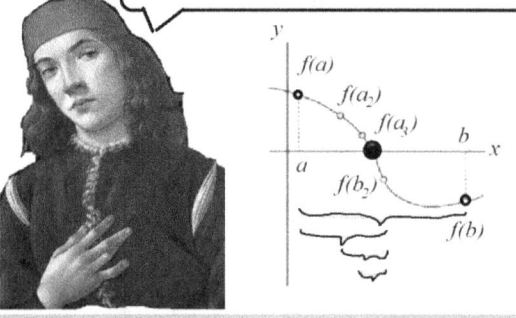

Renaissance

Figures.

If you have a function *y=f(x)* which is continuous on an interval *[a,b]*, the Bisection method to find a solution will work if you can find an interval where the function changes signs.

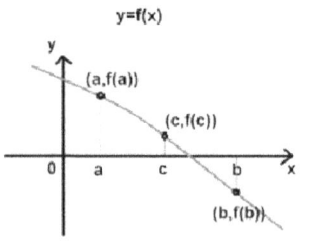

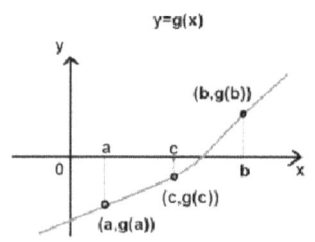

If $f(a) < f(b)$ then the function changes sign in the interval *(a,b)*, so an x-intercept exists for $x \in [a,b]$

If $f(b) < f(a)$ then the function changes sign in the interval *(a,b)*, so an x-intercept exists for $x \in [a,b]$

Act 2, Scene 2. The drama of finding solutions for $f(x) = 0$ **without technology..**

1. if *f(a)f(b) < 0*, halve the interval *[a,c],[c,b]*
2. if *f(a)f(c) < 0*, then function changes sign in the interval *[a,c]*, so an x-intercept exists for *x ∈ [a,c]*
3. if *f(c)f(b) < 0*, then function changes sign in the interval *[c,b]*, so an x-intercept exists for *x ∈ [c,b]*
4. Repeat the Bisection steps from step 1. until the level of accuracy has been reached

Renaissance

Figures.

An Example.

Let us use the Bisection Method to find the x-intercepts for *f(x)=x³+x-1*, that is solve *f(x)=0* for *x*.

Since *f(0)=-1* and *f(1)=1* , the *x-intercept* is between *x=0* and *x=1*. This will be the starting interval for the Bisection method to begin.

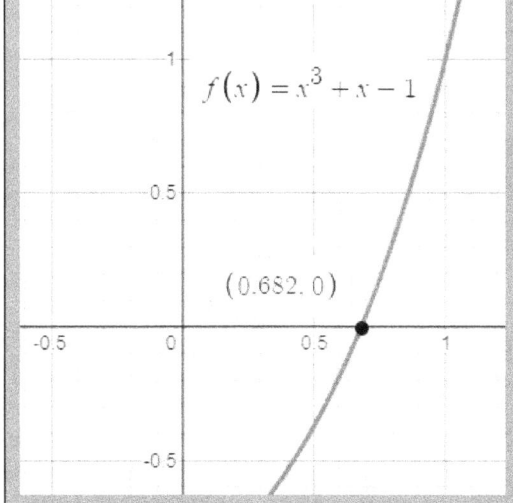

$f(x) = x^3 + x - 1$

$(0.682, 0)$

$f(0)f(1) < 0$, halve the interval *[0,1]* giving *c=0.5*

$f(0.5)f(1) < 0$, halve the interval *[0.5,1]* giving *c=0.75*

$f(0.5)f(0.75) < 0$, halve the interval *[0.5,0.75]* giving *c=0.625*

$f(0.625)f(0.75) < 0$, halve the interval *[0.625,0.75]* giving *c=0.6875*

$f(0.6875)f(0.75) > 0$, halve the interval *[0.625,0.6875]* giving *c=0.6563*

Continue the Bisection method until the required accuracy is met.

SUPER POWER FUNCTIONS SUMMARY:

Polynomial Functions

All the powers of x are positive integers.
- Linear functions $f(x) = mx + c$
- Quadratic functions $f(x) = ax^2 + bx + c$ in turning point form $f(x) = a(x-h)^2 + k$.
- Cubic functions $f(x) = ax^3 + bx^2 + cx + d$ in turning point form $f(x) = a(x-h)^3 + k$.
- Quartic functions $f(x) = ax^4 + bx^3 + cx^2 + dx + e$ in turning point form $f(x) = a(x-h)^4 + k$.

Null Factor Theorem

$$\frac{P(x)}{(x-a)} = Q(x) + \frac{R(x)}{(x-a)}$$

Rearranging the algebra gives factors,

$$P(x) = (x-a)Q(x) + R(x)$$

Negative Power Functions

The power or exponent is a negative integer.
- Hyperbola function: $f(x) = A\left(n(x-b)\right)^{-1} + c$
- Truncus function: $f(x) = A\left(n(x-b)\right)^{-2} + c$

Asymptotes are equations that represent places that the function does not exist, and can be determined by noting when a division by zero will occur and exploring extreme values.
As $x \to b^+, f(x) \to \infty$. As $x \to b^-, f(x) \to -\infty$
Extreme denominator exploration.
As $x \to \infty, f(x) \to 0^+$. As $x \to -\infty, f(x) \to 0^-$.

Quotient Power Functions

The power is a quotient or fraction.
- Square root: $f(x) = A\left(n(x-b)\right)^{1/2} + c$
- Cube root: $f(x) = A\left(n(x-b)\right)^{1/3} + c$
- General quotient power functions
$$f(x) = A\left(n(x-b)\right)^{p/q} + c$$

For even quotient powers when $x > b$. the function is undefined.
Inspect the quotient p/q.
- q is even, we cannot find real roots for negative values
- p is even, negative values to even powers become positive values.

Numerical Methods

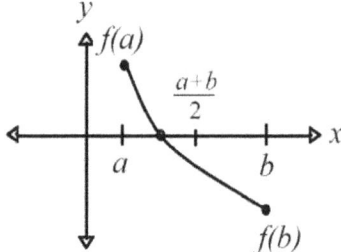

The Bisection Method is a repetitive approximation method that narrows down an interval that contains a solution to $f(x) = 0$.

- begins by guessing an initial interval $[a,b]$ that contains a solution for x, using the properties of the sign of the function $f(a)$ and $f(b)$.
- cuts the interval into 2 halves and check which half interval contains the solution by inspecting the sign of the function again.
- keep cutting the smaller interval in halves until the resulting interval is extremely small, the solution is then approximately equal to any value in the final (very small) interval.

SUPER POWER FUNCTIONS: Check your Understanding.

3.1 Explore the Polynomial functions.

a) Consider the following cubic graphs.

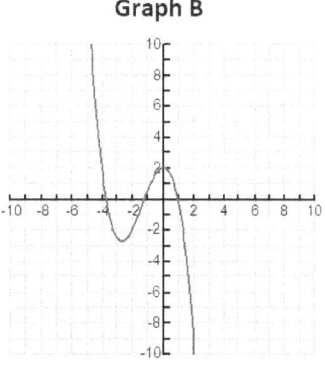

- i. How can you check to see if both graphs are functions?
- ii. How many x-intercepts do graphs A & B have?
- iii. What is the extreme value behavior as $x \to \pm\infty$ for each graph?
- iv. Which graph do you think has a positive leading coefficient? Why?
- v. Which graph do you think has a negative leading coefficient? Why?

b) Consider the following quartic graphs.

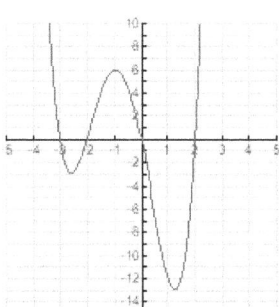 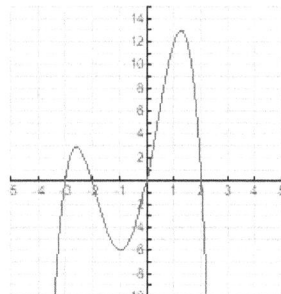

- i. How can you check to see if both graphs are functions?
- ii. How many x-intercepts do graphs A & B have?
- iii. What is the extreme value behavior as $x \to \pm\infty$ for each graph?
- iv. Which graph do you think has a positive leading coefficient? Why?
- v. Which graph do you think has a negative leading coefficient? Why?

c) A bit of polynomial long division. State the quotient and remainder of the following algebraic divisions.

    (i) $\dfrac{x^2 + 9x + 14}{x + 7}$ (ii) $\dfrac{3x^3 - 5x^2 + 10x - 3}{3x + 1}$

3.1 Explore the Polynomial functions. (continued)

d) Fun with transforming cubics. Fill in the blanks.

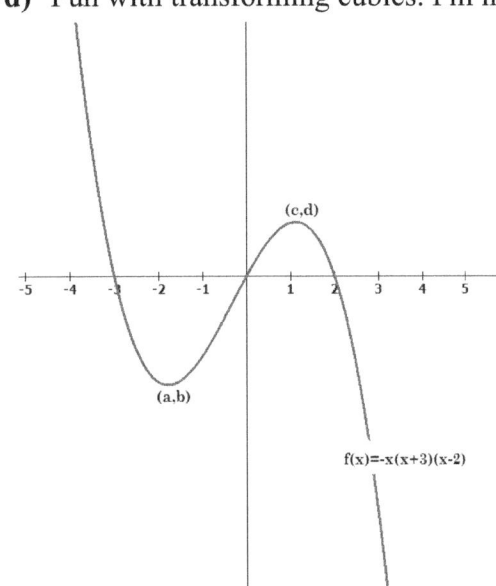

(c,d)

(a,b)

$f(x)=-x(x+3)(x-2)$

With x-intercepts at $\{-3,0,2\}$, the equation of the cubic shown could be:

$$f(x) = -(x+3)(x-0)(x-2)$$

Which simplifies to: $f(x) = -x(x+3)(x-2)$

A negative cubic: as $x \to \infty,\ y \to -\infty$

Stationary points:

minimum at (a, b), maximum at (c, d)

Starting with $f(x)$

- perform the transformations listed in i, ii, iii, iv and fill in the blanks below.
- complete information about changes to stationary points and extreme values.
- Sketch each graph for i, ii, iii, iv. Use your Technology to confirm the shape.

---

i.   After a dilation from x-axis by factor of 3, $y = 3f(x)$
  minimum (a, b) maps to point (,   )
  maximum (c, d) maps to point (,   )
  Extremes: as $x \to \infty,\ y \to -\infty$, as $x \to -\infty,\ y \to$

---

ii.   After a reflection in x-axis , $y = -f(x)$
  minimum (a, b) maps to point (,   )
  maximum (c, d) maps to point (,   )
  Extremes: as $x \to \infty,\ y \to$, as $x \to -\infty,\ y \to$

---

iii.   After a dilation from y-axis by a factor of 0.5, $y = f(2)$
  minimum (a, b) maps to point (,   )
  maximum (c, d) maps to point (,   )
  Extremes: as $x \to \infty,\ y \to$, as $x \to -\infty,\ y \to$

---

iv.   After a reflection in the y-axis $y = f(-x)$
  minimum (a, b) maps to point (,   )
  maximum (c, d) maps to point (,   )
  Extremes: as $x \to \infty,\ y \to$, as $x \to -\infty,\ y \to$

---

e) Find the intersections between the given cubic and quadratic.

$$f(x) = (x-2)(x+2)(x+5) \quad and \quad g(x) = \left(\frac{x}{2}\right)(x-2)$$

3.2 Explore the power functions for negative and quotient powers.

a) Negative powers. Find out if there is there a pattern for odd or even powers?

b) Quotient powers p/q. Explore patterns for p/q < 1
    i.   when p is even and q is odd
    ii.   when p is odd and q is even
    iii. when p is odd and q is odd.

c) Quotient powers p/q. Explore patterns for p/q > 1:
    i.   when p is even and q is odd
    ii.   when p is odd and q is even
    iii. when p is odd and q is odd.

d)

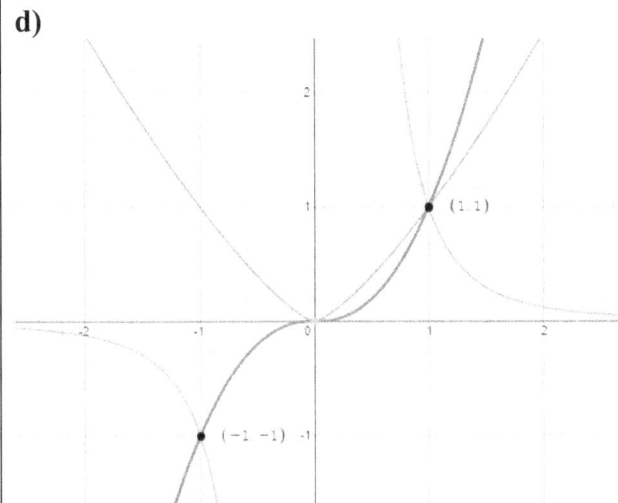

Pick the power functions when y=f(x) shown in the diagram.

$$f(x) = x^{4/3}$$

$$f(x) = x^{7/3}$$

$$f(x) = x^{-3}$$

3.3 Exploring the Power Functions for domain and range.

a) Using the points $x = \{0, 1, 4, 9\}$ find and plot the *(x,y)* coordinates for the function $f(x) = \sqrt{x} + 3$.

b) Use the resulting graph to determine the domain and range of this function.

c) Using the points $x = \{$ -2,-1,0, 1, 2$\}$ find and plot the *(x,y)* coordinates for the function $f(x) = x^3 - 3$

d) Use the resulting graph to determine the domain and range of this function.

e) Using the points $x = \{-4, -3, -2, -1, 0, 1, 2, 3, 4\}$ find and plot the *(x,y)* coordinates for the function

$$f(x) = \frac{2}{x - 3}$$

f) Use the resulting graph to determine the asymptotes, the domain and range.

g) Using the points $x = \{-4, -3, -2, -1, \frac{-1}{2}, 0, \frac{1}{2}, 1, 2, 3, 4\}$ find and plot the *(x,y)* coordinates for the

    function $f(x) = \frac{x^2}{x^2 - 1}$

h) Express $f(x) = \frac{x^2}{x^2 - 1}$ as $f(x) = A + \frac{B}{x^2 - 1}$

i) Use the resulting graph to determine the asymptotes, the domain and range.

3.4 Modelling with Power Functions.

a) Boyle's Law is an ideal gas law where at constant temperature, the volume of an ideal gas is inversely proportional to its absolute pressure. There are couple of ways of expressing the law as an equation. The most basic is: $PV = k$, or $V = \dfrac{k}{P}$.

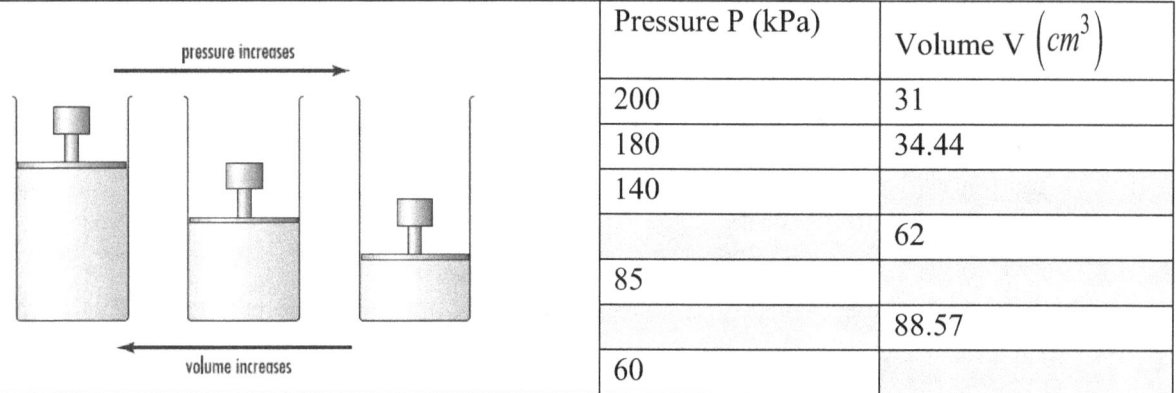

Pressure P (kPa)	Volume V $\left(cm^3 \right)$
200	31
180	34.44
140	
	62
85	
	88.57
60	

   i.   Complete the table for Boyle's Law. State the value of k for this particular ideal gas.
   ii.   Plot the points with Volume on the vertical axis and Pressure on the horizontal axis.
   iii.   Is it possible to have zero volume and infinite pressure in this model?

b) Numerical methods for solving equations. Use the Bisection method, find the approximation of $f(x) = 0$ for the cubic function which is defined by $f(x) = \dfrac{1}{2}x^3 - x^2 - 1$ using an accuracy of 2 decimal places, and starting in the interval $x \in [2,4]$.

c)

The population of a penguin colony is studied for many years to determine if they are under stress from overfishing.

The zoologist has recorded the following population information since the beginning of the observations.

Year	0	1	2	3	4	5	6	7	8	9	10
Population	250	270	310	375	410	395	335	290	290	320	325

   i.   Plot the points on a Cartesian plane.
   ii.   Can you fit a quadratic function to this information? Estimate the 11th year population using the quadratic function found.
   iii.   Can you fit a cubic function to this information? Estimate the 11th year population using the cubic function found.
   iv.   Can you fit a quartic function to this information? Estimate the 11th year population using the quartic function found.
   v.   Which type of function best fits the information? Can you justify your choice?

SUPER POWER FUNCTIONS SOLUTIONS.

3.1 Explore the Polynomial functions. (SOLUTIONS)

a)

i. How can you check to see if both graphs are functions?
Use the Vertical line test.

ii. How many x-intercepts do graphs A & B have?
Each has 3 x-intercepts.

iii. What is the extreme value behavior as $x \to \pm\infty$ for each graph?
Graph A: $as\ x \to -\infty, y \to -\infty\ as\ x \to \infty, y \to \infty$

Graph B: $as\ x \to -\infty, y \to \infty\ as\ x \to \infty, y \to -\infty$

iv. Which graph do you think has a positive leading coefficient? Why?
Graph A is a positive cubic due to the answer in part iii.

v. Which graph do you think has a negative leading coefficient? Why?
Graph B is a negative cubic due to the answer in part iii.

b)

i. How can you check to see if both graphs are functions?
Use the Vertical line test.

ii. How many x-intercepts do graphs A & B have?
Each has 4 x-intercepts.

iii. What is the extreme value behavior as $x \to \pm\infty$ for each graph?
Graph A: $as\ x \to -\infty, y \to \infty\ as\ x \to \infty, y \to \infty$

Graph B: $as\ x \to -\infty, y \to -\infty\ as\ x \to \infty, y \to -\infty$

iv. Which graph do you think has a positive leading coefficient? Why?
Graph A is a positive quartic due to the answer in part iii.

v. Which graph do you think has a negative leading coefficient? Why?
Graph B is a negative quartic due to the answer in part iii.

c)

$$x+7\overline{)\begin{array}{r} x+2 \\ x^2+9x+14 \end{array}}$$

$$\underline{^-x^2\,^-7x}$$

$$2x+14$$

$$\underline{^-2x^-14}$$

$$0$$

$$\frac{x^2+9x+14}{x+7} = x+2$$

$$3x+1\overline{)\begin{array}{r} x^2 -2x +4 \\ 3x^3 -5x^2 +10x -3 \end{array}}$$

$$\underline{^-3x^3\,^-1x^2}$$

$$-6x^2 +10x -3$$

$$\underline{^+6x^2 +\ 2x}$$

$$12x-3$$

$$\underline{^-12x^+4}$$

$$-7$$

$$\frac{3x^3-5x^2+10x-3}{3x+1} = x^2-2x+4-\frac{7}{3x+1}$$

d)

(i) Dilation from x-axis by factor of 3, $y = 3f(x)$

$y = -3x(x + 3)(x - 2)$

Intercepts unchanged.

(a, b) *maps to point* $(a, 3b)$

(c, d) *maps to point* $(c, 3d)$

Extremes: as $x \to \infty$, $y \to -\infty$, as $x \to -\infty$, $y \to -\infty$

Negative cubic

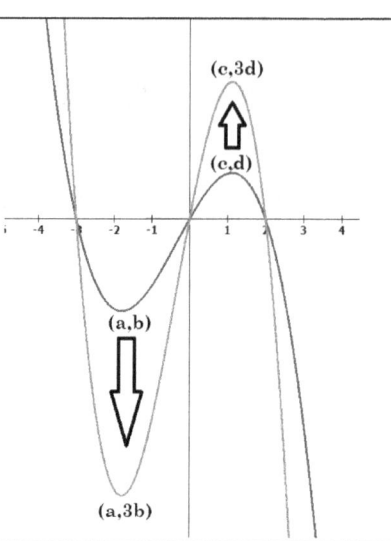

(ii)   Reflection in x-axis , $y = -f(x)$

$y = x(x + 3)(x - 2)$

x-intercepts at $\{-3, 0, 2\}$

minimum (a, b) *maps to* $(a, -b)$ *is now a maximum*

maximum (c, d) *maps to* $(c, -d)$ *is now a minimum*

Extremes: as $x \to \infty$, $y \to \infty$, as $x \to -\infty$, $y \to -\infty$

Positive cubic

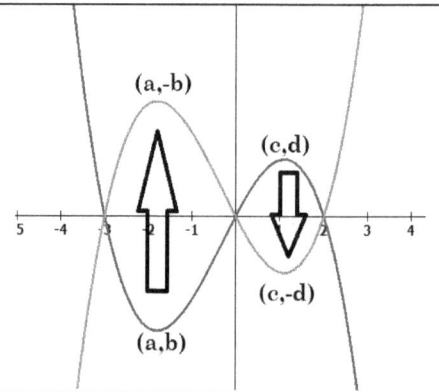

(iii)   Dilation from y-axis by a factor of 0.5 $y = f(2)$

$y = -(2x)(2x + 3)(2x - 2)$

x-intercepts at $\{-\frac{3}{2}, 0, 1\}$

minimum (a, b) *maps to* $\left(\frac{a}{2}, b\right)$

maximum (c, d) *maps to* $\left(\frac{c}{2}, d\right)$

Extremes: as $x \to \infty$, $y \to -\infty$, as $x \to -\infty$, $y \to -\infty$

Negative cubic

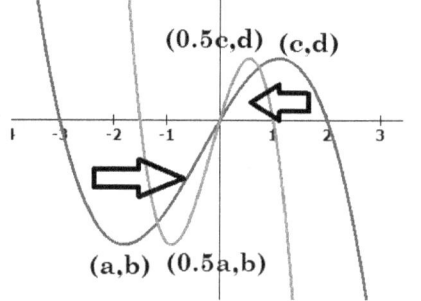

(iv)   Reflection in the y-axis $y = f(-x)$

$y = -- x(-x + 3)(-x - 2)$

$y = x(-)(x - 3)(-)(x + 2)$

$y = x(x - 3)(x + 2)$

x-intercepts at $\{-2, 0, 3\}$

minimum (a, b) *maps to* $(-a, b)$

maximum (c, d) *maps to* $(-c, d))$

Extremes: as $x \to \infty$, $y \to \infty$, as $x \to -\infty$, $y \to -\infty$

Positive cubic

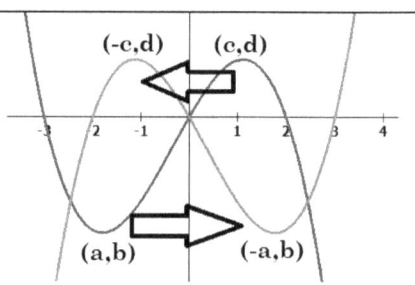

e) Find the intersections between the given cubic and quadratic.

$$f(x) = g(x)$$

$$(x-2)(x+2)(x+5) = \left(\frac{x}{2}\right)(x-2)$$

$$(x+2)(x+5) = \left(\frac{x}{2}\right)$$

$$x^2 + 7x + 10 = \left(\frac{x}{2}\right)$$

$$2x^2 + 14x + 20 = x$$

$$2x^2 + 13x + 20 = 0$$

quadratic formula $x = \dfrac{-b \pm \sqrt{b^2 - 4ac}}{2a}$

$$x = \frac{-13 \pm \sqrt{(13)^2 - 4(2)(20)}}{2(2)}$$

$$= \frac{-13 \pm \sqrt{169 - 160}}{4}$$

$$x = \frac{-13 - 3}{4} = -4,$$

$$or \ x = \frac{-13 + 3}{4} = -2.5$$

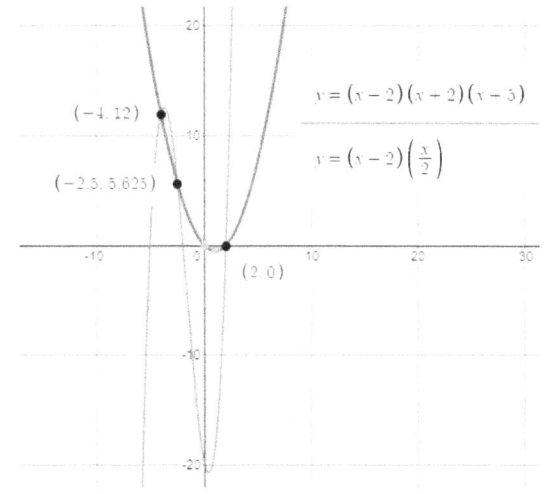

Substitute x values into $g(x)$ since it's simpler to find the y coordinates.

$$g(-4) = \left(\frac{-4}{2}\right)(-4-2)$$

$$= 12$$

$$\Rightarrow (-4, 12)$$

$$g(-2.5) = \left(\frac{-2.5}{2}\right)(-2.5-2)$$

$$= 5.625$$

$$\Rightarrow (-2.5, 5.625)$$

Classical Parabolas

3.2 Explore the power functions for negative and quotient powers. (SOLUTIONS)

a) Negative powers.

Odd negative powers		Even negative powers		Conclusions
-1	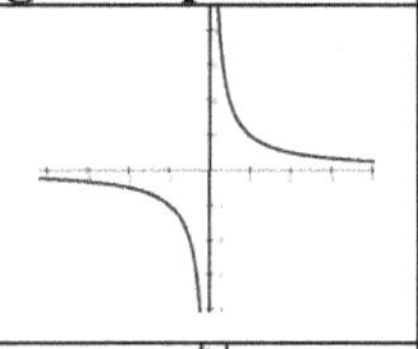 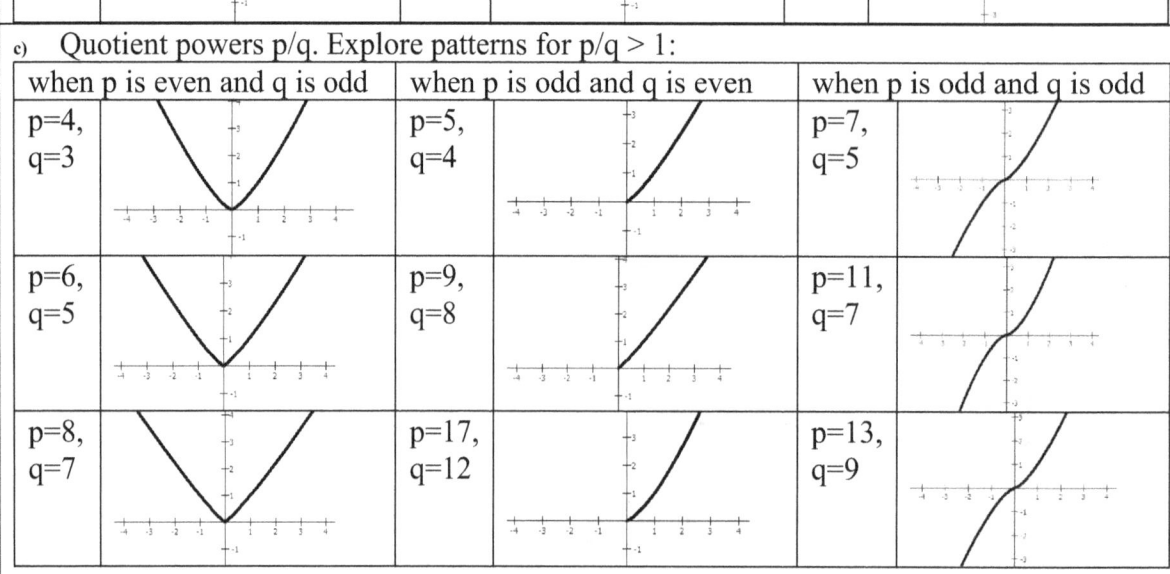	-2	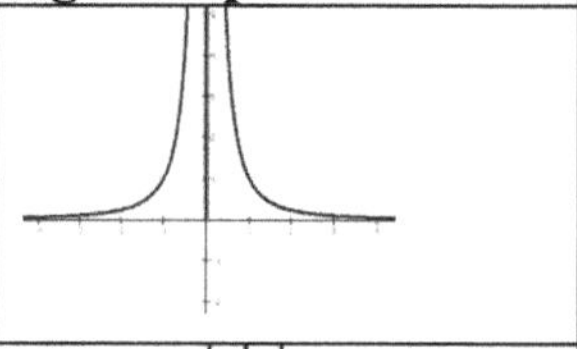	In summary it appears that:
-3	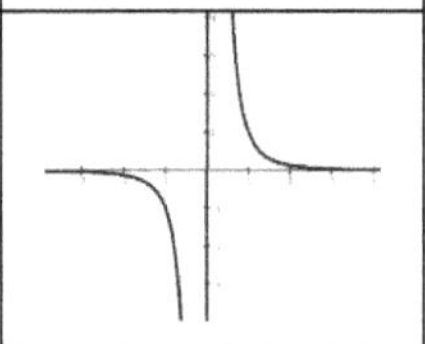	-4	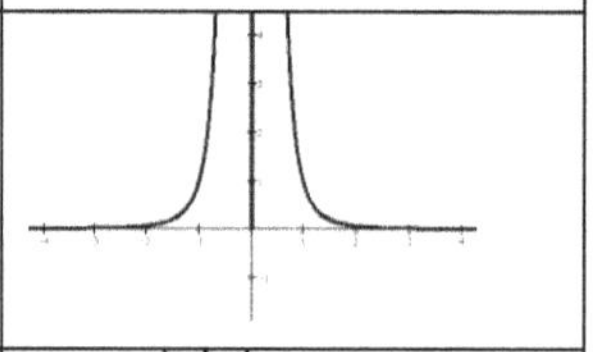	• Odd negative powers produce hyperbolic odd functions, which when reflected in the x-axis, followed by a reflection in the y-axis look exactly the same as the original function.
-5	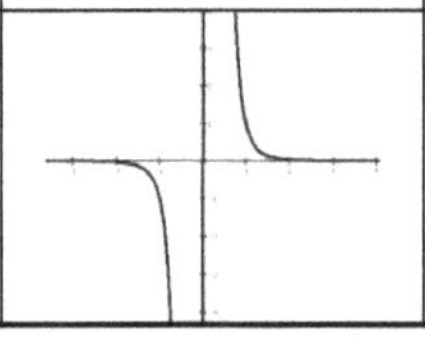	-6	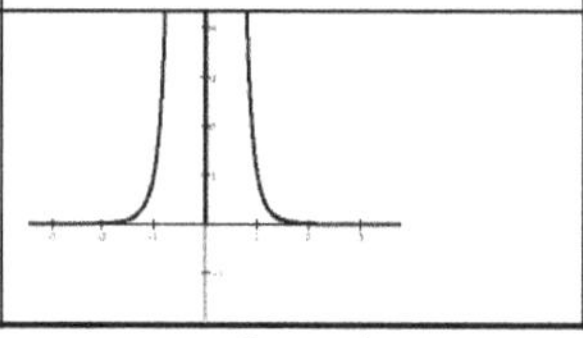	• Even negative powers produce even functions, which when reflected in the y-axis look exactly the same as the original function.

b) Quotient powers p/q. Explore patterns for p/q < 1

when p is even and q is odd		when p is odd and q is even		when p is odd and q is odd	
p=2, q=3	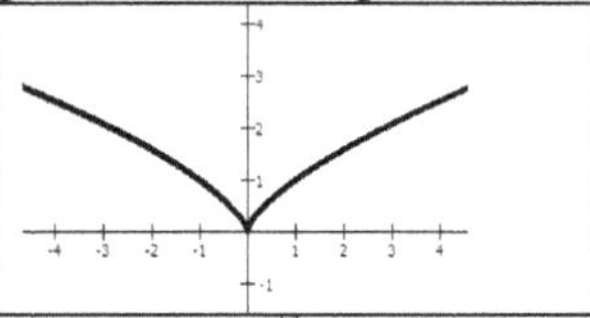	p=3, q=4	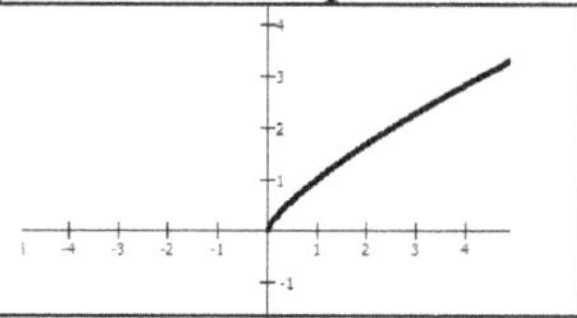	p=3, q=5	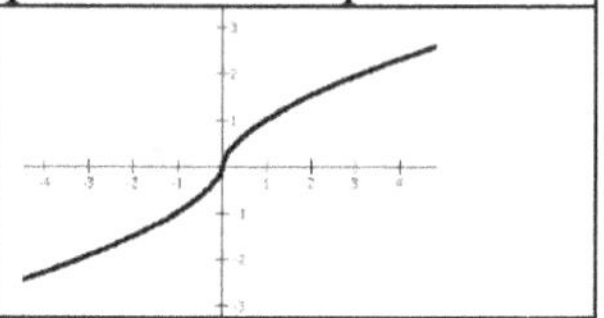
p=2, q=5	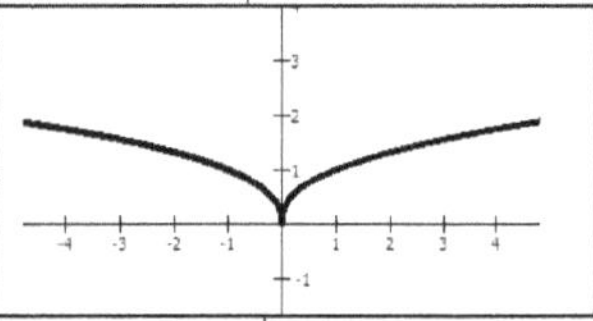	p=5, q=8	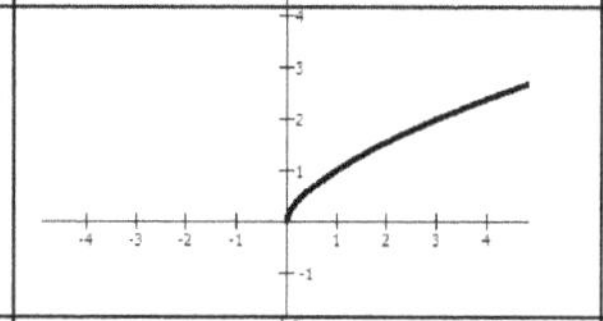	p=5, q=7	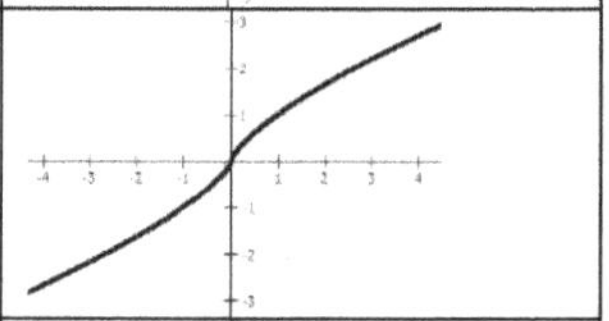
p=4, q=7	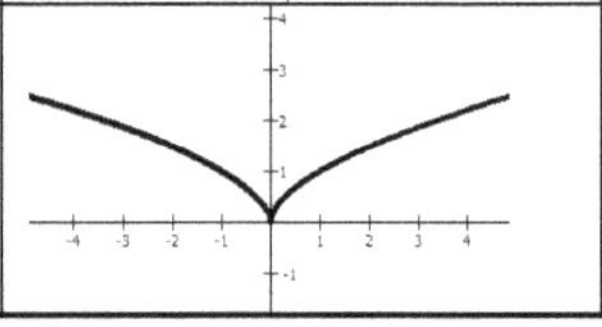	p=7, q=12	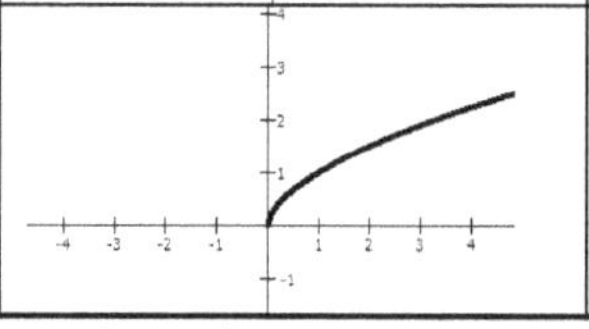	p=1, q=9	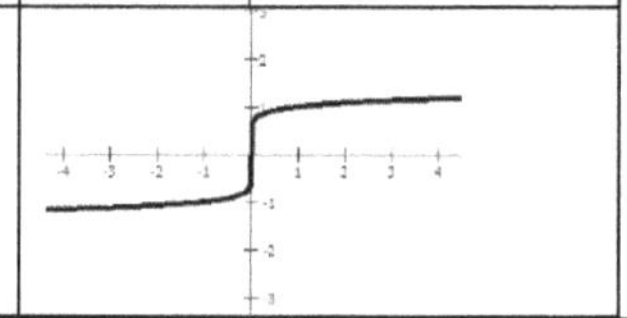

c) Quotient powers p/q. Explore patterns for p/q > 1:

when p is even and q is odd		when p is odd and q is even		when p is odd and q is odd	
p=4, q=3		p=5, q=4		p=7, q=5	
p=6, q=5		p=9, q=8		p=11, q=7	
p=8, q=7		p=17, q=12		p=13, q=9	

d) $f(x) = x^{\frac{4}{3}}$, $f(x) = x^{\frac{7}{3}}$, $f(x) = x^{-3}$ can be checked on a graphing calculator.

3.3 Exploring the Power Functions for domain and range. (SOLUTIONS)

a) Using the points $x = \{0, 1, 4, 9\}$ find and plot the *(x,y)* coordinates for the function
$f(x) = \sqrt{x} + 3$.

The coordinates are (0, 3), (1, 4), (4, 5), (9, 6)

Plot the points on a Cartesian plane. Join them with a smooth line.

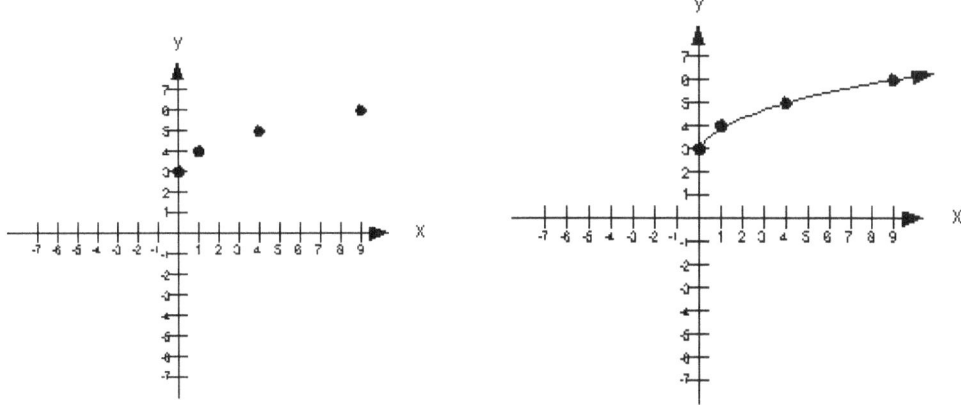

b) Use the resulting graph to determine the domain and range of this function.
To figure out the domain and range, look left and right, up and down.

- Domain: There is a left endpoint at $x = 0$, since we cannot take the square root of a negative number in the real number system, it can however go on forever for positive *x* values. $[0, \infty)$

- Range: For the range note how the lowest *y* value is 3, and then the graph in increasing. $[3, \infty)$

c) Using the points $x = \{-2, -1, 0, 1, 2\}$ find and plot the *(x,y)* coordinates for the function
$f(x) = x^3 - 3$.

The coordinates are (-2, -11), (-1, -4), (0, -3), (1, -2), (2, 5)

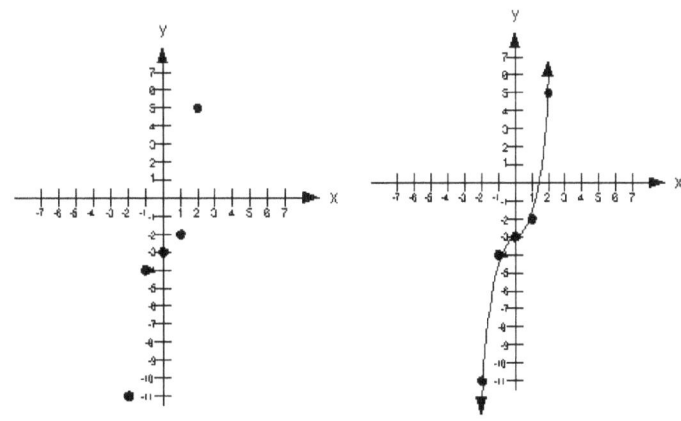

d) Use the resulting graph to determine the domain and range of this function.
To figure out the domain and range, look left and right, up and down. This graph can go on forever in both directions.

- Domain: You can cube any number and can go on forever for all *x* values. $(-\infty, \infty)$

- Range: Cubes can be any positive or negative real number, $(-\infty, \infty)$

e) For $x = \{-4, -3, -2, -1, 0, 1, 2, 3, 4\}$ the *(x,y)* coordinates for $f(x) = \dfrac{2}{x-3}$ are:

$$\left(-4, \frac{-2}{7}\right), \left(-3, \frac{-1}{3}\right), \left(-2, \frac{-2}{5}\right), \left(-1, \frac{-1}{2}\right), \left(0, \frac{-2}{3}\right), (1, -1), (2, -2), \left(3, \frac{2}{0}\right), (4, 2)$$

f) Use the resulting graph to determine the asymptotes, the domain and range.

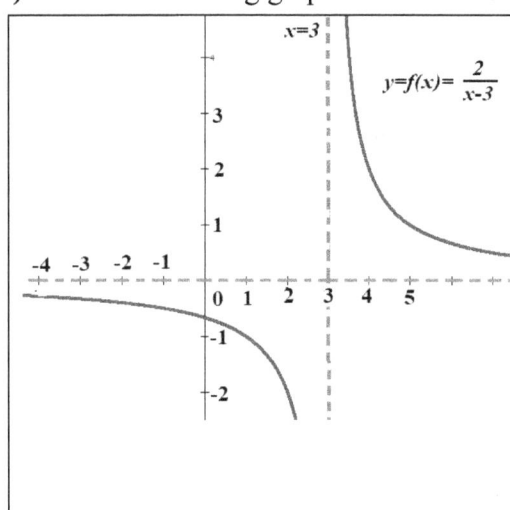

This graph can go on forever in both directions, but has one vertical asymptotes and one horizontal asymptote. You can never get $f(x) = \dfrac{2}{x-3} = 0$, because the numerator is 2!

- Asymptotes $x = 3, y = 0$
- Domain: Function does not exist at $x = 3$ but is ok for other x values.
$$(-\infty, 3) \cup (3, \infty)$$
- Range: There is no y value at $y = 0$, therefore the range is
$$(-\infty, 0) \cup (0, \infty)$$

g) For points $x = \{-4, -3, -2, -1, \dfrac{-1}{2}, 0, \dfrac{1}{2}, 1, 2, 3, 4\}$ the *(x,y)* coordinates for

$f(x) = \dfrac{x^2}{x^2 - 1}$, the coordinates are

$$\left(-4, \frac{16}{15}\right), \left(-3, \frac{9}{8}\right), \left(-2, \frac{4}{3}\right), \left(-1, \frac{1}{0}\right), \left(\frac{-1}{2}, \frac{-1}{3}\right), (0, 0), \left(\frac{1}{2}, \frac{-1}{3}\right), \left(1, \frac{1}{0}\right), \left(2, \frac{4}{3}\right), \left(3, \frac{9}{8}\right), \left(4, \frac{16}{15}\right)$$

Express $f(x) = \dfrac{x^2}{x^2 - 1}$ as $f(x) = A + \dfrac{B}{x^2 - 1}$

$$f(x) = \frac{x^2}{x^2 - 1} \text{ is } f(x) = 1 + \frac{-1}{x^2 - 1}$$

h) Use the resulting graph to determine the asymptotes, the domain and range.

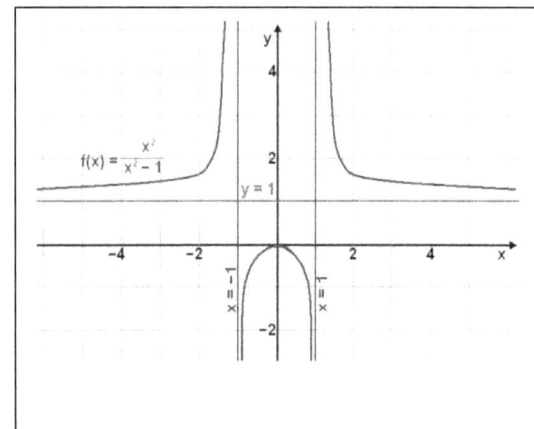

This graph can go on forever in both directions, but has 2 vertical asymptotes and one horizontal asymptote.

- Asymptotes $x = 1, x = -1, y = 1$
- Domain: Function does not exist at $x = 1, x = -1$ but is ok for other x values. $(-\infty, -1) \cup (-1, 1) \cup (1, \infty)$
- Range: There are no y values in the interval $(0, 1]$, therefore the range is
$$(-\infty, 0] \cup (1, \infty)$$

3.4 Modelling with Power Functions. (SOLUTIONS)

a) Boyle's Law

i.

Pressure P (kPa)	Volume V (cm^3)
200	31
180	34.44
140	44.29
100	62
85	72.94
70	88.57
60	103.33

k=6200

ii.

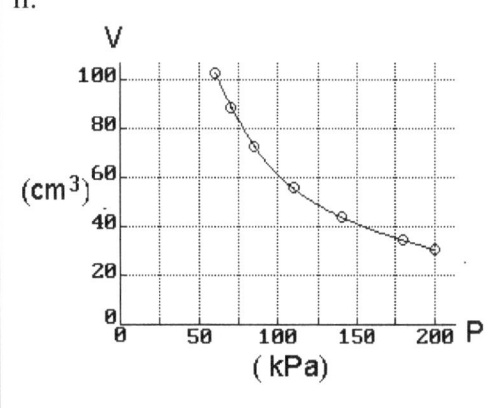

iii.   You can't defy the laws of physics.

b)

Bisection method, approximation of $f(x) = 0$ for $f(x) = \frac{1}{2}x^3 - x^2 - 1$ accurate to 2 decimal places, starting in the interval $x \in [2,4]$.

1	$f(2) = -1$	$f(4) = 15$
2	$f(2) = -1$	$f(3) = 3.5$
3	$f(2) = -1$	$f(2.5) = 0.56$
4	$f(2) = -1$	$f(2.25) = -0.37$
5	$f(2.25) = -0.37$	$f(2.37) = 0.04$

The approximate solution for $f(x) = 0$ is when x is in the interval $(2.34, 2.38)$

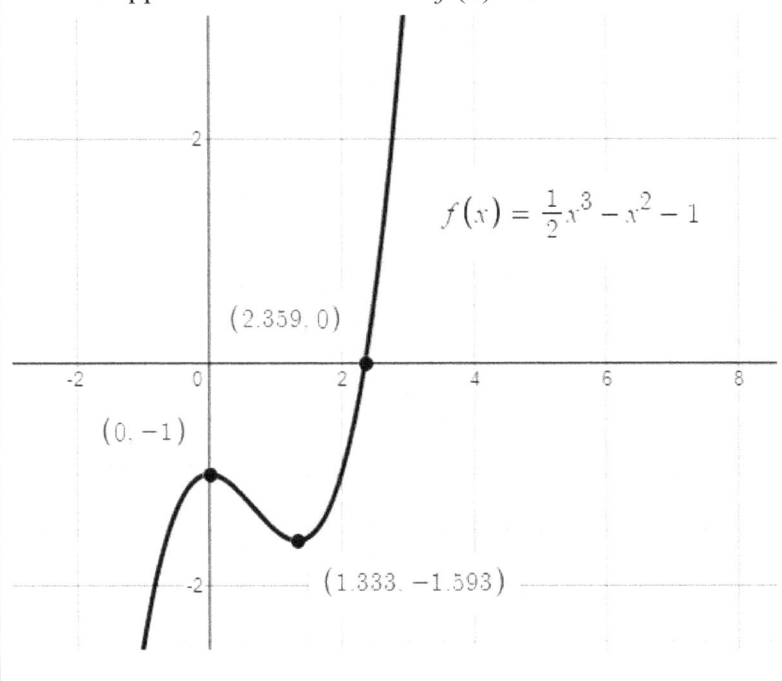

c)

The zoologist has recorded the following penguin population information since the beginning of the observations.

Year	0	1	2	3	4	5	6	7	8	9	10
Population	250	270	310	375	410	395	335	290	290	320	325

    i.   Plot the points on a Cartesian plane.

    ii.   Quadratic function model used to estimate the 11th year population.

Quadratic model using evenly spaced out 3 points $(0,250)$, $(5,395)$, $(10,325)$ fit to

$p(t) = at^2 + bt + c$ where a, b, c are real constants.

With the aid of technology solve the simultaneous equations for

model: $p(t) = \dfrac{-43}{10}t^2 + \dfrac{101}{2}t + 250$

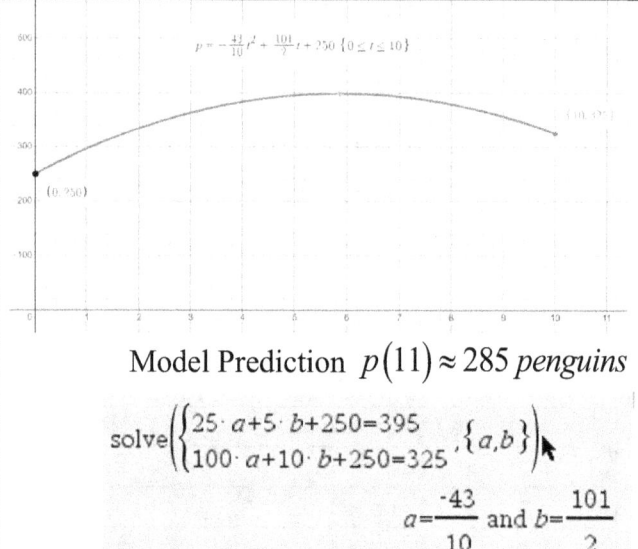

Model Prediction $p(11) \approx 285 \; penguins$

$$\text{solve}\left(\begin{cases} 25 \cdot a + 5 \cdot b + 250 = 395 \\ 100 \cdot a + 10 \cdot b + 250 = 325 \end{cases}, \{a, b\}\right)$$

$$a = \frac{-43}{10} \text{ and } b = \frac{101}{2}$$

    iii.   Cubic function model used to estimate the 11th year population.

Cubic model using evenly spaced out 4 points $(0,250)$, $(3,375)$, $(7,290)$, $(10,325)$ fit to

$p(t) = dt^3 + et^2 + ft + g$ where d, e, f, g are real constants.

With the aid of technology solve the simultaneous equations for model:

$p(t) = \dfrac{115}{84}t^3 + \dfrac{-635}{28}t^2 + \dfrac{2045}{21}t + 250$

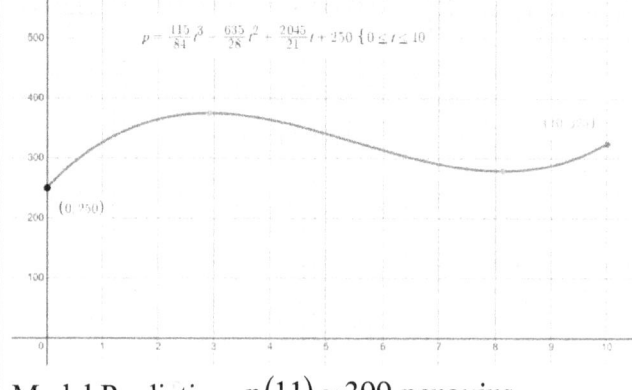

Model Prediction $p(11) \approx 399 \; penguins$

$$\text{solve}\left(\begin{cases} 27 \cdot d + 9 \cdot e + 3 \cdot f + 250 = 375 \\ 343 \cdot d + 49 \cdot e + 7 \cdot f + 250 = 290 \\ 1000 \cdot d + 100 \cdot e + 10 \cdot f + 250 = 325 \end{cases}, \{d, e\} \right.$$

$$d = \frac{115}{84} \text{ and } e = \frac{-635}{28} \text{ and } f = \frac{2045}{21}$$

c) continued

iv.   Quartic function model used to estimate the 11th year population.

Quartic model using evenly spaced out 5 points (0,250), (3,375), (5,395), (8,290), (10,325) fit to

$p(t) = at^4 + bt^3 + ct^2 + dt + e$ where

a, b, c, d, e are real constants.

With the aid of technology solve the simultaneous equations for model:

$p(t) = \dfrac{131}{420}t^4 + \dfrac{-559}{105}t^3 + \dfrac{8809}{420}t^2 + \dfrac{383}{21}t + 250$

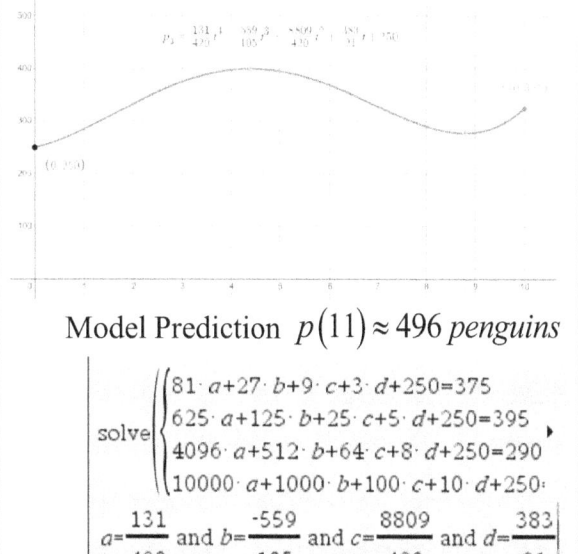

Model Prediction $p(11) \approx 496$ penguins

$$\text{solve}\begin{cases} 81 \cdot a + 27 \cdot b + 9 \cdot c + 3 \cdot d + 250 = 375 \\ 625 \cdot a + 125 \cdot b + 25 \cdot c + 5 \cdot d + 250 = 395 \\ 4096 \cdot a + 512 \cdot b + 64 \cdot c + 8 \cdot d + 250 = 290 \\ 10000 \cdot a + 1000 \cdot b + 100 \cdot c + 10 \cdot d + 250 \end{cases}$$

$a = \dfrac{131}{420}$ and $b = \dfrac{-559}{105}$ and $c = \dfrac{8809}{420}$ and $d = \dfrac{383}{21}$

v.   Which is the best model? The Quartic fits to more points with least difference to the actual observations. A higher order polynomial of order 10 could be used to fit to all 11 points observed.

However the 11th year observation can determine which model is best for this case. If the penguin colony is in decline then the quadratic model may be the closest.

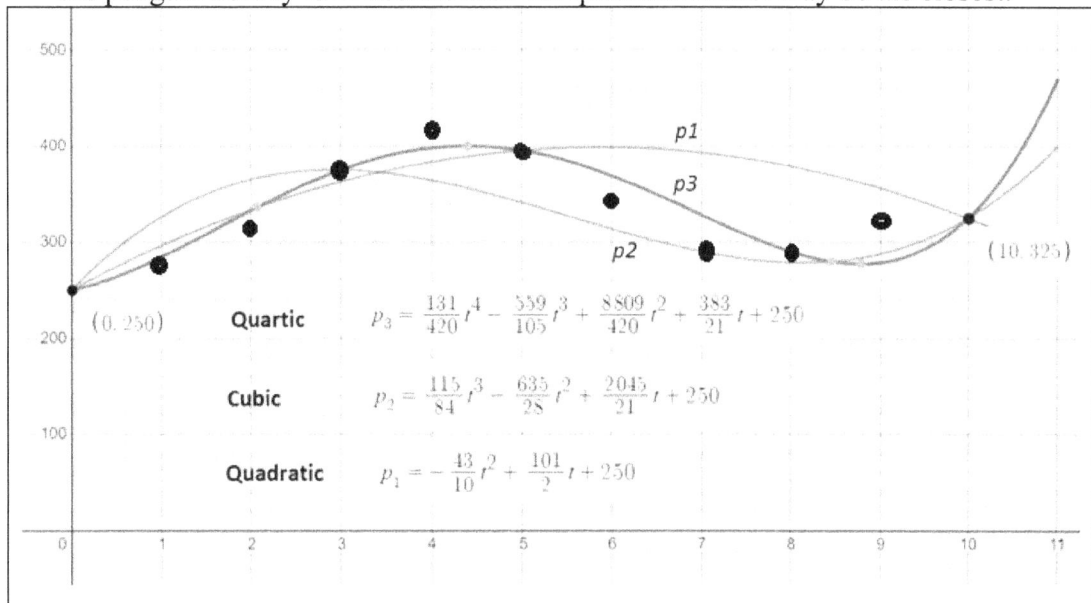

Quartic $p_3 = \dfrac{131}{420}t^4 - \dfrac{559}{105}t^3 + \dfrac{8809}{420}t^2 + \dfrac{383}{21}t + 250$

Cubic $p_2 = \dfrac{115}{84}t^3 - \dfrac{635}{28}t^2 + \dfrac{2045}{21}t + 250$

Quadratic $p_1 = -\dfrac{43}{10}t^2 + \dfrac{101}{2}t + 250$

Just for fun! Brain extenders. A selection of puzzles from Martin Gardner.

What angle is made by the two red lines drawn on the two sides of the cube, as shown in the illustration?

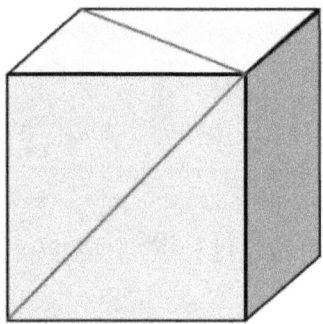

Cryptarithm In these problems, each letter corresponds to a single digit. For instance, can you figure out which digit each letter represents to make the sum at the right work?

```
      SEVEN
      SEVEN
      SEVEN
      SEVEN
      SEVEN
      SEVEN
  +   SEVEN
  ─────────
      FORTY9
```

You are to make one cut which needn't be straight that will divide the figure into two identical parts.

4 Fabulous Functions

Don't Forget: The functions that you need to know about in VCE Mathematical Methods 3&4.

Power functions in general form, $f(x) = A\left(n\left(x-b\right)\right)^p + c$ are discussed in Chapters 1,2 and 3.

Exponential functions with base of a and e in the general form, where $f(x) = Aa^{n(x-b)} + c$ or $f(x) = Ae^{n(x-b)} + c$.

Logarithmic functions with base of a and e in the general form, where $f(x) = A\log_a\left(n\left(x-b\right)\right) + c$ or $f(x) = A\log_e\left(n\left(x-b\right)\right) + c$.

Circular functions for anti-clockwise angle θ from the positive x-axis where A, n, b, c are constant reals.

- Sine function $f(\theta) = A\sin(n(\theta - b)) + c$.
- Cosine function $f(\theta) = A\cos(n(\theta - b)) + c$.
- Tangent function $f(\theta) = A\tan(n(\theta - b)) + c$.

Circular functions are used to study circular motion, and are sometimes also called **Trigonometric functions** because they are connected to **Trigonometry** the study of right angled triangles with the use of sin, cos and tan ratios. Do you know the mnemonic **SohCahToa**?

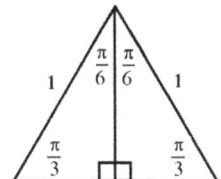 The sin, cos, tan ratios for $\dfrac{\pi}{6}$ and $\dfrac{\pi}{3}$ can be derived using an equilateral triangle of side length 1, by cutting one side in half, the height of the triangle is $\dfrac{\sqrt{3}}{2}$.

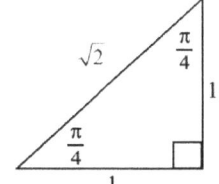 The sin, cos, tan ratios for $\dfrac{\pi}{4}$ can be derived using an isosceles triangle of side lengths $1, 1, \sqrt{2}$.

The sin, cos, tan ratios can also be used when studying linear functions and can be related to the gradient of a line, where all sorts of angles can also be found relative to the x and the y-axis.

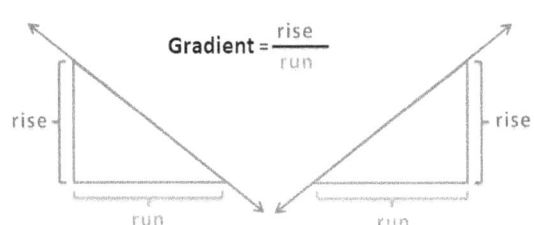

Fibonacci Blue/Flickr

89

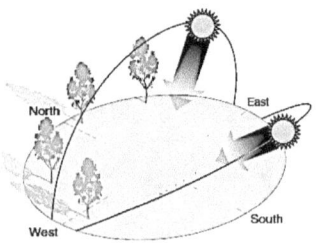

Circular functions

This study of Mathematics originated from observations of the sun rising in the east, and setting in the west for use in navigation. The model or abstract representation is a **unit circle** of radius 1, that is centred at (0,0) is used for the **derivation** of circular functions. The perimeter of a unit circle is equal to 2π units. The value of π is irrational and cannot be expressed exactly using fractions, but it is approximately equal to 3.14159....

The three basic circular functions are:

$$\sin(\theta) = \frac{opposite}{hypotenuse} = \frac{y}{1} = y$$

$$\cos(\theta) = \frac{adjacent}{hypotenuse} = \frac{x}{1} = x$$

$$\tan(\theta) = \frac{opposite}{adjacent} = \frac{y}{x} = \frac{\sin(\theta)}{\cos(\theta)}$$

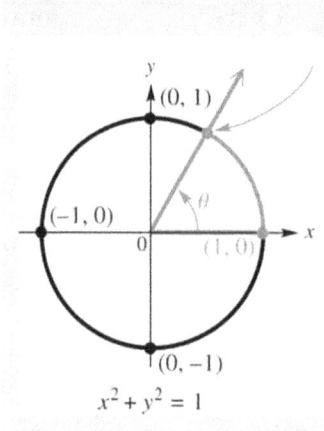

$x^2 + y^2 = 1$

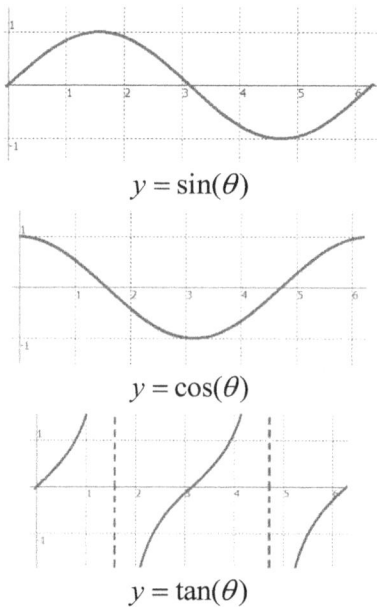

$y = \sin(\theta)$

$y = \cos(\theta)$

$y = \tan(\theta)$

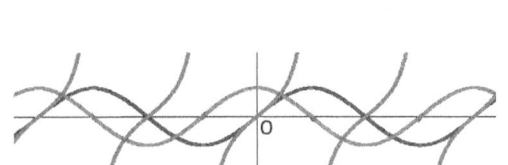

Angles are measured in **degrees** represented by the symbol $^\circ$ or **radians** represented by the symbol ^{c}. Imagine equating the perimeter of the unit circle which is 2π units and is equivalent to 360° that is the length $2\pi^c \equiv 360^\circ$ which simplifies to $\pi^c \equiv 180^\circ$, this means $\pi^c/2 \equiv 90^\circ$, $\pi^c/3 \equiv 60^\circ$, $\pi^c/4 \equiv 45$, $\pi^c/6 \equiv 30$.

Changing radians ^{c} to degrees $^\circ$ $\times \dfrac{180}{\pi}$	**Changing degrees $^\circ$ to radians ^{c}** $\times \dfrac{\pi}{180}$
Example: $4.2^c \times \dfrac{180}{\pi} \approx 240.64^\circ$	Example: $67.89^\circ \times \dfrac{\pi}{180} \approx 1.185^c$

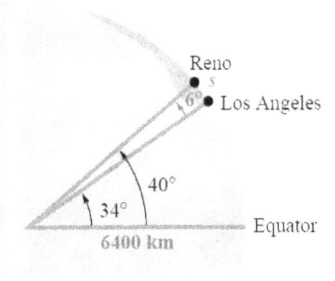

The Babylonians, some 5000 years ago developed a number system based on the number of days it takes for the earth to rotate around the sun. They measured each year as lasting 360 days, which is very close to the actual number of days of 365, and this is the reason there are 360 degrees in a circle in modern day mathematics.

When the circular functions are graphed, a repetitive **periodic** pattern of values is generated from repeatedly travelling around the perimeter of the **unit circle**.

Amplitude: Largest distance from the origin to the maximum.
Period: Interval of repetition of the graph shape.

The possible y values that can be generated by the **sine** (*sin*) function about the **unit circle** is in the interval *[-1,1]*.
The possible x values that can be generated by the **cosine** (*cos*) function about the **unit circle** is also in the interval *[-1,1]*.
The **tangent** (*tan*) function results in value in the interval *(-∞,∞)* due to being the ratio of sine divided by cosine.

$$\sin(\theta) = \frac{opposite}{hypotenuse} = \frac{y}{1} = y$$

$y = \sin(\theta)$ where the radius is rotated through the

angle $\theta \in [0, 2\pi^c]$

Amplitude=1, Period =2π

$$f(\theta) = \sin(\theta)$$

θ	0	$\frac{\pi}{2}$	π	$\frac{3\pi}{2}$	2π
$sin\ \theta$	0	1	0	-1	0

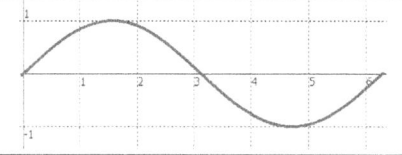

$$\cos(\theta) = \frac{adjacent}{hypotenuse} = \frac{x}{1} = x$$

$y = \cos(\theta)$ where the radius is rotated through the

angle $\theta \in [0, 2\pi^c]$

Amplitude=1, Period =2π

$$f(\theta) = \cos(\theta)$$

θ	0	$\frac{\pi}{2}$	π	$\frac{3\pi}{2}$	2π
$cos\ \theta$	1	0	-1	0	1

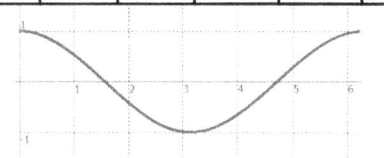

$$\tan(\theta) = \frac{opposite}{adjacent} = \frac{y}{x} = \frac{\sin(\theta)}{\cos(\theta)}$$

$\tan(\theta) = \dfrac{\sin(\theta)}{\cos(\theta)}$ where the radius is rotated through

the angle $\theta \in [0, 2\pi^c]$

Amplitude=∞, Period =π

$$f(\theta) = \tan(\theta)$$

θ	0	$\frac{\pi}{2}$	π	$\frac{3\pi}{2}$	2π
$tan\ \theta$	0	*undef*	0	*undef*	0

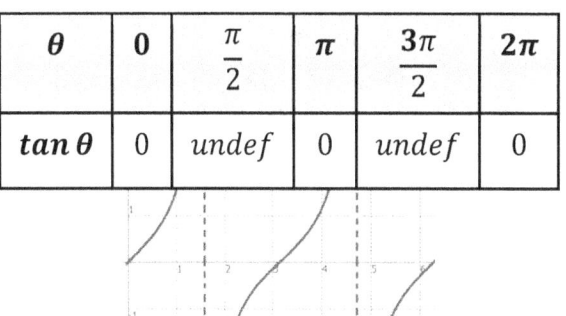

Coordinates $(x,y) \equiv (\cos\theta, \sin\theta)$ around the Unit Circle.

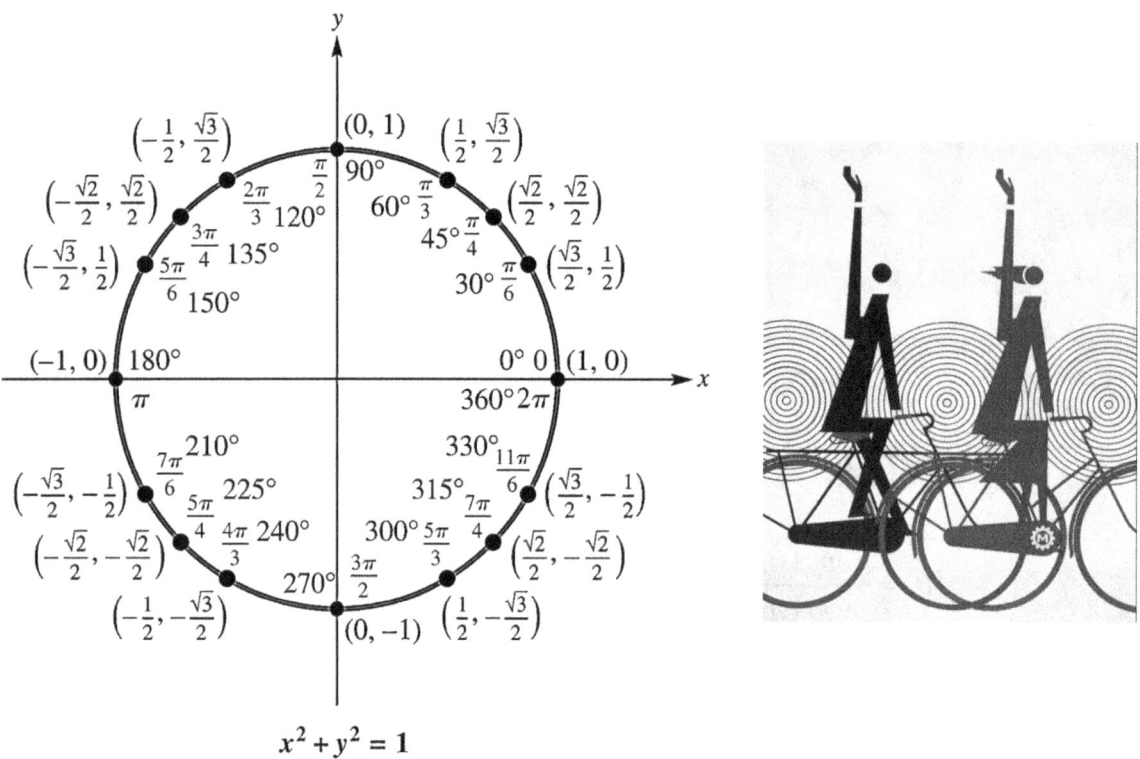

$$x^2 + y^2 = 1$$

Reference angles of rotation from the origin give exact values for sin, cos and tan

θ	0	$\dfrac{\pi}{6}$ 30°	$\dfrac{\pi}{4}$ 45°	$\dfrac{\pi}{3}$ 60°	$\dfrac{\pi}{2}$ 90°
$\sin\theta$	$\dfrac{\sqrt{0}}{2}$	$\dfrac{\sqrt{1}}{2}$	$\dfrac{\sqrt{2}}{2}$	$\dfrac{\sqrt{3}}{2}$	$\dfrac{\sqrt{4}}{2}$
$\cos\theta$	$\dfrac{\sqrt{4}}{2}$	$\dfrac{\sqrt{3}}{2}$	$\dfrac{\sqrt{2}}{2}$	$\dfrac{\sqrt{1}}{2}$	$\dfrac{\sqrt{0}}{2}$
$\tan\theta$	0	$\dfrac{\sqrt{3}}{3}$	1	$\sqrt{3}$	undefined

Quadrant I – All sin, cos, tan positive
Quadrant IV – Cosine positive
Quadrant III – Tangent positive
Quadrant II – Sine positive

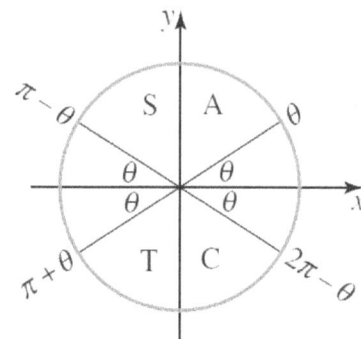

Solving Circular Equations, thinking about the inverse functions and the **Reference angles** helps.

ratio	0	$\dfrac{\sqrt{1}}{2}$	$\dfrac{\sqrt{2}}{2}$	$\dfrac{\sqrt{3}}{2}$	$\dfrac{\sqrt{4}}{2}$
$\sin^{-1}(ratio)$	0	$\dfrac{\pi}{6}$	$\dfrac{\pi}{4}$	$\dfrac{\pi}{3}$	$\dfrac{\pi}{2}$
$\cos^{-1}(ratio)$	$\dfrac{\pi}{2}$	$\dfrac{\pi}{3}$	$\dfrac{\pi}{4}$	$\dfrac{\pi}{6}$	0
ratio	0	$\dfrac{1}{\sqrt{3}}$	1	$\dfrac{\sqrt{3}}{1}$	$undef$
$\tan^{-1}(ratio)$	0	$\dfrac{\pi}{6}$	$\dfrac{\pi}{4}$	$\dfrac{\pi}{3}$	$\dfrac{\pi}{2}$

$$\cos(x) = a$$
$$\sin(x) = a$$
$$\tan(x) = a$$

(Where a is a constant)
Use the inverse function.

$$x = \cos^{-1}(a)$$
$$x = \sin^{-1}(a)$$
$$x = \tan^{-1}(a)$$

Without a calculator this can be solved if we use the properties of the unit circle to find the values corresponding to the **Reference angles.** Take care with the sign of "a" for the correct quadrants, and use the symmetry properties of the unit circle to find all relevant solutions.

Find **all solutions** in the given domain. If the domain is larger than the period of the circular function, then you will need to add/subtract the period to each value found until the entire domain is covered.

If the input value to the circular function is transformed (**$nx + b$**) then a simple substitution allows us to find the solutions using a similar method to the one already described.

$\cos(nx+b) = a$ $\sin(nx+b) = a$ $\tan(nx+b) = a$ Where the pronumerals (a, b and n are constants)	Let $= \mathbf{nx + b}$, this substitution allows the simplification of the method. Solve for θ, $$\cos(\theta) = a$$ $$\sin(\theta) = a$$ $$\tan(\theta) = a$$	$\theta = \cos^{-1}(a)$ $\theta = \sin^{-1}(a)$ $\theta = \tan^{-1}(a)$	Finally we need to solve for x $$(nx+b) = \cos^{-1}(a)$$ $$(nx+b) = \sin^{-1}(a)$$ $$(nx+b) = \tan^{-1}(a)$$

- Use the properties of the unit circle to find all the values for the **Reference angle(s)** taking note of the sign of "a", and ensure that you find **all solutions** in the given domain for x.

Example 4.1: Solving equations with circular functions and algebra.

Solve for x where $\cos(x) = -\dfrac{\sqrt{3}}{2}$, $x \in [0, 2\pi]$, note the restricted domain for providing a solution.

Step 1. Determine the **reference angle** for the **cosine** ratio provided.

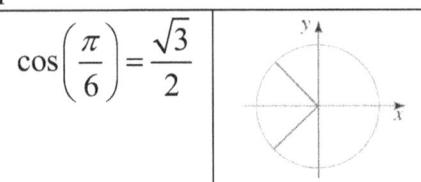

$$\cos\left(\frac{\pi}{6}\right) = \frac{\sqrt{3}}{2}$$

Step 2. Consider the sign of the ratio and determine the quadrants where the solution lies.

Cosine is negative in the 2nd and 3rd Quadrants of the Cartesian plane.

Step 3. Determine the solutions in the required domain.

$$x = \left\{ \pi - \frac{\pi}{6}, \pi + \frac{\pi}{6} \right\}$$

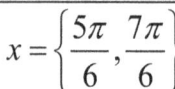

$$x = \left\{ \frac{5\pi}{6}, \frac{7\pi}{6} \right\}$$

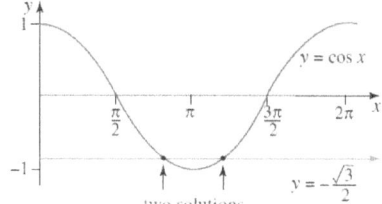

two solutions

Solve for x where $\sin(x) = -\dfrac{\sqrt{3}}{2}$, $x \in [0, 4\pi]$, note the restricted domain for providing a solution.

Step 1. Determine the **reference angle** for the sine ratio provided.

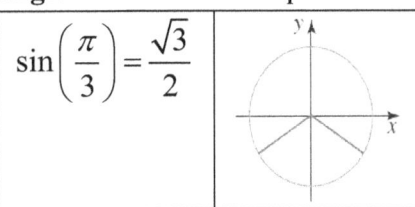

$$\sin\left(\frac{\pi}{3}\right) = \frac{\sqrt{3}}{2}$$

Step 2. Consider the sign of the ratio and determine the quadrants where the solution lies.
Sine is negative in the 3rd and 4th Quadrants of the Cartesian plane.

Step 3. Determine the solutions in the required domain.
All the solutions can be found by adding the period of 2π.

$$x = \left\{ \begin{array}{l} \pi + \dfrac{\pi}{3}, 2\pi - \dfrac{\pi}{3}, \\ 3\pi + \dfrac{\pi}{3}, 4\pi - \dfrac{\pi}{3} \end{array} \right\}$$

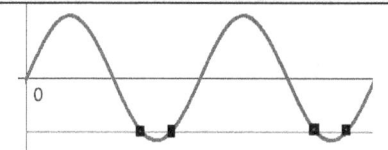

Simplify the final answers.

$$x = \left\{ \frac{4\pi}{3}, \frac{5\pi}{3}, \frac{10\pi}{3}, \frac{11\pi}{3} \right\}$$

Solve for x where $\sin(3x) = -\dfrac{\sqrt{3}}{2}$, $x \in [0, 2\pi]$, note the coefficient of x and restricted domain.

Step 1. Determine the **reference angle** for the sine ratio provided, same as example above.
Step 2. Consider the sign of the ratio and determine the quadrants where the solution lies.
Sine is negative in the 3rd and 4th Quadrants of the Cartesian plane. $3x = \left\{ \pi + \dfrac{\pi}{3}, 2\pi - \dfrac{\pi}{3} \right\}$

Step 3. Determine the solutions in the required domain.
All the solutions can be found by adding the value of 2π two more times since the co-efficient of x is 3, which means there are three cycles of sin within the interval $x \in [0, 2\pi]$.

$$3x = \left\{ \begin{array}{l} \pi + \dfrac{\pi}{3}, 2\pi - \dfrac{\pi}{3}, \\ 3\pi + \dfrac{\pi}{3}, 4\pi - \dfrac{\pi}{3}, \\ 5\pi + \dfrac{\pi}{3}, 6\pi - \dfrac{\pi}{3} \end{array} \right\} \rightarrow 3x = \left\{ \begin{array}{l} \dfrac{4\pi}{3}, \dfrac{5\pi}{3}, \\ \dfrac{10\pi}{3}, \dfrac{11\pi}{3}, \\ \dfrac{16\pi}{3}, \dfrac{17\pi}{3} \end{array} \right\} \rightarrow x = \left\{ \begin{array}{l} \dfrac{4\pi}{9}, \dfrac{5\pi}{9}, \\ \dfrac{10\pi}{9}, \dfrac{11\pi}{9}, \\ \dfrac{16\pi}{9}, \dfrac{17\pi}{9} \end{array} \right\}$$

If the domain is not specified for a system of circular equations, then there are *infinite* solutions.

Equation	Infinite Solutions exist	Graphically
$\cos(x) = \dfrac{1}{2}$	$x = \{.., -\dfrac{5\pi}{3}, -\dfrac{\pi}{3}, \dfrac{\pi}{3}, \dfrac{5\pi}{3}, ...\}$ Reference angle(s) $-\dfrac{\pi}{3}, \dfrac{\pi}{3}$ add $\pm 2\pi$ forever to find all the solutions. Use the symmetry of this **Even** function.	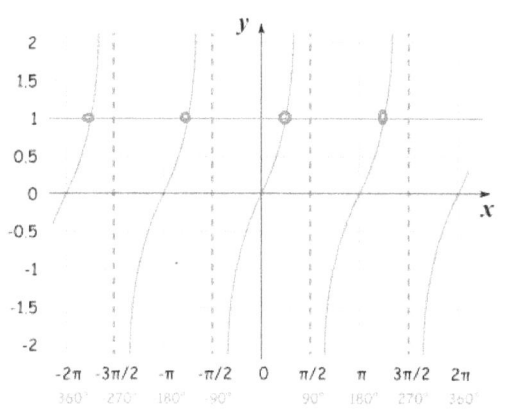
$\sin(x) = \dfrac{1}{2}$	$x = \{.., -\dfrac{11\pi}{6}, -\dfrac{7\pi}{6}, \dfrac{\pi}{6}, \dfrac{5\pi}{6}, \dfrac{13\pi}{6}...\}$ Reference angle(s) $\dfrac{\pi}{6}, \dfrac{5\pi}{6}$ add $\pm 2\pi$ forever to find all the solutions. Use the rotational symmetry of this **Odd** function.	
$\tan(x) = 1$	$x = \{.., -\dfrac{7\pi}{4}, \dfrac{\pi}{4}, \dfrac{5\pi}{4}, ...\}$ Reference angle $\dfrac{\pi}{4}$ add $\pm \pi$ forever to find all the solutions.	

To generalize this behavior for circular functions we have a general solution for an undefined domain equation.

Circular Function	General Solution: *Z is the set of integers, $k \in Z$*
$\cos(x) = a$	$x = 2k\pi \pm \cos^{-1}(a), k \in Z, a \in [-1,1]$
$\sin(x) = a$	$x = \{2k\pi + \sin^{-1}(a),\ (2k+1)\pi - \sin^{-1}(a)\}\ k \in Z, a \in [-1,1]$
$\tan(x) = a$	$x = k\pi + \tan^{-1}(a), k \in Z$

Example 4.2: Solving equations with circular functions and algebra.

Solve for x where $\tan(x) = 1$, $x \in [0, 2\pi]$, note the restricted domain for providing a solution.

Step 1. Determine the **reference angle** for the **tangent** ratio provided.

$$\tan\left(\frac{\pi}{4}\right) = 1$$

Step 2. Consider the sign of the ratio and determine the quadrants where the solution lies.

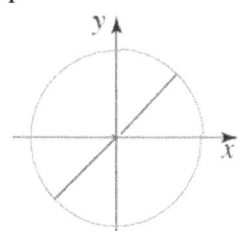

Tangent is positive in the 1st and 3rd Quadrants of the Cartesian plane.

Step 3. Determine the solutions in the required domain.

$$x = \left\{\frac{\pi}{4}, \pi + \frac{\pi}{4}\right\}$$

$$x = \left\{\frac{\pi}{4}, \frac{5\pi}{4}\right\}$$

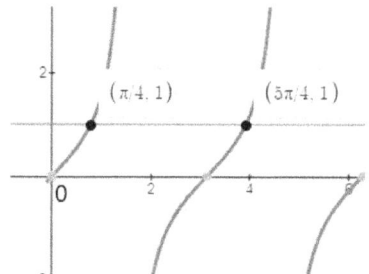

Solve for x where $\tan\left(2\left(x - \frac{\pi}{8}\right)\right) = 1$, $x \in [0, 2\pi]$, note the coefficient of x and the restricted domain for providing a solution.

Step 1. Determine the **reference angle** for the tangent ratio provided.

$$\tan\left(\frac{\pi}{4}\right) = 1$$

Step 2. Consider the sign of the ratio and determine the quadrants where the solution lies.

Let $\theta = 2\left(x - \frac{\pi}{8}\right)$

Tangent is positive in the 1st and 3rd Quadrants of the Cartesian plane.

$$\theta = \left\{\frac{\pi}{4}, \pi + \frac{\pi}{4}\right\}$$

Step 3. Determine the solutions in the required domain.

All the solutions can be found by adding the value of 2π one more time to each basic solution, since the co-efficient of x is 2, which means there are two more solutions of tan within the interval $x \in [0, 2\pi]$

$$\theta = \left\{\begin{array}{l} \frac{\pi}{4}, \pi + \frac{\pi}{4}, \\ \frac{\pi}{4} + \pi, \pi + \frac{\pi}{4} + \pi \end{array}\right\} \rightarrow \theta = \left\{\begin{array}{l} \frac{\pi}{4}, \frac{5\pi}{4}, \\ \frac{9\pi}{4}, \frac{13\pi}{4} \end{array}\right\}$$

$$2\left(x - \frac{\pi}{8}\right) = \left\{\begin{array}{l} \frac{\pi}{4}, \frac{5\pi}{4}, \\ \frac{9\pi}{4}, \frac{13\pi}{4} \end{array}\right\} \rightarrow \left(x - \frac{\pi}{8}\right) = \left\{\begin{array}{l} \frac{\pi}{8}, \frac{5\pi}{8}, \\ \frac{9\pi}{8}, \frac{13\pi}{8} \end{array}\right\}$$

$$x = \left\{\begin{array}{l} \frac{\pi}{8} + \frac{\pi}{8}, \frac{5\pi}{8} + \frac{\pi}{8}, \\ \frac{9\pi}{8} + \frac{\pi}{8}, \frac{13\pi}{8} + \frac{\pi}{8} \end{array}\right\} \rightarrow x = \left\{\begin{array}{l} \frac{\pi}{4}, \frac{3\pi}{4}, \\ \frac{5\pi}{4}, \frac{7\pi}{4} \end{array}\right\}$$

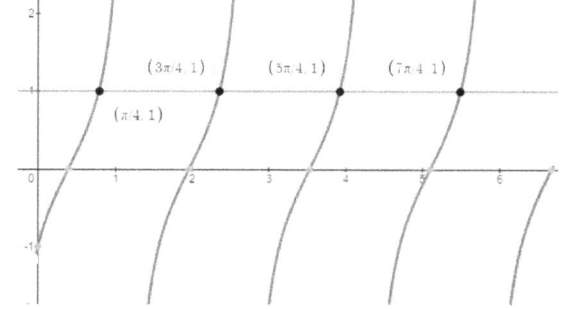

For acoustic waves in air, the peaks represent high air pressure, and the troughs, low air pressure. At any point in space, the air pressure varies or *oscillates* from high to low with a frequency set by the *pitch* of the sound.

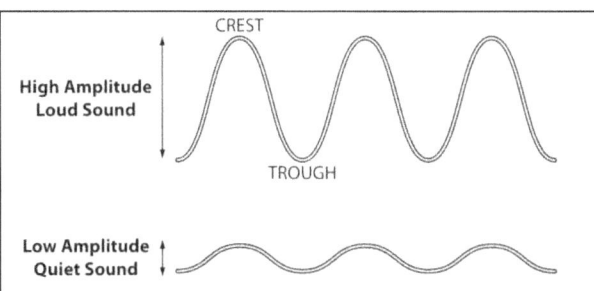

The Amplitude of the sound wave is a measure of how much energy the sound has. The larger the Amplitude the louder the sound.

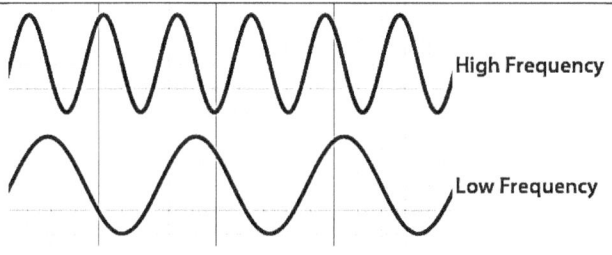

Frequency is how many waves there are per second. The higher the frequency, the more quickly air particles vibrate and the higher the pitch. In sound waves, a high pitch means a high note and a low pitch is a low note.

Comparing the sound waves 1, 2, 3 and 4. Wave 1 has the highest frequency and therefore the highest pitch and the same loudness as Wave 2, since the amplitudes are the same, and three times as loud as Wave 3 and 4.

Wave 2 and 3 have the same frequency and pitch, while Wave 2 is louder with the higher amplitude.

Wave 4 has the lowest frequency and therefore pitch, it has the same loudness as Wave 3.

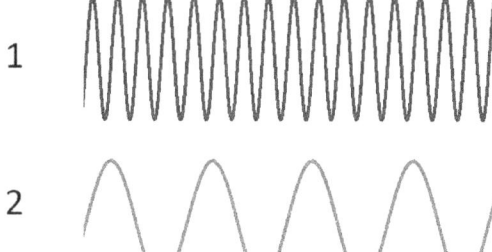
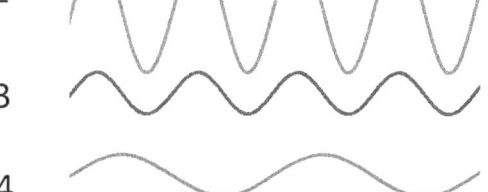

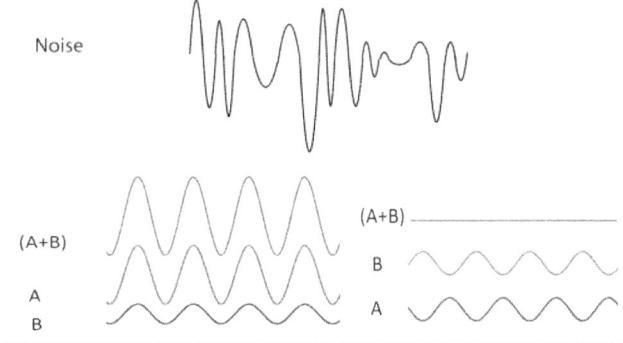

A noise is the addition of many circular functions of varying amplitudes and periods.

Sound waves can combined to become even louder or can cancel each other out to become silent.

Some properties of the unit circle, the Pythagorean Identity.

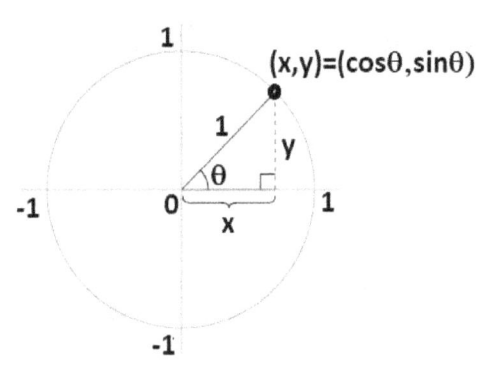

The relation $x^2 + y^2 = 1$ defines a unit circle with the centre at $(0,0)$ and a radius of 1.

Since $x = \cos(\theta)$ and $y = \sin(\theta)$, for the radius=1, when θ is the angle made relative the positive x-axis.

We get the "Pythagorean Identity" $\cos^2(\theta) + \sin^2(\theta) = 1$ since the hypotenuse squared is equal to the sum of the squares of the other two sides.

Example 4.3: Solving equations with circular functions using quadratic methods.

Solve for x where $2\cos^2(x) = \cos(x)$ where $0 \le x \le 2\pi$

Let $c = \cos(x)$ and we get a quadratic.

$2c^2 = c$

$2c^2 - c = 0$

$c(2c - 1) = 0$

$c = 0 \ or \ (2c - 1) = 0$

Replace $c = \cos(x)$

$\cos(x) = 0 \ or \ 2\cos(x) - 1 = 0$

Solve for $\cos(x) = 0 \ or \ \cos(x) = \dfrac{1}{2}$

$x = \left\{\dfrac{\pi}{2}, \dfrac{3\pi}{2}\right\}$ or $x = \left\{\dfrac{\pi}{2}, \dfrac{3\pi}{2}\right\}$

Solutions are: $x = \left\{\dfrac{\pi}{3}, \dfrac{\pi}{2}, \dfrac{3\pi}{2}, \dfrac{5\pi}{3}\right\}$

Solve for x where $2\cos^2(x) - \sin(x) = 1$ where $0 \le x \le 2\pi$

Use the Pythagorean identity

$$\cos^2(x) + \sin^2(x) = 1$$
$$\cos^2(x) = 1 - \sin^2(x)$$

Replace $\cos^2(x)$ with $1 - \sin^2(x)$

$$2(1 - \sin^2(x)) - \sin(x) = 1$$
$$2 - 2\sin^2(x) - \sin(x) = 1$$
$$-2\sin^2(x) - \sin(x) + 1 = 0$$
$$2\sin^2(x) + \sin(x) - 1 = 0$$

Let $s = \sin(x)$ and we get a quadratic.

$$2s^2 + s - 1 = 0$$

Factorise and solve for s.

$$(2s - 1)(s + 1) = 0$$
$$s = \frac{1}{2} \ or \ s = -1$$

Replace $s = \sin(x)$

$$\sin(x) = \frac{1}{2} \ or \ \sin(x) = -1$$

$$x = \left\{\dfrac{\pi}{6}, \dfrac{5\pi}{6}\right\} \ or \ x = \left\{\dfrac{3\pi}{2}\right\}$$

Solutions are: $x = \left\{\dfrac{\pi}{6}, \dfrac{3\pi}{2}, \dfrac{5\pi}{6}\right\}$

Exponential functions

This study of Mathematics originated from observations of the growth and decay in the natural world. Exponential functions are fast changing functions. $f(x) = a^x, a > 1$ and $f(x) = a^{-x}, a > 1$ are some of the forms of these functions.

The value a is referred to as the base of the function, and it is always a positive value. Why? Try evaluating values for $f(x) = a^x$ with a negative base value of a. What is the conclusion?

Exponential growth	Exponential decay
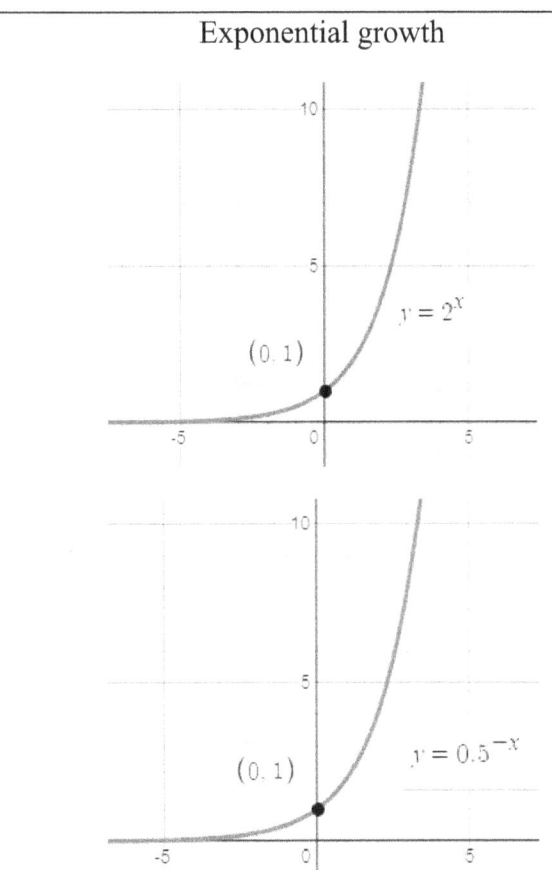	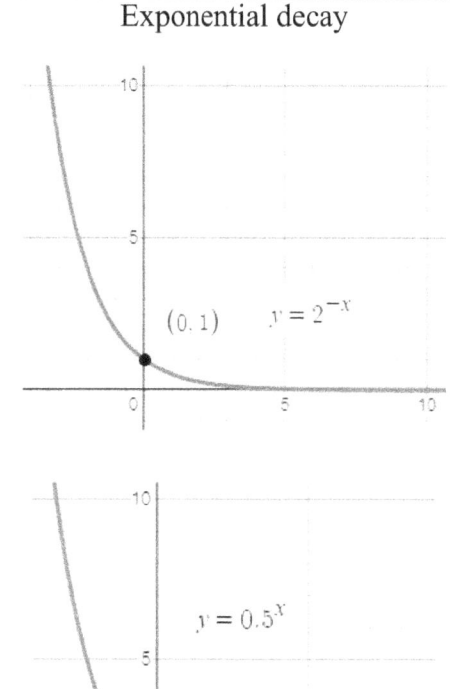
$f(x) = a^x, a > 1$	$f(x) = a^{-x}, a > 1$
$f(x) = a^{-x}, 0 < a < 1$	$f(x) = \dfrac{1}{a^x}, a > 1$
$f(x) = \dfrac{1}{a^x}, 0 < a < 1$	$f(x) = a^x, 0 < a < 1$
The *x-axis* is a horizontal asymptote	The *x-axis* is a horizontal asymptote
• As $x \to \infty, f(x) \to \infty$	• As $x \to \infty, f(x) \to 0^+$
• As $x \to -\infty, f(x) \to 0^+$	• As $x \to -\infty, f(x) \to \infty$
• domain $f = R^+$	• domain $f = R^+$
• range $f = R^+$	• range $f = R^+$
• $f(0) = 1$	• $f(0) = 1$

Rumour spreading.	Car depreciation.

If 5 people hear a rumour, and each spread it to 5 new people every day, the spread can be modelled by an exponential function.

New cars start losing their value as soon as they leave the showroom. If a car loses its value at a rate of 10% per year, it retains 90% of its current value each year. The loss of value called depreciation can be modelled by an exponential function.

Day	People who heard the rumour
0	5^0=1
1	5^1=5
2	5^2=25
3	5^3=125
4	5^4=625

Year	Car Value
0	$30000(0.9^0)$=30000
1	$30000(0.9^1)$=27000
2	$30000(0.9^2)$=24300
3	$30000(0.9^3)$=21870
4	$30000(0.9^4)$=19683

$$f(x) = 5^x$$

$$f(x) = 30000(0.9)^x$$

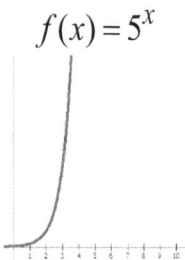

The world's population in 2017 was 7.53 billion people. At the rate of rumour spread modelled here, everyone in the world will hear this rumour after approximately 21 days.

Why doesn't everyone in the world hear this particular rumour? What is the domain of the rumour spreading function?

What is the domain of the depreciation? Maybe if you keep your old car long enough and keep it in reasonable condition, it may become a collectable vintage car.

Vintage cars may start appreciating again.

Ferraris manufactured between 1950 and 1970 have increased in value at a rate that is greater than the rate of inflation and many other conventional investments.

Growth rate of compound interest and Euler's number.

Compound interest Frequency	Calculation	Amount
Twice a year	$\left(1+\dfrac{1}{2}\right)^2$	$2.25
Every month	$\left(1+\dfrac{1}{12}\right)^{12}$	$2.61304
Every week	$\left(1+\dfrac{1}{52}\right)^{52}$	$2.6926
Every day	$\left(1+\dfrac{1}{365}\right)^{365}$	$2.71457
Every hour	$\left(1+\dfrac{1}{8760}\right)^{8760}$	$2.71813
Every minute	$\left(1+\dfrac{1}{525600}\right)^{525600}$	$2.71828
Every second	$\left(1+\dfrac{1}{31536000}\right)^{31536000}$	$2.71828

As the interval of calculating compound interest becomes smaller, we're approaching a limit, and earning less additional interest. The Swiss Mathematician Leonhard Euler found that if you accrue interest continually, you end up at the end of the year with 2.718281828459045… now known as Euler's number or the "Natural Exponential".

Euler's number e, can be calculated using $e = \lim\limits_{n\to\infty}\left(1+\dfrac{1}{n}\right)^n$, so for $n=10000$

$$e = \left(1+\dfrac{1}{10000}\right)^{10000} \approx 2.718$$

Euler's number e, is irrational and cannot be represented as a fraction, e is often used as the base for exponential and logarithmic functions $f(x)=e^x$ or $g(x)=\log_e x$.

Example 4.4: Exponential function $M(t)$ showing Radioactive decay.

Radioactive decay – the rate at which a radioactive substance decays is proportional to the mass (M) of the substance remaining, and can be modelled using **Euler's number** as the base of an exponential function.

$$M(t) = M_0 e^{-kt}$$

M is the mass of substance at time t

M_0 is the starting mass at $t=0$

Magnesium-27 decays exponentially and has a half-life of 9.45 minutes

$$M(t) = 10e^{-0.07335t}, \text{ where } t \text{ is minutes, } M_0=10g$$

$f(x) = 10e^{-0.07335x}$

Euler's number *e*, can also be determined by a series summation

$$e = \frac{1}{0!} + \frac{1}{1!} + \frac{1}{2!} + \frac{1}{3!} + ... + \frac{1}{n!} \approx 2.718........$$

2.718281828459045235360287
471352662497757247093 6999
595749669672   07663035
4759457132   8   6642742
746639192   030   1817413
5966290   290   42952605
073813   3490763233
829880   952510190115738
341879   21540   4993488
4167509   6   668082264
800168477   5537423454424
371075390777 4499206955170
6133138458300075204...

As an example, here is the computation of *e* to 10 decimal places:

1/0! =	1/1 =	1.0000000000
1/1! =	1/1 =	1.0000000000
1/2! =	1/2 =	0.5000000000
1/3! =	1/6 =	0.1666666667
1/4! =	1/24 =	0.0416666667
1/5! =	1/120 =	0.0083333333
1/6! =	1/720 =	0.0013888889
1/7! =	1/5040 =	0.0001984127
1/8! =	1/40320 =	0.0000248016
1/9! =	1/362880 =	0.0000027557
1/10! =	1/3628800 =	0.0000002756
1/11! =	1/39916800 =	0.0000000251
1/12! =	1/479001600 =	0.0000000021
1/13! =	1/6227020800 =	0.0000000002
1/14! =	1/87178291200 =	0.0000000000

The sum of the values in the right column is 2.7182818285 which is approximately "*e*."

Example 4.5: Using numerical methods to find $e^x = \frac{x^0}{0!} + \frac{x^1}{1!} + \frac{x^2}{2!} + \frac{x^3}{3!} + ... + \frac{x^n}{n!}$

$$e^{0.3} = \frac{0.3^0}{0!} + \frac{0.3^1}{1!} + \frac{0.3^2}{2!} + \frac{0.3^3}{3!} + \frac{0.3^4}{4!} + .$$
$$= 1 + 0.3 + 0.045 + 0.0045 + 0.0003375 \approx 1.3498$$

Compare this with finding $e^{0.3}$ on your calculator.

Alternatively you can use a Scientific calculator to work it out. Scientific calculator buttons for finding exponents.

x^y	y^x	a^y
$x^\blacksquare$	$\wedge$	a^x

Exponential Growth.

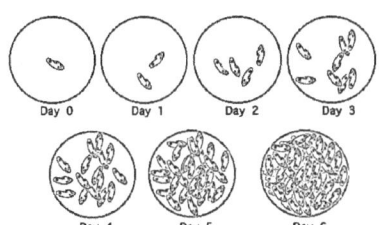

Logarithmic functions

Logarthmic functions are the inverse functions of the Exponential functions, and in contrast are slow changing functions. $f(x) = \log_a x, a > 1$ and $f(x) = \log_e x$ are some of the forms of these functions.

$\log_a N = x \Leftrightarrow N = a^x$

$\log_e N = x \Leftrightarrow N = e^x$

$\log_e N \equiv \ln N \quad natural \log$

As for exponential functions, the value a is referred to as the base of the function, and it is always a positive value. Why? Try evaluating values for $f(x) = \log_a x$ with a negative base value of a. What is te conclusion?

Logarithms are very useful in solving exponential equations being their inverse functions, they naturally complement each other. Logarithms are used to allow simpler calculations for extreme sized measurements and metrics, which can be converted to logarithms.

Logarithmic growth	Inverse function of the Exponential function.
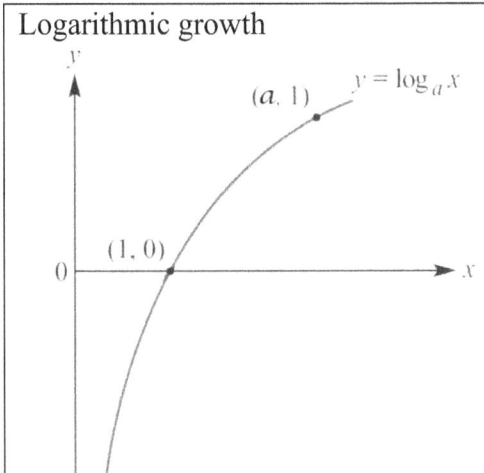	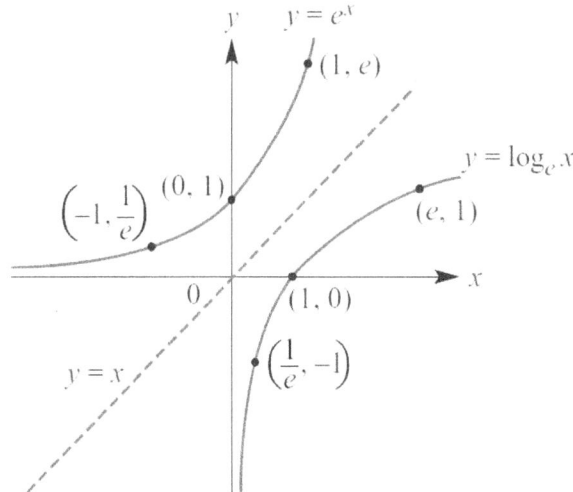
$f(x) = \log_a(x)$ is a 1:1 function	$f(x) = \log_a(x)$ is the inverse function for
The *y-axis* is a vertical asymptote	$g(x) = a^x$
As $x \to 0^+$, $f(x) \to -\infty$	
Domain f is R^+, Range f is R	Mirror image in the line $y = x$
$f(a) = \log_a(a) = 1$	Composing the functions results in x
$f(1) = \log_a(1) = 0$	• $f(g(x)) = \log_a(a^x) = x$
$f\left(\dfrac{1}{a}\right) = \log_a\left(\dfrac{1}{a}\right)$	• $g(f(x)) = a^{\log_a x} = x$
$\quad = \log_a(a^{-1})$	If $f(x) = \log_2(x)$ and $g(x) = 2^x$
$\quad = (-1)\log_a(a)$	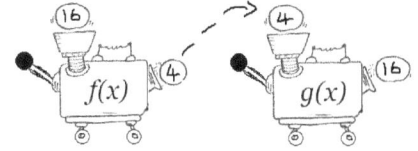
$\quad = -1$	

Example 4.6: Logarithms and exponents are useful for arithmetic with extremely large or extremely small numbers.

Large numbers	Small Numbers
$L = 100000 \times 1000000 = 100000000000$	$S = 0.001 \div 10000000 = 0.0000000001$
$L = 10^5 \times 10^6 = 10^{11}$	$S = 10^{-3} \div 10^7 = 10^{-10}$
The same arithmetic with logarithms.	The same arithmetic with logarithms.
$\log_{10} L = \log_{10} 100000 + \log_{10} 1000000$	$\log_{10} S = \log_{10} 0.001 - \log_{10} 10000000$
$= 5 + 6$	$= -3 - 7$
$= 11$	$= -10$
$\log_{10} L = 11 \Leftrightarrow L = 10^{11}$	$\log_{10} S = -10 \Leftrightarrow S = 10^{-10}$

Logarithmic laws mirror the Index laws for doing arithmetic.

Logarithm Law	For identical bases.	Related Index Law
Logarithm of a product	$\log_a m + \log_a n = \log_a mn$	$a^m \times a^n = a^{m+n}$
Logarithm of a quotient	$\log_a m - \log_a n = \log_a(\frac{m}{n})$	$a^m \div a^n = a^{m-n}$
Logarithm of a power	$\log_a m^n = n\log_a m$	$(a^m)^n = a^{mn}$
Logarithm of the base	$\log_a a = 1$	$a^1 = a$
Logarithm of one	$\log_a 1 = 0$	$a^0 = 1$

Example 4.7: Using the logarithm laws to solve for unknown variables or to expand or simplify.

For identical log bases.	For identical log bases.	
$\log (x + 3) = \log 14$	$\log x + \log 3 = \log 15$	$\log_a \dfrac{A\sqrt{C}}{B^2}$
$x + 3 = 14$	$\log 3x = \log 15$	$\log_a A\sqrt{C} - \log_a B^2$
$x = 11$	$3x = 15$	$\log_a A + \log_a \sqrt{C} - 2\log_a B$
	$x = 5$	$\log_a A + \dfrac{1}{2}\log_a C - 2\log_a B$

Proof of Quotient Logarithm Law. $\log_a\left(\dfrac{m}{n}\right) = \log_a(m) - \log_a(n)$

Let $x = \log_a(m)$ and $y = \log_a(n)$ then $a^x = m$ and $a^y = n$

$$\log_a\left(\frac{m}{n}\right) = \log_a\left(\frac{a^x}{a^y}\right) = \log_a\left(a^{x-y}\right) = x - y = \log_a(m) - \log_a(n)$$

Logarithmic bases can be any positive real value.

- e^x is an exponent to base e and *ln* is called a natural logarithm to base e.
- 10^x is an exponent to base 10 , $\log_{10}$ is called a logarithm to base 10.
- 2^x is an exponent to base 2 , $\log_2$ is called a logarithm to base 2.

You can change the base $\log_a x = \dfrac{\log_b x}{\log_b a}$, where **b** can be **10** or **e** if using scientific calculator

For example: $\log_2 7 = \dfrac{\log_e 7}{\log_e 2} = \dfrac{\log_{10} 7}{\log_{10} 2} \approx 2.8074$ *therefore* $2^{2.8074} \approx 7$, give it a try.

Finding the logarithm of any real positive number using technology.

Scientific calculators have function keys to find the logarithm of numbers to the base of 10 or of e.

A scientific calculator can be used to find the logarithm of any number x to any base a.

$$\log_a x = \frac{\log_{10} x}{\log_{10} a} \text{ or } \log_a x = \frac{\log_e x}{\log_e a}$$

	0	1	2	3	4	5	6	7	8	9	1 2 3	4 5 6	7 8 9
1·0	·0000	0043	0086	0128	0170	0212	0253	0294	0334	0374	4 8 12	17 21 25	29 33 37
1·1	·0414	0453	0492	0531	0569	0607	0645	0682	0719	0755	4 8 11	15 19 23	26 30 34
1·2	·0792	0828	0864	0899	0934	0969	1004	1038	1072	1106	3 7 10	14 17 21	24 28 31
1·3	·1139	1173	1206	1239	1271	1303	1335	1367	1399	1430	3 6 10	13 16 19	23 26 29
1·4	·1461	1492	1523	1553	1584	1614	1644	1673	1703	1732	3 6 9	12 15 18	21 24 27
1·5	·1761	1790	1818	1847	1875	1903	1931	1959	1987	2014	3 6 8	11 14 17	20 22 25
1·6	·2041	2068	2095	2122	2148	2175	2201	2227	2253	2279	3 5 8	11 13 16	18 21 24
1·7	·2304	2330	2355	2380	2405	2430	2455	2480	2504	2529	2 5 7	10 12 15	17 20 22
1·8	·2553	2577	2601	2625	2648	2672	2695	2718	2742	2765	2 5 7	9 12 14	16 19 21
1·9	·2788	2810	2833	2856	2878	2900	2923	2945	2967	2989	2 4 7	9 11 13	16 18 20
2·0	·3010	3032	3054	3075	3096	3118	3139	3160	3181	3201	2 4 6	8 11 13	15 17 19

Source: http://abitofauldmaths.org/2013/08/how-to-use-log-tables/

To calculate the value of $\log_{10}(1.4)$

using an old fashioned method of looking up tables, travel down the leftmost column to 1.4 and read off the digits next to this to get the value

$\log_{10}(1.4) = 0.1461$, which is the value correct to 4 decimal places.

In the same row $\log_{10}(1.44) = 0.1584$.

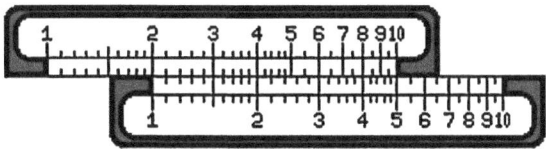

Using a slide rule to multiply by 2, align the two slide rule scales so, 1 was placed under the 2; therefore, 2 is under the 4, 3 is under the 6, 4 is under the 8, and 5 is under the 10.

---

https://en.wikipedia.org/wiki/Logarithm#Calculation

A classical method for positive numbers N larger than 2 is to find the $\log_2 N$, by repeatedly dividing N by 2. The answer found can be converted to any other base by $\log_a x = \dfrac{\log_b x}{\log_b a}$.

There are several algorithms for working out logarithms from first principles if you have the interest and the time you can do a bit of research.

QUESTION 1: Is it possible for a Logarithm to have a negative base?

If you graph $f(x) = (+2)^x$ you get a smooth continuous curve. It's inverse $f^{-1}(x) = \log_2 x$ also gives a smooth continuous logarithm curve.

If you try graphing $f(x) = (-2)^x$, the result **will not** be a smooth continuous curve.

- When x is an odd integer, $f(x)$ is negative. When x is an even integer, $f(x)$ is positive.
- When x is a fraction with an odd denominator, $f(x)$ will be positive or negative, depending on the numerator.
- When x is a fraction with even denominator, $f(x)$ is undefined in the Real number system (because it requires you to take an even root of -2).
- When x is not a fraction or an integer (like $(-2)^\pi$ or $(-2)^{7.865}$), we cannot tell whether $f(x)$ is positive or negative.
- The result is a discontinuous set of points jumping from positive to negative, and undefined for most real numbers.
- Its inverse $f^{-1}(x) = \log_{(-2)} x$ is mostly undefined in the Real system of numbers.

ANSWER: Not in the Real Number System.

Unreal.

QUESTION 2: Is it possible to find the Logarithm of a negative number?

If we follow on our argument from QUESTION 1 above and we cannot have a logarithm with a negative base in the real number system. Then the answer to QUESTION 2 will also be NO, since the base will always be positive real number.

Consider the real values of variables N, x, y where $N > 0$, and $\log_N y = x$, since $N > 0$, then $y = N^x$ will always be positive, and this is still true if x is negative.

If you think about this with N=5, then $y = 5^2$ and $y = 5^{-2}$ are positive.

Logarithmic Functions in real life.

Logarithms were introduced by the Scottish Mathematician John Napier in the early 17th century as a means to simplify calculations.

They were rapidly adopted by navigators, scientists, engineers, surveyors and others to perform high-accuracy computations more easily. (source Wikipedia)

Logarithms are very useful in solving exponential equations being their inverse functions, they naturally complement each other.

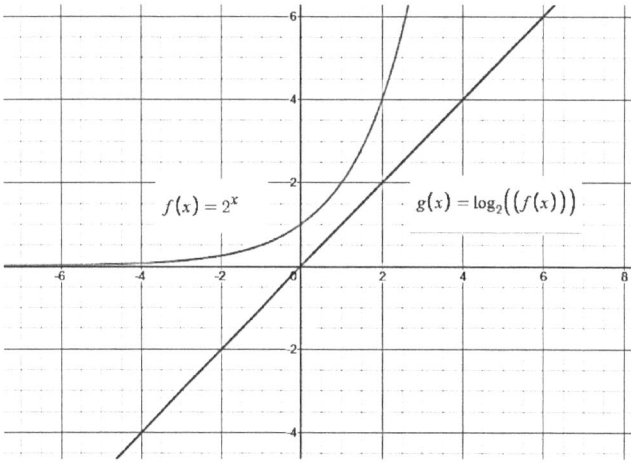

In real life logarithmic functions are used to represent levels of sound in decibels, the brightness of stars, the acidity and pH balance of chemicals and the scale of earthquakes, using the Richter scale.

Logarithms are used to allow simpler calculations for these measurements and metrics, since these measurements grow exponentially they become difficult to apply calculations to as the scales are either extremely enormous or extremely tiny.

Richter scale for measuring the magnitude of earthquakes is a logarithmic scale.

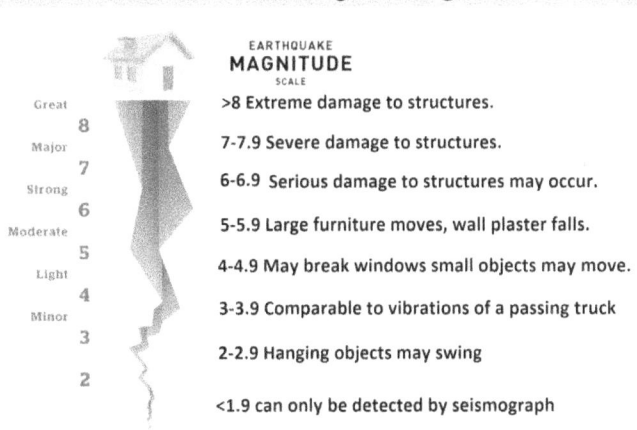

EARTHQUAKE
MAGNITUDE
SCALE

Great
8
Major
7
Strong
6
Moderate
5
Light
4
Minor
3
2

>8 Extreme damage to structures.

7-7.9 Severe damage to structures.

6-6.9 Serious damage to structures may occur.

5-5.9 Large furniture moves, wall plaster falls.

4-4.9 May break windows small objects may move.

3-3.9 Comparable to vibrations of a passing truck

2-2.9 Hanging objects may swing

<1.9 can only be detected by seismograph

The Richter scale measurements are logarithmic base 10, which means that an earthquake of magnitude 9.0 would be 10 times as strong as an earthquake of magnitude 8.0

Similarly, an earthquake of magnitude 9.0 would be 10^4 times as powerful as an earthquake that measured 5.0

Decibel (dB) scale for measuring noise is a logarithmic scale.

Typical Sound Levels (dBA)

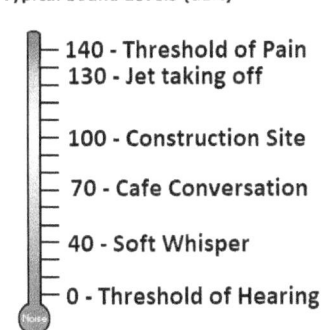

140 - Threshold of Pain
130 - Jet taking off

100 - Construction Site

70 - Cafe Conversation

40 - Soft Whisper

0 - Threshold of Hearing

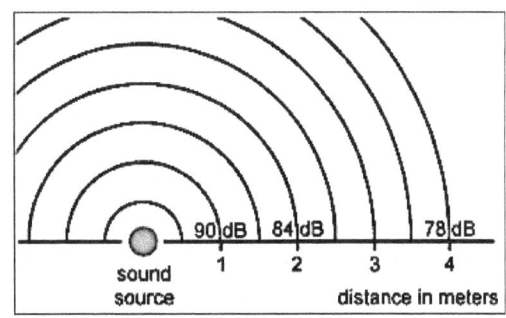

90 dB 84 dB 78 dB
sound   1   2   3   4
source   distance in meters

A small change in the number of decibels indicates a huge change in the amount of noise.

The decibel scale compresses sound pressures important to human hearing 0dB to 140dB into a manageable scale.

FABULOUS FUNCTIONS SUMMARY:

Circular Functions

Sine function $f(\theta) = A\sin(n(\theta - b)) + c$.

Cosine function $f(\theta) = A\cos(n(\theta - b)) + c$.

Tangent function $f(\theta) = A\tan(n(\theta - b)) + c$

Solving Equations.

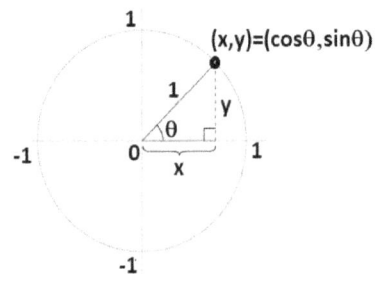

$cos(x) = \theta$ $x = \cos^{-1}(\theta)$

$\sin(x) = \theta$ $x = \sin^{-1}(\theta)$

$\tan(x) = \theta$ $x = \tan^{-1}(\theta)$

Reference angles of rotation from the origin give exact values for sin, cos and tan

θ	0	$\dfrac{\pi}{6}$ 30°	$\dfrac{\pi}{4}$ 45°	$\dfrac{\pi}{3}$ 60°	$\dfrac{\pi}{2}$ 90°
$\sin\theta$	$\dfrac{\sqrt{0}}{2}$	$\dfrac{\sqrt{1}}{2}$	$\dfrac{\sqrt{2}}{2}$	$\dfrac{\sqrt{3}}{2}$	$\dfrac{\sqrt{4}}{2}$
$\cos\theta$	$\dfrac{\sqrt{4}}{2}$	$\dfrac{\sqrt{3}}{2}$	$\dfrac{\sqrt{2}}{2}$	$\dfrac{\sqrt{1}}{2}$	$\dfrac{\sqrt{0}}{2}$
$\tan\theta$	0	$\dfrac{\sqrt{3}}{3}$	1	$\sqrt{3}$	undefined

Pythagorean Identity: $\cos^2(\theta) + \sin^2(\theta) = 1$

Exponential Functions

Exponential Growth	Exponential Decay
$f(x) = a^x, a > 1$	$f(x) = a^{-x}, a > 1$
$f(x) = a^{-x}, 0 < a < 1$	$f(x) = \dfrac{1}{a^x}, a > 1$
$f(x) = \dfrac{1}{a^x}, 0 < a < 1$	$f(x) = a^x, 0 < a < 1$

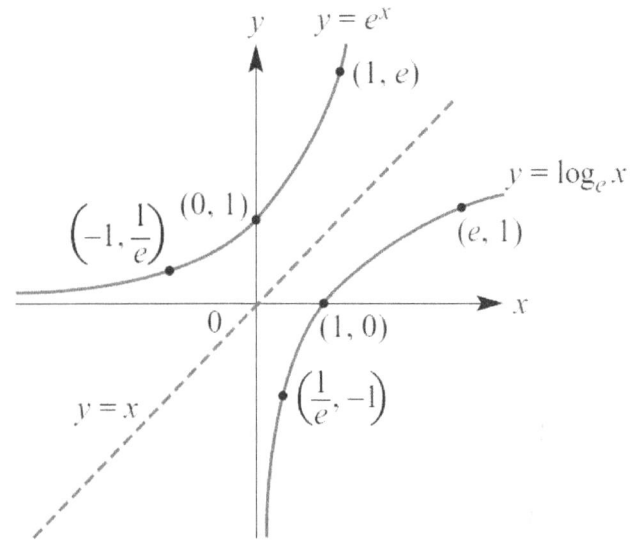

Logarithmic Functions

$\log_a N = x \Leftrightarrow N = a^x$

$\log_e N = x \Leftrightarrow N = e^x$

$\log_e N \equiv \ln N$ *natural* log

$f(x) = \log_a(x)$ is the inverse function for

$g(x) = a^x$

Mirror image in the line $y = x$

Composing the functions results in x

- $f(g(x)) = \log_a(a^x) = x$

- $g(f(x)) = a^{\log_a x} = x$

For identical bases.	Related Index Law
$\log_a m + \log_a n = \log_a mn$	$a^m \times a^n = a^{m+n}$
$\log_a m - \log_a n = \log_a(\dfrac{m}{n})$	$a^m \div a^n = a^{m-n}$
$\log_a m^n = n\log_a m$	$(a^m)^n = a^{mn}$
$\log_a a = 1$	$a^1 = a$
$\log_a 1 = 0$	$a^0 = 1$

FABULOUS FUNCTIONS: Check your Understanding.

4.1 Explore the Circular Functions.

a) Describe the transformations in function and matrix notation and show the mappings for $y = \tan(x)$

to become $y^* = 3\tan\left(\dfrac{x}{2} - \dfrac{\pi}{4}\right)$ and sketch both functions for $x \in [0, 4\pi]$

b) The Ferris Wheel at Melbourne Show rotates clockwise at a constant speed completing 15 rotations every hour.

The wheel has a diameter of 90 metres and the bottom of the wheel is 6 metres above the ground.

The Wheel has twelve *swing style suspended* cabins similar to the ones shown in the picture, that are evenly spaced around the wheel.

A cabin starts at the bottom of the wheel with the top of the cabin 6 metres above the ground.

   i.   Draw a rough diagram of the Ferris Wheel and the cabins, labelling all the given measurements.

   ii.   Find the greatest height of the top of any cabin reaches as the wheel rotates.

After t minutes, the height $h(t)$ metres above the ground of the top of a cabin is given by the function $h(t) = 51 - a\cos(bt)$, where a, b are real constants.

   iii.   Find the period of $h(t)$, find the value of b, find the value of a.

   iv.   Find the height of all the other cabins at this moment.

   v.   Sketch the graph of $h(t)$, for one full rotation of the wheel for that cabin.

   vi.   In one rotation of the wheel, find the probability that a randomly selected seat is at least 70 metres above the ground. Give your answer to two decimal places.

c) Helen the hula hoop lady works in the Circus next to the Ferris Wheel.

In one show she spins 7 hula hoops with the following periodic functions.

$h_1(x) = 2\cos(2x)$

$h_2(x) = -2\cos(4(x - \pi))$

$h_3(x) = 2\cos\left(3\left(x - \dfrac{\pi}{2}\right)\right)$

$h_4(x) = 3\sin(4x - \pi)$

$h_5(x) = -3\sin(2x)$

$h_6(x) = \cos(3(x - \pi)) + 1$

$h_7(x) = -\cos(3x) - 1$

   i.   Find the value(s) of x where $h_3(x) = 0$

   ii.   Find the value(s) of x where $h_2(x) = 1$

   iii.   Graph some of the solutions for $h_2(x) = 1$ about $x = 0$

   iv.   Find the value(s) of x where $h_6(x) = 1$

   v.   Using technology, or otherwise find the general solution for $h_2(x) = h_6(x) = 1$

   vi.   Using technology find and state the range of

        $y = h_1(x) + h_2(x) + h_3(x) + h_4(x) + h_5(x) + h_6(x) + h_7(x)$ to the accuracy of 2 decimal places.

4.2 Explore the Exponential function.

a) Take the function(s) $f(x) = 5e^x$ and $g(x) = 0.1e^{-x}$ and apply the transformations in sequence of Reflection, Dilation and Translations as described. Graph the final results.

   i.   **Reflection in the x-axis** $(x, y) \rightarrow (x, -y)$

   ii.   **Followed by Dilations by a factor of 4 from the y-axis** $(x, y) \rightarrow (4x, y)$

   iii.   **Followed by Translations +2 units horizontally +1 units vertically**

      $(x, y) \rightarrow (x + 2, y + 1)$

b) Find the equation of the form $f(x) = ae^x + b$, where $a, b \in R$ for the graph shown below: We are given the points $(0, 1)$ and $(\log_e 2, 0)$.

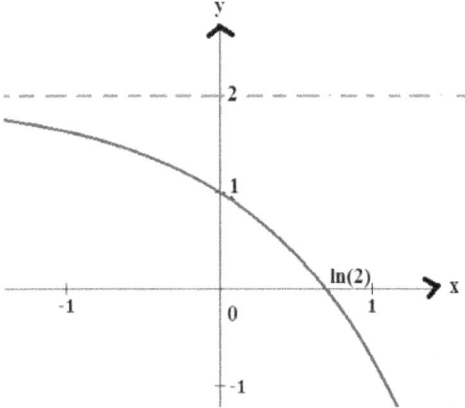

c) On the same axes sketch $f(x) = e^x + 2$ and $g(x) = e^{-x}$ in the domain $[-3, 3]$ by the addition of ordinates sketch

    $h(x) = f(x) + g(x)$

       $= e^x + 2 + e^{-x}$

d) The population of a town is estimated to increase by 15% per year. The population today is 20000. The function describing the population P as a function of time t is of the form

    $P(t) = A(b)^t, \text{ where } A, b \in R$

   i.   Find *A, b* for the function *P(t)*

   ii.   Complete the following table for

t	$P(t)$
-5	$P(-5) =$
0	$P(0) =$
5	$P(5) =$
10	$P(10) =$

   iii.   Does a negative value of *t* make sense?

   iv.   Make a graph of the population function

   v.   What is the population in ten years from now?

4.3 Explore the logarithmic function.

a)

Find the equation of the form $y = A\ln(x+b)+B$

where $A,b,B \in R$

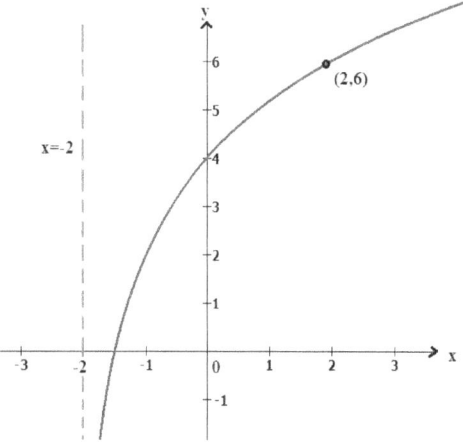

Use the given points $(0,4)$ and $(2,6)$ and the

asymptote $x = -2$

b) Caffeine stays in our system for some time after drinking a cup. It leaves the body at around 12% per hour, giving it a half-life of approximately 5 hours, with variations for individuals based on metabolism, tolerance, genetics, diet and body mass.

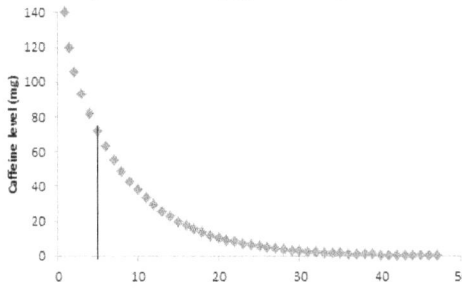

Cal drank a cup of coffee with 140mg of caffeine at 11am, and had 70mg of caffeine remaining around 4pm.

Ann, Bob, Don and Enid had slightly different reductions in caffeine over time, as shown in the table below.

Person	Caffeine from 1 cup of Coffee serving in mg	After 5 hours in mg	Caffeine C as a function of time t in hours.
Ann	140	30	$C_A(t) = 140(2)^{-at}, a = ?$
Bob	140	50	$C_B(t) = 140(2)^{-bt}, b = ?$
Cal	140	70	$C_C(t) = 140(2)^{-ct}, c = ?$
Don	140	90	$C_D(t) = 140(2)^{-dt}, d = ?$
Enid	140	110	$C_E(t) = 140(2)^{-et}, e = ?$

i. Complete the functions $C_X(t)$ for each person $X=\{Ann, Bob, Cal, Don, Enid\}$ using the information in the table and determine the value of the constants a,b,c,d,e to 4 decimal places.

ii. Plot the amount of Caffeine remaining for each person for 24 hours on a Cartesian plane. Is there a pattern that emerges?

iii. Using algebra find how long does it take for each person to have < 10mg of caffeine in their system in hours to 4 decimal places.

iv. For each person find the inverse function of Caffeine in their system over time, graph the results on the same axis. What is the interpretation of the inverse functions?

FABULOUS FUNCTIONS SOLUTIONS.

4.1 Explore the Circular Functions. (SOLUTIONS)

a) By function recognition

$$y^* = 3\tan\left(\frac{x}{2} - \frac{\pi}{4}\right) = 3\tan\left(\frac{1}{2}\left(x - \frac{\pi}{2}\right)\right)$$

Let $f(x) = \tan(x)$ then

$$y^* = 3f\left(\frac{1}{2}\left(x - \frac{\pi}{2}\right)\right)$$

- Dilation by 3 from x-axis
- Dilation by 2 from y-axis
- Translation $+\frac{\pi}{2}$ horizontally

Period of $y^* = \frac{\pi}{\frac{1}{2}} = 2\pi$

By mapping $(x, y) \to (x', y')$

rearrange the y^*

$$\frac{y^*}{3} = \tan\left(\frac{x}{2} - \frac{\pi}{4}\right)$$

$$y = \frac{y'}{3} \qquad x = \frac{x'}{2} - \frac{\pi}{4}$$

$$y' = 3y \qquad x' = 2x + \frac{\pi}{2}$$

$$(x, y) \to \left(2x + \frac{\pi}{2}, 3y\right)$$

b) Ferris Wheel Drawings

 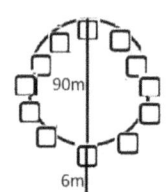

Many possible Ferris Wheels can be drawn (Not to scale)

i. Cabins are swing style and suspended.
ii. 96 metres
iii. With 15 rotations per hour, each period is 4 minutes. Therefore the dilation factor b of the regular cosine function needs to be
$$\frac{2\pi}{b} = 4 \Rightarrow b = \frac{2\pi}{4} = \frac{\pi}{2}.$$
The amplitude a is equal to 45, which is the radius of the Ferris wheel. $h(t) = -45\cos\left(\frac{\pi}{2}t\right) + 51$

iv. Ferris Wheel circle equation is $x^2 + (y - 51)^2 = 45^2$ and the cabins are evenly spaced with an angle of $\frac{\pi}{6}$ between them relative to the centre of the Ferris Wheel circle.

coordinates cabins

$$(0,6), \left(\frac{45\sqrt{3}}{2}, -\frac{45}{2} + 51\right), \left(\frac{45}{2}, -\frac{45\sqrt{3}}{2} + 51\right),$$

$$(45,51), \left(\frac{45}{2}, \frac{45\sqrt{3}}{2} + 51\right), \left(\frac{45\sqrt{3}}{2}, \frac{45}{2} + 51\right),$$

$$(0,96), \left(\frac{45\sqrt{3}}{2}, \frac{45}{2} + 51\right), \left(\frac{45}{2}, \frac{45\sqrt{3}}{2} + 51\right),$$

$$(-45,51), \left(-\frac{45}{2}, -\frac{45\sqrt{3}}{2} + 51\right), \left(-\frac{45\sqrt{3}}{2}, -\frac{45}{2} + 51\right)$$

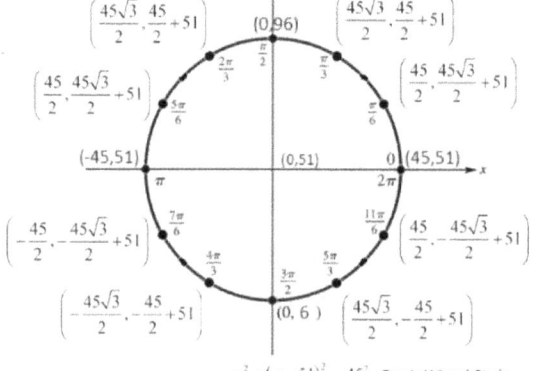

$x^2 + (y - 51)^2 = 45^2$ Ferris Wheel Circle

1 cabin at height 6m,
1 cabin at height 96 m,
2 cabins at height 51m,
2 cabins at height 73.5 m,
2 cabins at height $\approx$89.97m,
2 cabins at height 28.5 m,
1 cabins at height $\approx$12.03m

v. Ferris Wheel cabin 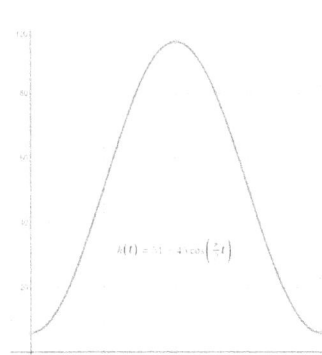	vi. Each cabin is above 70m for approximately 2.723-1.277=1.446 minutes. 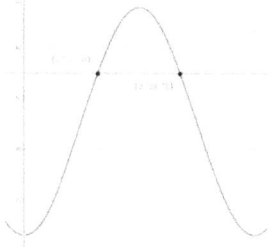 Therefore in a full rotation $$\Pr(cabin > 70m) = \frac{1.446}{4} \approx 0.3615$$

c) Hula hoops

i.   Find the value(s) of x where $h_3(x) = 0$ as a general solution since the domain is unrestricted.

$$2\cos\left(3\left(x - \frac{\pi}{2}\right)\right) = 0$$

$$let\ \theta = 3\left(x - \frac{\pi}{2}\right)$$

$$2\cos(\theta) = 0$$

$$\cos(\theta) = 0$$

$$\theta = \left\{\frac{\pi}{2}, \frac{3\pi}{2}\right\}$$

$$\theta = \frac{\pi}{2} + \pi k, k \in Z$$

$$3\left(x - \frac{\pi}{2}\right) = \frac{\pi}{2} + \pi k$$

$$\left(x - \frac{\pi}{2}\right) = \frac{\pi}{6} + \frac{\pi}{3}k$$

$$x = \frac{\pi}{6} + \frac{\pi}{2} + \frac{\pi}{3}k$$

$$x = \frac{4\pi}{6} + \frac{\pi}{3}k$$

$$x = \frac{2\pi}{3} + \frac{\pi}{3}k$$

ii.   Find the value(s) of x where $h_2(x) = 1$ as a general solution since the domain is unrestricted.

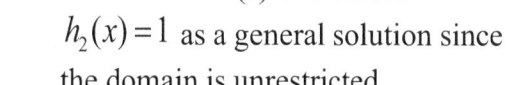

$$-2\cos(4(x - \pi)) = 1$$

$$\cos(4(x - \pi)) = -\frac{1}{2}$$

$$let\ \theta = 4(x - \pi)$$

$$\cos(\theta) = -\frac{1}{2}$$

$$\theta = \left\{\frac{2\pi}{3} + 2k\pi, \frac{4\pi}{3} + 2k\pi\right\}, k \in Z$$

$$4(x - \pi) = \left\{\frac{2\pi}{3} + 2k\pi, \frac{4\pi}{3} + 2k\pi\right\}$$

$$(x - \pi) = \left\{\frac{2\pi}{12} + \frac{2k\pi}{4}, \frac{4\pi}{12} + \frac{2k\pi}{4}\right\}$$

$$\pi = \frac{12\pi}{12}$$

$$x = \left\{\frac{2\pi}{12} + \frac{12\pi}{12} + \frac{2k\pi}{4}, \frac{4\pi}{12} + \frac{12\pi}{12} + \frac{2k\pi}{4}\right\}$$

$$x = \left\{\frac{14\pi}{12} + \frac{2k\pi}{4}, \frac{16\pi}{12} + \frac{2k\pi}{4}\right\}$$

$$x = \left\{\frac{7\pi}{6} + \frac{k\pi}{2}, \frac{4\pi}{3} + \frac{k\pi}{2}\right\}$$

c) continued

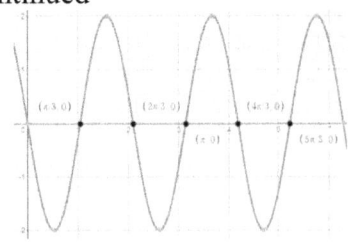

k	-3	-2	-1	0	1	2	3
$\left\{\dfrac{7\pi}{6}+\dfrac{k\pi}{2}\right\}$	$\dfrac{-\pi}{3}$	$\dfrac{\pi}{6}$	$\dfrac{2\pi}{3}$	$\dfrac{7\pi}{6}$	$\dfrac{5\pi}{3}$	$\dfrac{13\pi}{6}$	$\dfrac{8\pi}{3}$
$\left\{\dfrac{4\pi}{3}+\dfrac{k\pi}{2}\right\}$	$\dfrac{-\pi}{6}$	$\dfrac{\pi}{3}$	$\dfrac{5\pi}{6}$	$\dfrac{4\pi}{3}$	$\dfrac{11\pi}{6}$	$\dfrac{7\pi}{3}$	$\dfrac{17\pi}{6}$

iii.   Find the value(s) of x where $h_2(x)=1$, graph some solutions about $x=0$

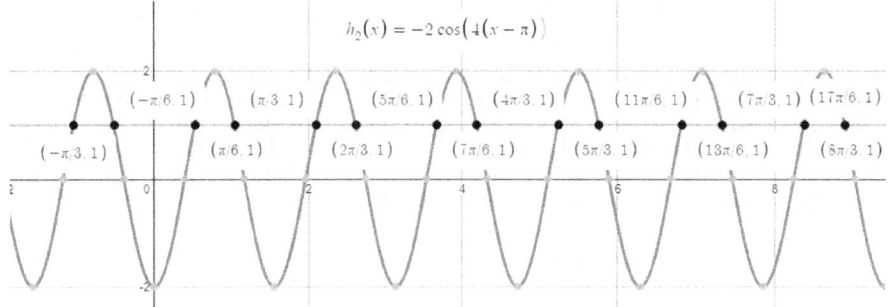

iv.   Find the value(s) of x where $h_6(x)=1$ as a general solution since the domain is unrestricted.

$$\cos\left(3\left(x-\pi\right)\right)+1=1$$
$$\cos\left(3\left(x-\pi\right)\right)=0$$
$$\text{let } \theta = 3\left(x-\pi\right)$$
$$\cos\left(\theta\right)=0$$
$$\theta=\left\{\frac{\pi}{2},\frac{3\pi}{2}\right\}$$
$$\theta=\frac{\pi}{2}+\pi k, k\in Z$$

$$3\left(x-\pi\right)=\left\{\frac{\pi}{2}+k\pi\right\}$$
$$\left(x-\pi\right)=\left\{\frac{\pi}{6}+\frac{k\pi}{3}\right\}$$
$$\pi=\frac{6\pi}{6}$$
$$x=\left\{\frac{\pi}{6}+\frac{6\pi}{6}+\frac{2k\pi}{6}\right\}$$
$$x=\left\{\frac{7\pi+2k\pi}{6}\right\}$$

k	-3	-2	-1	0	1	2	3
$\left\{\dfrac{7\pi}{6}+\dfrac{2k\pi}{6}\right\}$	$\dfrac{\pi}{6}$	$\dfrac{\pi}{2}$	$\dfrac{5\pi}{6}$	$\dfrac{7\pi}{6}$	$\dfrac{3\pi}{2}$	$\dfrac{11\pi}{6}$	$\dfrac{13\pi}{3}$

114

4.1 Explore the Circular Functions….continued.. (SOLUTIONS)

v.   Pattern 1

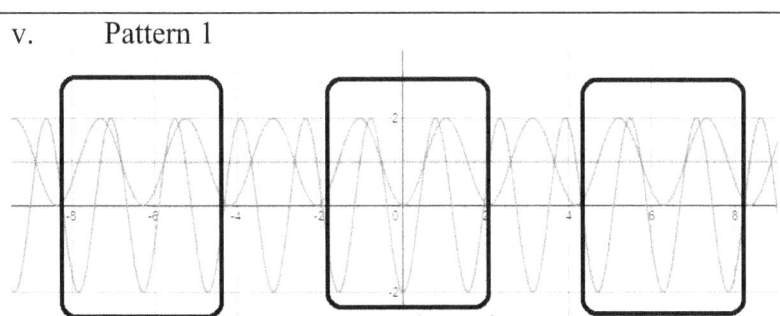

$$h_2(x) = -2\cos(4(x-\pi))$$
$$h_6(x) = \cos(3(x-\pi))-1$$
$$y = 1$$

$$\left\{......\frac{-13\pi}{6},\frac{-11\pi}{6},\frac{-\pi}{6},\frac{\pi}{6},\frac{11\pi}{6},\frac{13\pi}{6},......\right\} \ pattern 1 \ \left\{\frac{-\pi}{6}+2k\pi,\frac{\pi}{6}+2k\pi\right\}$$

Pattern 2

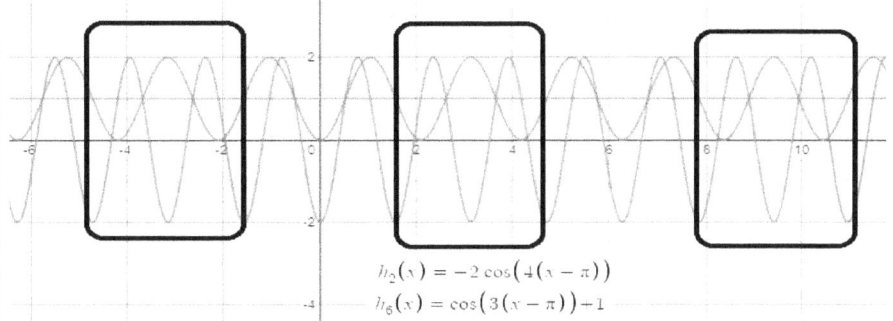

$$h_2(x) = -2\cos(4(x-\pi))$$
$$h_6(x) = \cos(3(x-\pi))-1$$
$$y = 1$$

$$\left\{......\frac{-7\pi}{6},\frac{-5\pi}{6},\frac{5\pi}{6},\frac{7\pi}{6},\frac{17\pi}{6},\frac{19\pi}{6},......\right\} \ pattern 2 \ \left\{\frac{-5\pi}{6}+2k\pi,\frac{5\pi}{6}+2k\pi\right\}$$

vi.   The range is approximately [-6.43,9.43]

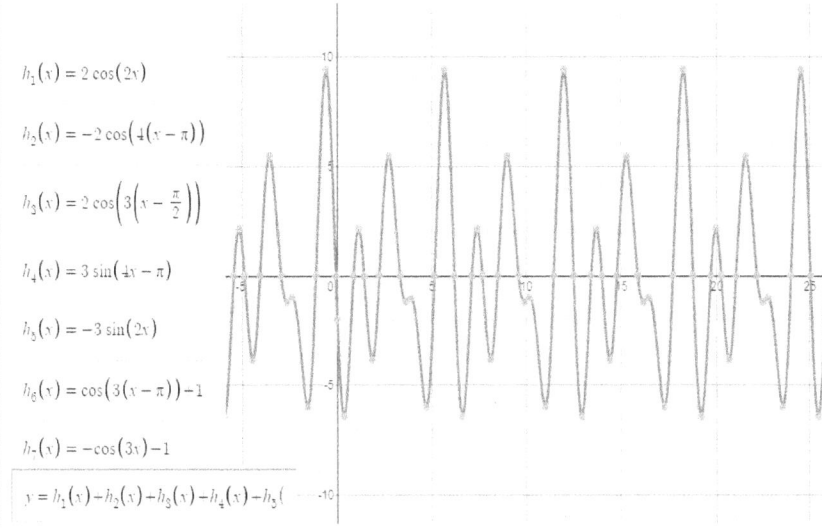

$$h_1(x) = 2\cos(2x)$$
$$h_2(x) = -2\cos(4(x-\pi))$$
$$h_3(x) = 2\cos\left(3\left(x-\frac{\pi}{2}\right)\right)$$
$$h_4(x) = 3\sin(4x-\pi)$$
$$h_5(x) = -3\sin(2x)$$
$$h_6(x) = \cos(3(x-\pi))-1$$
$$h_7(x) = -\cos(3x)-1$$
$$y = h_1(x) - h_2(x) + h_3(x) - h_4(x) - h_5($$

a) Take the function(s) $f(x) = 5e^x$ and $g(x) = 0.1e^{-x}$.

    i.   **Reflection in the x-axis** $(x, y) \rightarrow (x, -y)$

$$-f(x) = -5e^x, \quad -g(x) = -0.1e^{-x}$$

    ii.   **Followed by Dilations by a factor of 4 from the y-axis** $(x, y) \rightarrow (4x, y)$

$$-f\left(\frac{x}{4}\right) = -5e^{\left(\frac{x}{4}\right)}, \quad -g\left(\frac{x}{4}\right) = -0.1e^{-\left(\frac{x}{4}\right)}$$

    iii.   **Followed by Translations +2 units horizontally +1 units vertically**
$(x, y) \rightarrow (x+2, y+1)$

$$-f\left(\frac{1}{4}(x-2)\right) = -5e^{\left(\frac{1}{4}(x-2)\right)} + 1, \quad -g\left(\frac{1}{4}(x-2)\right) = -0.1e^{-\left(\frac{1}{4}(x-2)\right)} + 1$$

b) Find the equation of the form $f(x) = ae^x + b$, where $a, b \in R$ for the graph shown below:

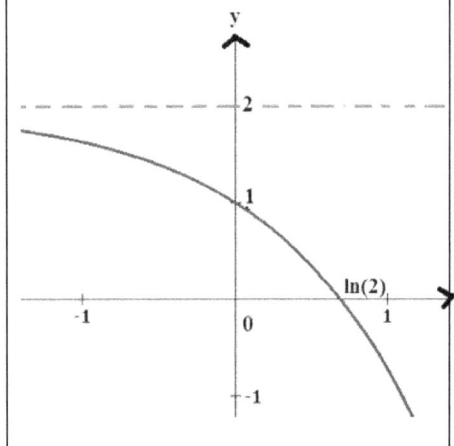

given points are $(0,1)$ *and* $(\log_e 2, 0)$

Using the x-intercept $(\log_e 2, 0)$
$$0 = ae^{\log_e 2} + b$$
$$\Rightarrow \quad 2a + b = 0 \quad eqn. 1$$

Using the y-intercept $(0,1)$
$$ae^0 + b = 1$$
$$\Rightarrow \quad a + b = 1 \quad eqn. 2$$

$eqn. 1 - eqn. 2 \qquad a = -1 \quad \therefore b = 2$

Equation is: $f(x) = -e^x + 2$

c) On the same axes sketch $f(x) = e^x + 2$ and $g(x) = e^{-x}$ in the domain $[-3, 3]$ by the addition of ordinates sketch
$$h(x) = f(x) + g(x)$$
$$= e^x + 2 + e^{-x}$$

x	-3	-2	-1	0	1	2	3
f(x)	2.05	2.13	2.35	3	4.75	9.39	22.28
g(x)	20.13	7.41	2.72	1	0.37	0.14	0.05
h(x)	22.18	9.54	5.07	4	5.12	9.53	22.32

Do your best to sketch this addition of ordinates by hand and compare it to a graph generated by technology.

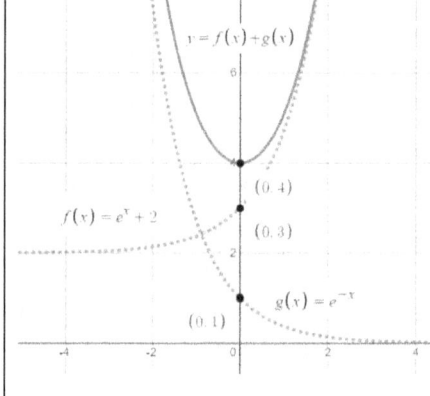

4.2 Explore the Exponential functions…continued.. (SOLUTIONS)

d) Population of a town increases by 15% per year. The population today is 20000.

   i. P is the population as a function of time t, A is the initial population, so $A=20000$. We must find what b is. Each year, the population increases by 15% of A, or $0.15A$. To find the total population for the following year, we must add the current population to the increase in population. In other words, $A+0.15A=1.15A$. This means that the base of the exponential is $b=1.15$. The formula that describes this problem is $P(t) = 20000(1.15)^{t}$.

   ii.   Complete the table of values.

t	$P(t) = 20000(1.15)^{t}$
-5	$P(-5) = 20000(1.15)^{-5}$
0	$P(0) = 20000(1.15)^{0} = 20000$
5	$P(5) = 20000(1.15)^{5} = 40200$
10	$P(10) = 20000(1.15)^{10} = 80911$

   iii.   Yes; negative time can represent time in the past. For example, $t=-5$ in this problem represents the population from five years ago.

   iv.   Graph the function.

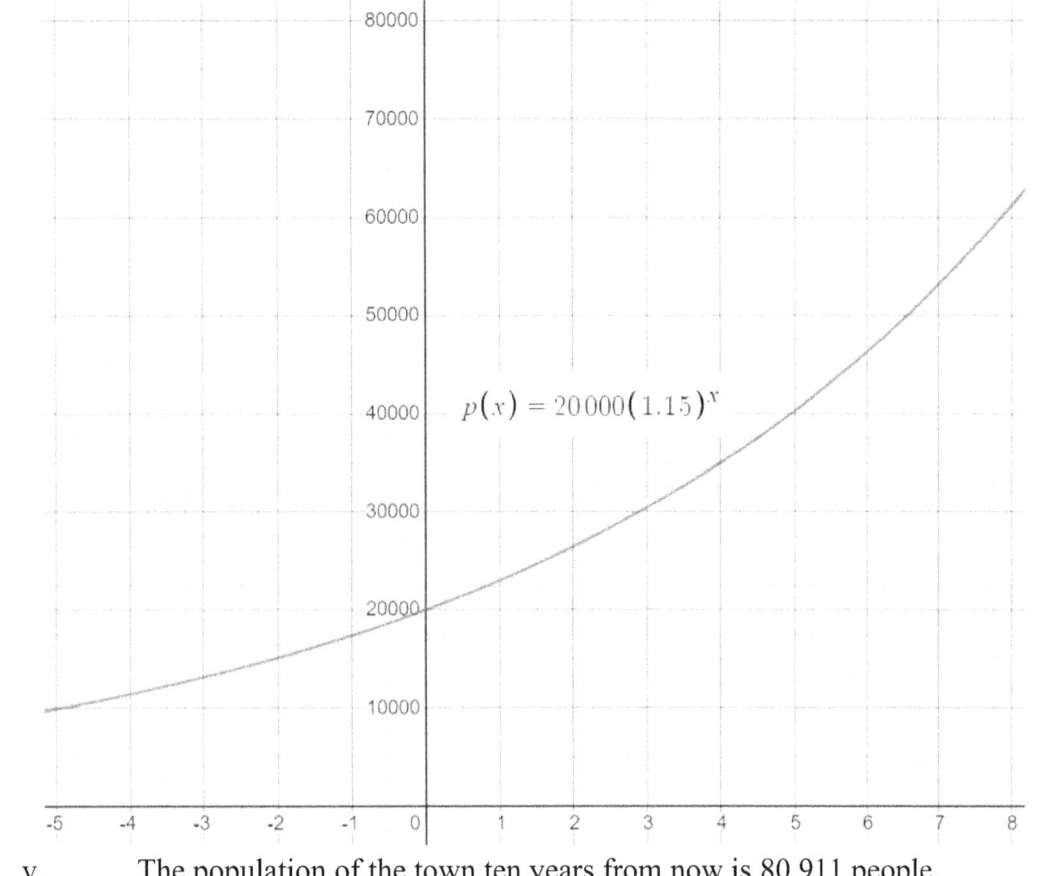

$$p(x) = 20000(1.15)^{x}$$

   v.   The population of the town ten years from now is 80,911 people.

a) Find $y = A\ln(x+b) + B$ where $A, b, B \in R$ points $(0,4)$ and $(2,6)$ and the asymptote $x = -2$

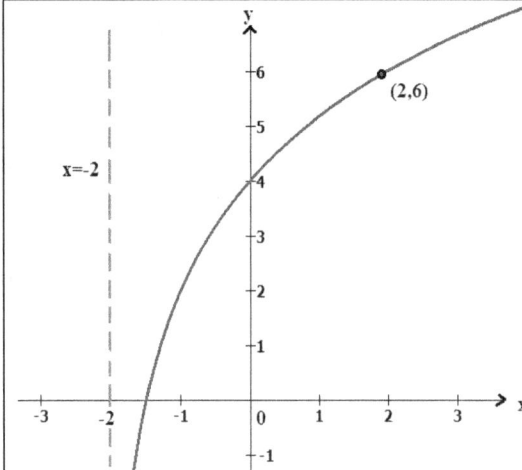

$x + b > 0 \quad \therefore x > -b$

Using the asymptote $x = -2 \Rightarrow b = 2$

Using the y-intercept $(0,4)$ and the point

$(2,6)$

$$4 = A\ln(2) + B \qquad (1)$$
$$6 = A\ln(4) + B \qquad (2)$$
$$(2) - (1)$$
$$2 = A\ln(4) - A\ln(2)$$
$$2 = A\ln(2)$$
$$A = \frac{2}{\ln(2)}$$

Substitute the <u>exact value of A</u> into the first equation

$$4 = \frac{2}{\ln(2)} \times \ln(2) + b$$

$4 = 2 + B \quad \therefore B = 2$

Equation is $y = \dfrac{2}{\ln(2)}\ln(x+2) + 2$

b) Caffeine decay modelling

i.

Ann: $C_A(0) = 140$,	Bob: $C_B(0) = 140$,	Don: $C_D(0) = 140$,
$C_A(5) = 30$	$C_B(5) = 50$	$C_D(5) = 90$
$30 = 140(2)^{-5a}$	$50 = 140(2)^{-5b}$	$90 = 140(2)^{-5d}$
$\dfrac{30}{140} = 2^{-5a}$	$b = \dfrac{1}{5}\log_2\left(\dfrac{14}{5}\right)$	$d = \dfrac{1}{5}\log_2\left(\dfrac{14}{9}\right)$
$\Rightarrow \dfrac{140}{30} = 2^{5a}$	≈ 0.2971	≈ 0.1275
$2^{5a} = \dfrac{14}{3}$	Cal: $C_C(0) = 140$,	Enid: $C_E(0) = 140$,
$\Rightarrow \log_2\left(2^{5a}\right) = \log_2\left(\dfrac{14}{3}\right)$	$C_C(5) = 70$	$C_E(5) = 110$
	$70 = 140(2)^{-5c}$	$110 = 140(2)^{-5e}$
$5a = \log_2\left(\dfrac{14}{3}\right)$	$c = \dfrac{1}{5}\log_2\left(\dfrac{14}{7}\right)$	$e = \dfrac{1}{5}\log_2\left(\dfrac{14}{11}\right)$
$a = \dfrac{1}{5}\log_2\left(\dfrac{14}{3}\right)$	$= 0.2$	≈ 0.0695
≈ 0.4445		

4.3 Explore the logarithmic function…continued.. (SOLUTIONS)

b) Continued….

Person	Caffeine from 1 cup of Coffee serving in mg	After 5 hours in mg	Caffeine C as a function of time t in hours.
Ann	140	30	$C_A(t) = 140(2)^{-at}, a = 0.4445$
Bob	140	50	$C_B(t) = 140(2)^{-bt}, b = 0.2971$
Cal	140	70	$C_C(t) = 140(2)^{-0.2t}, c = 0.2$
Don	140	90	$C_D(t) = 140(2)^{-dt}, d = 0.1275$
Enid	140	110	$C_E(t) = 140(2)^{-et}, e = 0.0695$

ii.

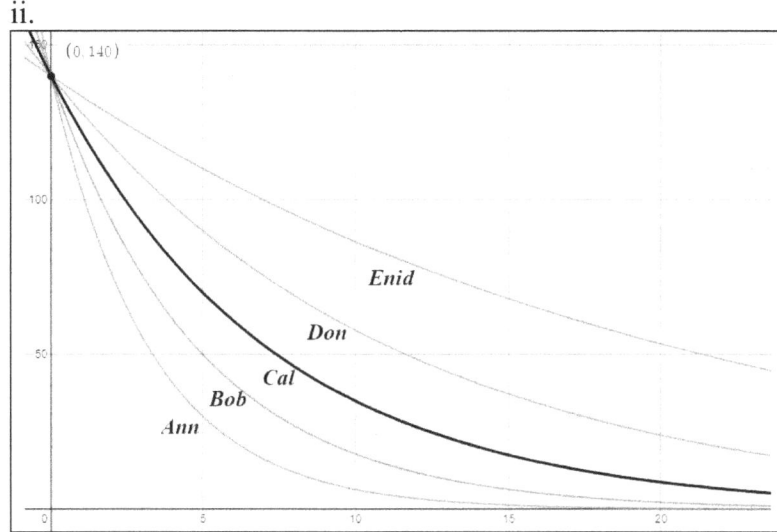

iii.

$Ann: C_A(t) = 10$

$10 = 140(2)^{-0.4445t}$

$2^{-0.4445t} = \dfrac{140}{10}$

$2^{0.4445t} = \dfrac{10}{140}$

$0.4445t = \log_2\left(\dfrac{1}{14}\right)$

$t \approx 8.5655 \ hours$

iii.

$Bob: C_B(t) = 10$	$Cal: C_C(t) = 10$	$Don: C_D(t) = 10$	$Enid: C_E(t) = 10$
$10 = 140(2)^{-0.2971t}$	$10 = 140(2)^{-0.2t}$	$10 = 140(2)^{-0.1275t}$	$10 = 140(2)^{-0.0695t}$
$2^{-0.2971t} = \dfrac{140}{10}$	$2^{-0.2t} = \dfrac{140}{10}$	$2^{-0.1275t} = \dfrac{140}{10}$	$2^{-0.0695t} = \dfrac{140}{10}$
$2^{0.2971t} = \dfrac{10}{140}$	$2^{0.2t} = \dfrac{10}{140}$	$2^{0.1275t} = \dfrac{10}{140}$	$2^{0.0695t} = \dfrac{10}{140}$
$0.2971t = \log_2\left(\dfrac{1}{14}\right)$	$0.2t = \log_2\left(\dfrac{1}{14}\right)$	$0.1275t = \log_2\left(\dfrac{1}{14}\right)$	$0.0695t = \log_2\left(\dfrac{1}{14}\right)$
$t \approx 12.8151 \ hours$	$t \approx 19.0368 \ hours$	$t \approx 29.8616 \ hours$	$t \approx 54.7821 \ hours$

The graphs forms are dilations of each other. Enid shouldn't drink coffee!

b) Continued…..

iv.

Person	Time t hours as a function of Caffeine C.
Ann	$t = \dfrac{1}{a}\log_2\left(\dfrac{140}{C_A}\right), a = 0.4445 \Rightarrow t = 2.2497\log_2\left(\dfrac{140}{C_A}\right)$
Bob	$t = \dfrac{1}{b}\log_2\left(\dfrac{140}{C_B}\right), a = 0.2971 \Rightarrow t = 3.3659\log_2\left(\dfrac{140}{C_B}\right)$
Cal	$t = \dfrac{1}{c}\log_2\left(\dfrac{140}{C_C}\right), c = 0.2 \Rightarrow t = 5\log_2\left(\dfrac{140}{C_C}\right)$
Don	$t = \dfrac{1}{d}\log_2\left(\dfrac{140}{C_D}\right), a = 0.1275 \Rightarrow t = 7.8431\log_2\left(\dfrac{140}{C_D}\right)$
Enid	$t = \dfrac{1}{e}\log_2\left(\dfrac{140}{C_E}\right), e = 0.0695 \Rightarrow t = 14.3885\log_2\left(\dfrac{140}{C_E}\right)$

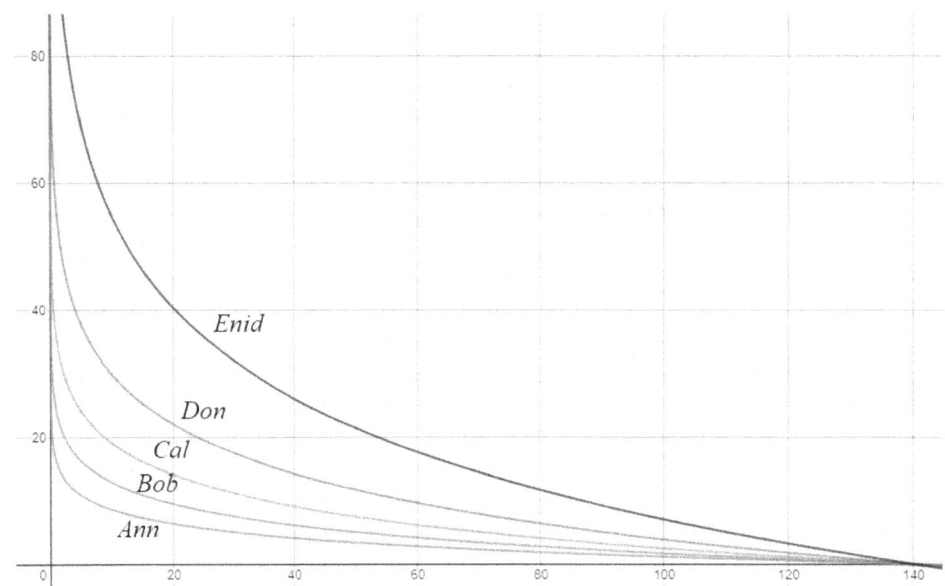

120

Deriving Derivatives.

Functions $f(x)$ change as x changes. We measure the changes in a function $f(x)$ by finding the derivative, which is a measure of change and is also known as the gradient function $f'(x) = \dfrac{dy}{dx}$.

This branch of mathematics is called differential calculus, and is used to work out the rate of change of quantities represented by functions.

The **average rate of change** for a function $f(x)$ between the coordinates $(x_1, f(x_1))$ and $(x_2, f(x_2))$ is given by the gradient of the line connecting the two points $$\frac{rise}{run} = \frac{f(x_2) - f(x_1)}{x_2 - x_1}$$	To measure the **instantaneous rate of change** for a function $f(x)$ from first principles, the change in y, represented by the symbol Δy, divided by the change in x, represented by the symbol Δx is found when the change in x is approaching zero, $\Delta x \to 0$. $$f'(x) = \lim_{h \to 0} \frac{f(x+h) - f(x)}{h} \qquad (notice\ h \to 0)$$
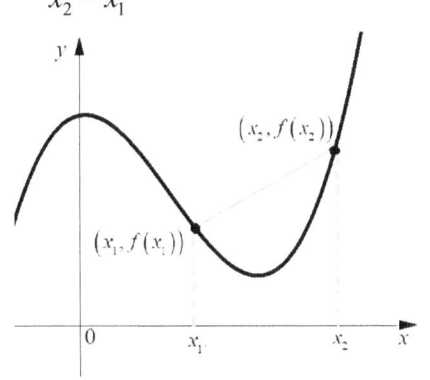	

Newton

$$f(x) = x^3 + x + 5$$
$$f'(x) = 3x^2 + 1$$

Leibniz

$$y = x^3 + x + 5$$
$$\frac{dy}{dx} = 3x^2 + 1$$

An important milestone in the history of mathematics, Newton and Leibniz independently discover differential calculus, and integral calculus at about the same time. Snap!

Act 1, Scene 1. Deriving Derivatives. The drama of finding the instant rate of change of a function.

The derivative is also represented with the symbol dy/dx.
So f'(x)=dy/dx

The derivative f'(x) is also called the gradient function, and this is because not only is it the instantaneous rate of change, it is also the instantaneous gradient of the function.

Finding f'(x) is like finding the rise/run when the run is approaching zero.

The derivative f'(x) is a function itself, and can be used to find change.

Amazing

Algebrains.

$$f'(x) = \lim_{h \to 0} \frac{f(x+h) - f(x)}{h}$$

Given $f(x) = 3x^2 + x + 8$ find $f'(x)$ from first principles.

$$f'(x) = \lim_{h \to 0} \frac{\left(3(x+h)^2 + (x+h) + 8\right) - \left(3x^2 + x + 8\right)}{h}$$

$$f'(x) = \lim_{h \to 0} \frac{\left(3\left(x^2 + 2xh + h^2\right) + (x+h) + 8\right) - \left(3x^2 + x + 8\right)}{h}$$

$$f'(x) = \lim_{h \to 0} \frac{\left(3x^2 + 6xh + 3h^2 + x + h + 8\right) - \left(3x^2 + x + 8\right)}{h}$$

$$f'(x) = \lim_{h \to 0} \frac{\left(6xh + 3h^2 + h\right)}{h}$$

$$f'(x) = \lim_{h \to 0} \frac{\cancel{h}\left(6x + 3h + 1\right)}{\cancel{h}} \qquad (h \text{ is cancelled})$$

$$f'(x) = \lim_{h \to 0} \left(6x + 3h + 1\right)$$

$$f'(x) = 6x + 1$$

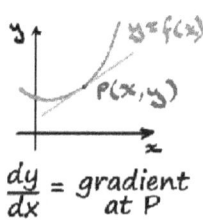

$\frac{dy}{dx} = $ gradient at P

The gradient function can be used to find out how fast quantities change.

Act 1, Scene 2. The drama of finding the rate of change of a function.

Graphing f(x) and f'(x) together shows their relationship. For any point (p,f(p)), examine the gradient of the tangent line at that point.

If (p,f(p)) has a positive gradient line, then f'(p) is a positive value,

If (p,f(p)) has a negative gradient line, then f'(p) is a negative value

If (p,f(p)) has a zero gradient line, then f'(p) is equal to zero.

The relationship between f(x) and f'(x).

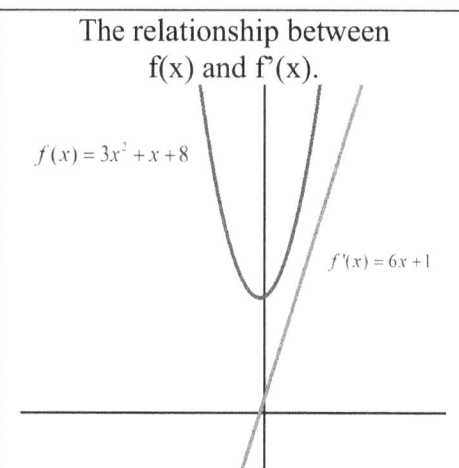

$f(x) = 3x^2 + x + 8$

$f'(x) = 6x + 1$

Amazing

Algebrains.

Using technology, if we take a quartic function f(x) and find its derivative f'(x), we see a cubic. Take the derivative of the cubic f''(x) we get a quadratic, take the derivative of the quadratic, f'''(x) we get a linear function.

QUARTIC

$y = ax^4 + bx^3 + cx^2 + dx + e$

when differentiated becomes:

...a **CUBIC**

$y = fx^3 + gx^2 + hx + i$

when differentiated becomes:

...a **QUADRATIC**

$y = jx^2\ kx + L$

when differentiated becomes:

...a **LINE**

$y = mx + c$

When we calculate the derivative function for a power function $f(x)$, the pattern observed is that $f'(x)$ is of an order one less than the original function.

The action of finding the derivative is called differentiation.

Common notation for derivatives:

First derivative	$f'(x)$	$\dfrac{dy}{dx}$	y'
Second derivative	$f''(x)$	$\dfrac{d^2y}{dx^2}$	y''
Third derivative	$f'''(x)$	$\dfrac{d^3y}{dx^3}$	y'''
nth derivative	$f^n(x)$	$\dfrac{d^ny}{dx^n}$	y^n

Application 5.1: Filling the vases, and thinking about the height changing.

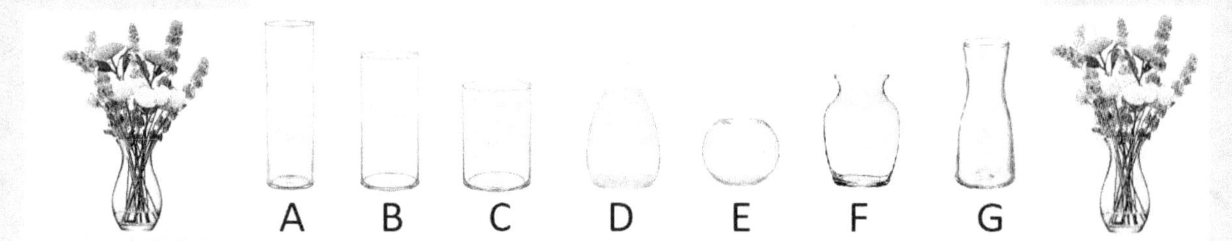

In a controlled experiment where the flow of water is a slow constant rate, we are filling these initially empty glass vases of various heights, shapes and dimensions. How is the height of water changing with time for each vase?

Here are some Height vs Time graphs.
Which ones best match the vases?

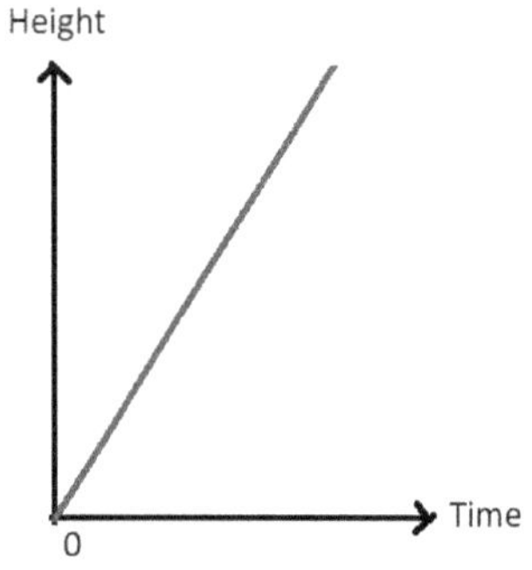

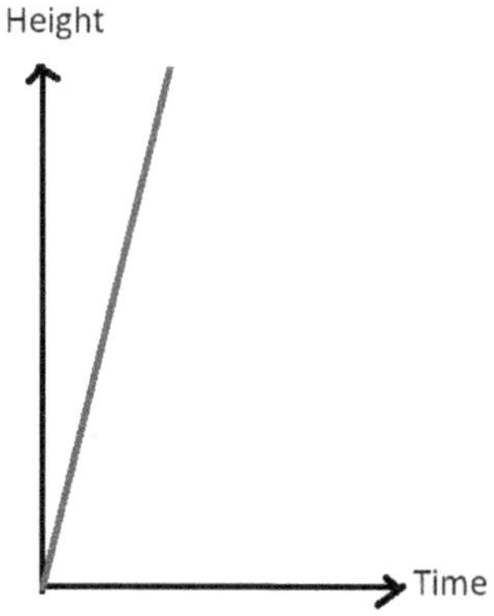

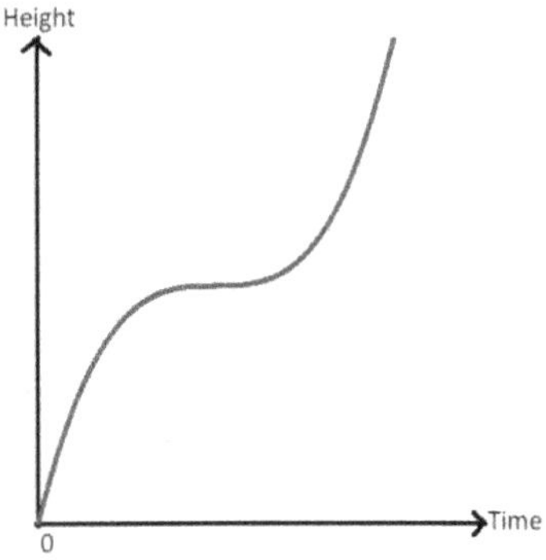

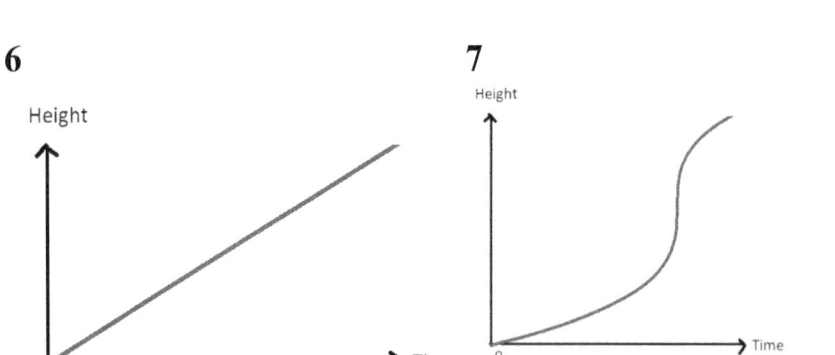

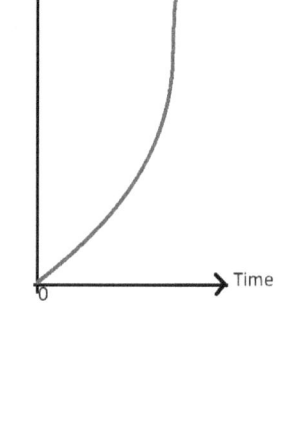

Do you agree with the pairings A2, B1, C6, D4, E3, F7, G5?

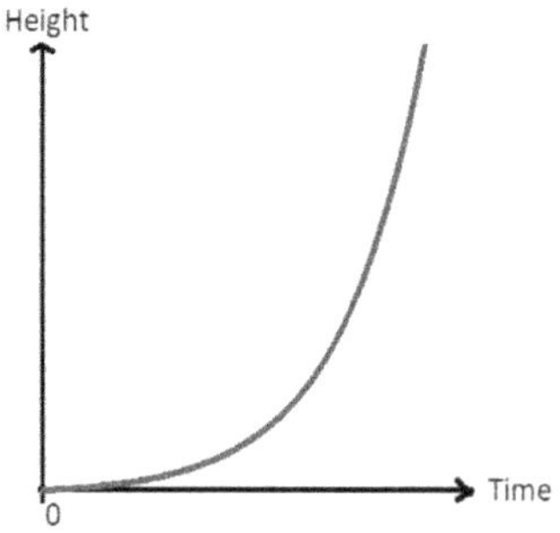

124

Differentiation by rule.

By observation of the patterns of applying differentiation (deriving the derivative) using first principles, several rules have been discovered to make finding the derivative a lot easier.

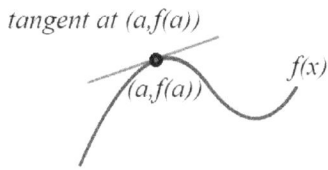

tangent at (a,f(a))

f(x)

(a,f(a))

Rewrite all roots of x as indices

eg. $\sqrt{x} = x^{\frac{1}{2}}$

Rewrite all fractions of x as indices

eg. $\dfrac{1}{x^2} = x^{-2}$

The **Power Rule** can be used on power functions for quick differentiation (deriving the derivative). It is applied to each power term in the function.

$$f(x) = x^n \Rightarrow f'(x) = nx^{n-1}$$

$$y = x^n \Rightarrow \frac{dy}{dx} = nx^{n-1}$$

$f(x) = x^2 - 4x + 1 \quad \Rightarrow f'(x) = 2x - 4$

$f(x) = 2x^3 + x^{-3} \quad \Rightarrow f'(x) = 6x^2 - 3x^{-4}$

$y = 9x - 5x^{\frac{1}{2}} \quad \Rightarrow \dfrac{dy}{dx} = 9 - \dfrac{5}{2}x^{-\frac{1}{2}}$

The **Product Rule** can be used on functions that are or can be factorised for quick differentiation (deriving the derivative).

$$f(x) = g(x)h(x)$$

$$\Rightarrow f'(x) = g(x)h'(x) + g'(x)h(x)$$

$$y = uv \Rightarrow \frac{dy}{dx} = u\frac{dv}{dx} + v\frac{du}{dx}$$

$f(x) = x^2(x+1)$

$Let\ g(x) = x^2,\ h(x) = (x+1)$

$\Rightarrow g'(x) = 2x,\ h'(x) = 1$

$f'(x) = g(x)h'(x) + g'(x)h(x)$

$\quad = x^2(1) + 2x(x+1)$

$\quad = x^2 + 2x^2 + 2x$

$\quad = 3x^2 + 2x$

$f'(x) = 3x^2 + 2x$

The **Quotient Rule** can be used on functions in rational or fraction form for quick differentiation (deriving the derivative).

$$f(x) = \frac{g(x)}{h(x)}$$

$$\Rightarrow f'(x) = \frac{h(x)g'(x) - h'(x)g(x)}{[h(x)]^2}$$

$$y = \frac{u}{v} \Rightarrow \frac{dy}{dx} = \frac{v\frac{du}{dx} - u\frac{dv}{dx}}{v^2}$$

$y = \dfrac{x^2}{(x+1)}$

$Let\ u = x^2,\ v = (x+1)$

$\Rightarrow \dfrac{du}{dx} = 2x,\ \dfrac{dv}{dx} = 1$

$\dfrac{dy}{dx} = \dfrac{v\frac{du}{dx} - u\frac{dv}{dx}}{v^2}$

$\dfrac{dy}{dx} = \dfrac{(x+1)(2x) - (x^2)(1)}{(x+1)^2}$

simplify

$\dfrac{dy}{dx} = \dfrac{x^2 + 2x}{(x+1)^2}$

Continuous and smooth functions.

The derivative at any point $f'(x_0)$ can be found only if the function is **continuous and smooth** at the point $(x_0, f(x_0))$.

Continuous functions on an interval $x \in [a,b]$ can be drawn without lifting a pen from the paper from $x = a$ to $x = b$. If $x = x_0$ is in the interval $x \in [a,b]$ then the function is continuous at $x = x_0$ where the limit from the left of $x = x_0$ is equal to the limit on the right of $x = x_0$. $\lim_{x \to x_0^-} f(x) = \lim_{x \to x_0^+} f(x) = f(x_0)$

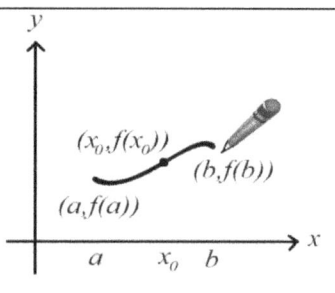

Smooth functions have no corners or sharp turns, so that the *left* derivative at the point $x = x_0$ equals the *right* derivative.

$$f'(x_0) = \lim_{h \to 0} \frac{f(x_0 - h) - f(x_0)}{h} = \lim_{h \to 0} \frac{f(x_0 + h) - f(x_0)}{h}$$

Special care needs to be taken when defining the domain of the derivative when differentiating:

- Functions with restricted domains, with holes, asymptotes
- End points do not fit the definition of smoothness
- Functions with sharp corners

Not Continuous.

Following the line of the function in the interval $x \in [a,b]$ in which $P(x,y)$ exists cannot be drawn without lifting the pen.

The discontinuity can occur at an asymptote or a gap in the function and is excluded from the domain of the derivative function as shown.

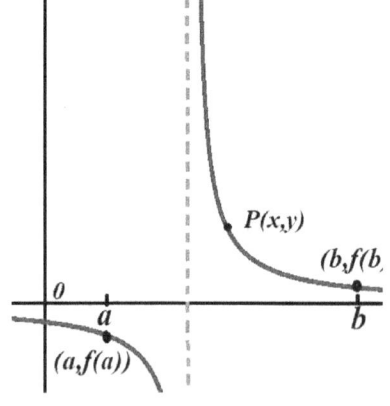

Asymptote in interval $x \in [a,b]$

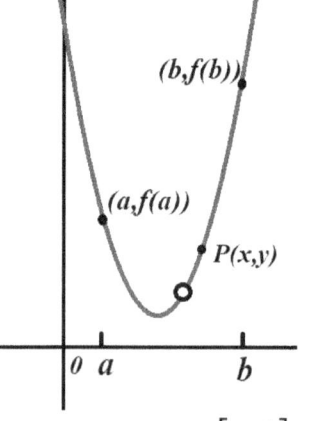

Gap in interval $x \in [a,b]$

Not Smooth.

The gradient to the left of P is not the same as gradient to the right of P as the change in x approaches zero either side of *P(x,y)*.

Therefore $f'(x)$ doesn't exist at point P, and point P is explicitly excluded from the domain of the derivative function.

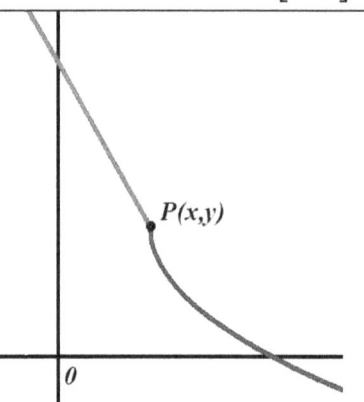

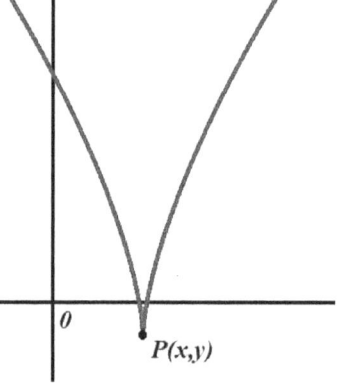

| **More rules for differentiation. You just need to know them.** |

$$\frac{d}{dx}\left(x^n\right)=nx^{n-1} \qquad \frac{d}{dx}\left(e^{ax}\right)=ae^{ax} \qquad \frac{d}{dx}\left(\sin\left(ax\right)\right)=a\cos\left(ax\right)$$

$$\frac{d}{dx}\left(\left(ax+b\right)^n\right)=an\left(ax+b\right)^{n-1} \qquad \frac{d}{dx}\left(\log_e\left(ax\right)\right)=\frac{a}{x} \qquad \frac{d}{dx}\left(\cos\left(ax\right)\right)=-a\sin\left(ax\right)$$

$$\frac{d}{dx}\left(\tan\left(ax\right)\right)=\frac{a}{\cos^2\left(ax\right)}=a\sec^2\left(ax\right)$$

| **Chain Rule: Differentiation of Composite Functions** $f(x)=f(g(x))$ |

The Chain rule $y=u(x),\dfrac{dy}{dx}=\dfrac{dy}{du}\times\dfrac{du}{dx}$ also stated as $f'(x)=f'(g(x))g'(x)$ "chains" the derivatives of each composite function together. Working from the inside function to the outside function we are finding… "the derivative of the outside function times derivative of the inside function".

Example 5.1: Applying the Chain rule to composite functions. Step by Step.

Find the derivative of	Derivative of the outside function	Derivative of the inside function	Chaining the derivatives together.
$y=3\left(4-x^2\right)^5$ Let $u(x)=\left(4-x^2\right)$ or expressed more simply $u=\left(4-x^2\right)$ therefore $y=3u^5$	$\dfrac{dy}{du}=15u^4$	$\dfrac{du}{dx}=-2x$	$y=u(x),\dfrac{dy}{dx}=\dfrac{dy}{du}\times\dfrac{du}{dx}$ $\dfrac{dy}{dx}=15u^4\times-2x$ $=15\left(4-x^2\right)^4\times-2x$ $=-30x\left(4-x^2\right)^4$
$y=\sin\left(5\left(4-x^2\right)\right)$ Let $u=\left(4-x^2\right)$ therefore $y=\sin(5u)$	$\dfrac{dy}{du}=5\cos(5u)$	$\dfrac{du}{dx}=-2x$	$y=u(x),\dfrac{dy}{dx}=\dfrac{dy}{du}\times\dfrac{du}{dx}$ $\dfrac{dy}{dx}=5\cos(5u)\times-2x$ $=5\cos\left(5\left(4-x^2\right)\right)\times-2x$ $=-10x\cos\left(5\left(4-x^2\right)\right)$
$y=e^{-\left(4-x^2\right)}$ Let $u=\left(4-x^2\right)$ therefore $y=e^{-u}$	$\dfrac{dy}{du}=-e^{-u}$	$\dfrac{du}{dx}=-2x$	$y=u(x),\dfrac{dy}{dx}=\dfrac{dy}{du}\times\dfrac{du}{dx}$ $\dfrac{dy}{dx}=-e^{-u}\times-2x$ $=-e^{-\left(4-x^2\right)}\times-2x$ $=2xe^{-\left(4-x^2\right)}$

The derivative is the gradient function. By substitution you can use the derivative $f'(x)$ to find the gradient anywhere along the curve of the function $f(x)$.

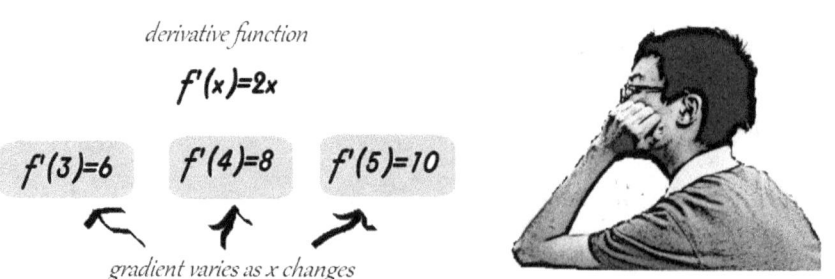

derivative function

$f'(x)=2x$

$f'(3)=6$ $f'(4)=8$ $f'(5)=10$

gradient varies as x changes

Example 5.2: The "Rickety Rollercoaster" track, is a hybrid function composed of several functions. The ride is rickety and thrilling.

The profile of the Rollercoaster in one section has a curve described by the following hybrid aka piecewise function.

$$f(x) = \begin{cases} 4\sin(x)+8, & 0 \le x < 2\pi \\ 2\cos(x)+6, & 2\pi \le x < 3\pi \\ e^{-3x}+4, & 3\pi \le x < 12.158 \\ -\tan\left(\dfrac{x+6}{10}\right), & 12.158 \le x \le 16 \end{cases}$$

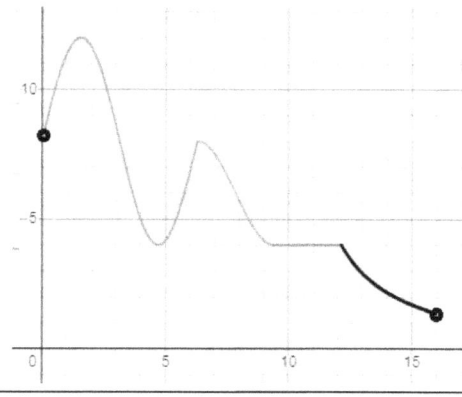

To find the derivative or the gradient function $f'(x)$ we take the derivative of each piece of the function using the rules for differentiation.

$$f'(x) = \begin{cases} 4\cos(x), & 0 < x < 2\pi \\ -2\sin(x), & 2\pi < x < 3\pi \\ -3e^{-3x}, & 3\pi < x < 12.158 \\ -0.1\sec^2\left(\dfrac{x+6}{10}\right), & 12.158 < x < 16 \end{cases}$$

Due to lack of **smoothness**, there are holes in the gradient function $f'(x)$ at the ends of the sub-intervals for the hybrid function $f(x)$.

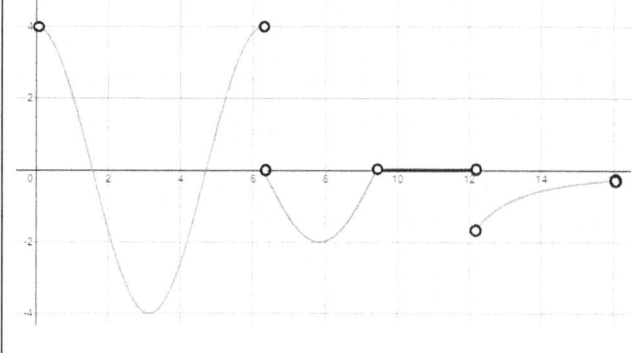

Find where the gradient of the rollercoaster is equal to -2 taking care of the sub-domain restrictions.

$4\cos(x) = -2 \Rightarrow \cos(x) = \dfrac{-1}{2} \Rightarrow x \in \left\{ \dfrac{2\pi}{3}, \dfrac{4\pi}{3} \right\}$

$-2\sin(x) = -2 \Rightarrow \sin(x) = 1 \Rightarrow x \in \left\{ \dfrac{5\pi}{2} \right\}$

$-3e^{-3x} = -2 \Rightarrow e^{-3x} = \dfrac{2}{3} \Rightarrow$ no solution in interval

$-0.1\sec^2\left(\dfrac{x+6}{10}\right) = -2 \Rightarrow$ no solution in interval

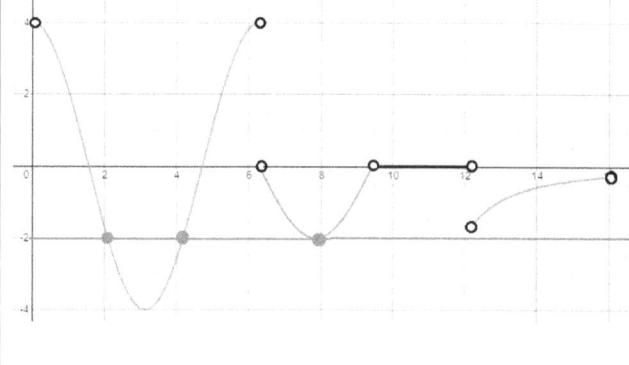

128

Many children have enjoyed the application of gradient functions in their local playgrounds in the form of slides, with steep sections for going down fast and the flatter sections for slowing down.

Vintage slide.	Extra long slide, Parramatta, NSW. Australia.

Adventure park slide.

Finding Stationary points for a function, which exist when the derivative equals zero, $f'(x) = 0$.

Stationary points have a gradient of zero. They can be local minimums, maximums or points of inflection. Their nature or type can be determined by exploring their neighbouring x values.

Type of Stationary point $f'(x_0) = 0$	Just before $f'(x_0^-)$	$f'(x_0) = 0$	Just after $f'(x_0^+)$
Minimum	Negative gradient	▬	Positive gradient
Maximum	Positive gradient	▬	Negative gradient
Point of Inflection	Positive gradient	▬	Positive gradient
Point of Inflection	Negative gradient	▬	Negative gradient

Act 2, Scene 1. Derivative functions. The drama of stationary points and finding a tangent.

Let's find the stationary points for
$f(x) = 2x^3 - 5x^2 + 7$
and determine their nature
by exploring their neighbourhood.

So we can find the derivative
using the power rule $f'(x) = 6x^2 - 10x$
And then find the stationary points by solving
for $f'(x) = 0$

Amazing

Algebrains

Nature of Stationary point $f'(x_0) = 0$	Just before $f'(x_0^-)$	$f'(x_0) = 0$	Just after $f'(x_0^+)$
Minimum	\	—	/
Maximum	/	—	\
Point of Inflection	/	—	/
Point of Inflection	\	—	\

For this case we solve $6x^2 - 10x = 0$,
factorised $2x(3x - 5) = 0$
We get stationary points at $x = 0$ and $x = \frac{5}{3}$

Just before	$f'(x) = 0$	Just after
$f'(-\frac{1}{5}) = \frac{6}{25} + 2$	$x = 0$	$f'(-\frac{1}{5}) = \frac{6}{25} - 2$
> 0	—	< 0
/		\
$f'(\frac{3}{3}) = 6(1)^2 - 10(1)$	$x = \frac{5}{3}$	$f'(\frac{6}{3}) = 6(2)^2 - 10(2)$
< 0		> 0
\	—	/

Act 2, Scene 2. Derivative functions. The drama of stationary points and finding a tangent.

The gradients of the neighbouring values of the stationary points shows that

$x = 0$ is a maximum and

$x = \frac{5}{3}$ is a minimum.

A quick graph using technology confirms this result.

Find the coordinates of the stationary points.

$f(x) = 2x^3 - 5x^2 + 7$

$f(0) = 7 \quad (0,7) \text{ max}$

$f\left(\frac{5}{3}\right) = \frac{64}{27} \quad \left(\frac{5}{3}, \frac{64}{27}\right) \text{ min}$

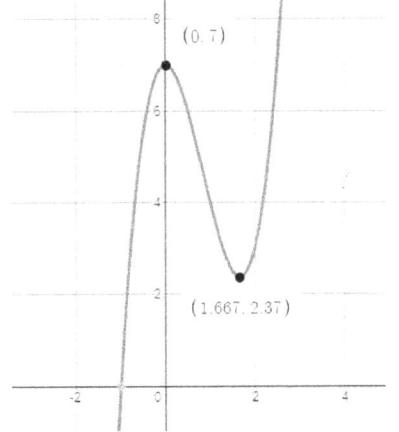

(0.7)

(1.667, 2.37)

Amazing

Algebrains

Let's find the tangent line for $x = 1$ now.

$f(1) = 4$, the line has to pass through point $(1, 4)$

The gradient is $f'(1) = -4$

Using the linear equation form $m = \frac{y - y_1}{x - x_1}$

we get $-4 = \frac{y - 4}{x - 1}$

Which gives us the line $-4(x - 1) = y - 4$

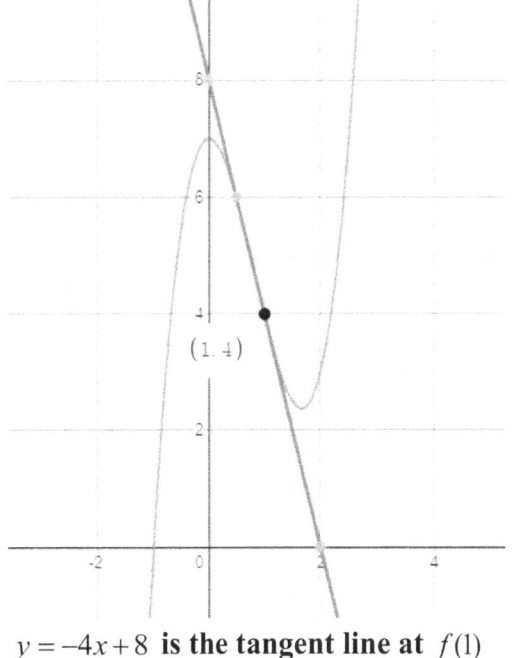

(1, 4)

$y = -4x + 8$ **is the tangent line at** $f(1)$

131

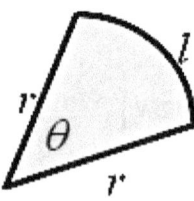

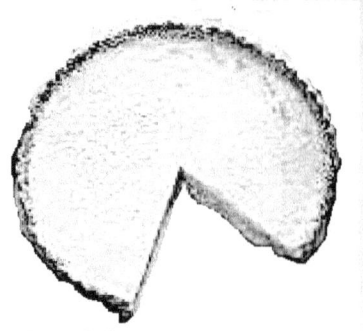

A sector slice of delicious lemon tart with a base area of 16 cm² is cut from a whole tart which has a radius of r.

$A_S = 16cm^2$

$P_S = 2r + l$

Find the minimum Perimeter P of the slice.

Show that the perimeter P_S of this sector slice that is cut from the centre of this tart is given by:

$P_S(r) = 2\left(r + \dfrac{16}{r}\right)$, then we need to find the minimum value of the perimeter P_S in cm. How

big in area is that sector slice relative to the whole uncut tart area? What is the fraction?

We need to relate the perimeter P_S to the radius r, since we are given the area of the tart.

- The area of sector slice $A_S = \dfrac{\theta^o}{360}\pi r^2$ where

  θ is the angle of sector sliced, is a fraction of the whole circular tart.

- Arc length of sector $l = \dfrac{\theta^o}{360^o}(2\pi r)$ is a

  fraction of the circumference of the whole circular tart.

$\dfrac{l}{2\pi r} = \dfrac{\theta^o}{360^o} = \dfrac{A_S}{\pi r^2}$

$\dfrac{l}{2\pi r} = \dfrac{16}{\pi r^2}$

$l = \dfrac{(16)(2\pi r)}{\pi r^2}$

$l = \dfrac{32}{r}$

$P_S(r) = 2r + l$

$P_S(r) = 2r + \dfrac{32}{r}$

$P_S(r) = 2\left(r + \dfrac{16}{r}\right)$

Shown

To find the minimum value of $P_S(r)$

- Re-express $P_S(r)$ with negative indices to use the power rule
- Find the derivative function
- Find the stationary points by solving $P_S'(r) = 0$

$P_S(r) = 2\left(r + \dfrac{16}{r}\right)$

$P_S(r) = 2r + 32r^{-1}$

$P_S'(r) = 2 - 32r^{-2}$

$P_S'(r) = 0$

$2 - 32r^{-2} = 0$

$2 - \dfrac{32}{r^2} = 0$

$2 = \dfrac{32}{r^2}$

$2r^2 = 32$

$r^2 = 16$

$r = \pm 4$

Discard the negative radius, imaginary pies are not allowed.

When $r = 4$

$P_S(4) = 2\left(4 + \dfrac{16}{4}\right)$

$= 16cm$

Since $r = 4$ the whole uncut tart has an area of 16π, the slice therefore $\dfrac{16}{16\pi}$

or $\dfrac{1}{\pi}$ of the whole tart area.

Strictly Increasing continuous functions

A function that is continuous and has a positive gradient, $f'(x) > 0$ that is increasing in size as x is increasing in a defined interval or sub-domain, that is the gradient is becoming increasingly steeper, can be said to be a strictly increasing function for that interval or sub-domain.

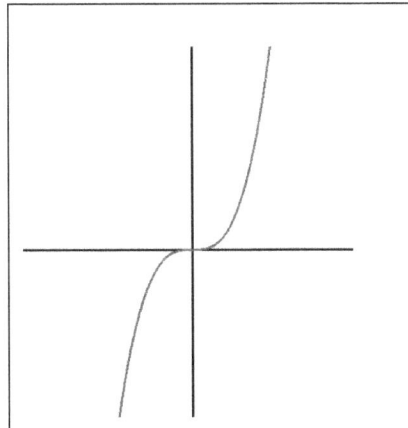

The function to the left is continuous and has a positive gradient, $f'(x) > 0$ everywhere except the origin, which looks like a stationary point.

The function is getting steeper and more steeper, **strictly increasing** over the interval $x > 0$.

For the interval $x < 0$ the gradient is still positive, however as the values of x get larger in this interval, $f'(x)$ approaches smaller gradient values eventually reaching $f'(x) = 0$ which is the stationary point. This function is **not** strictly increasing where $x < 0$.

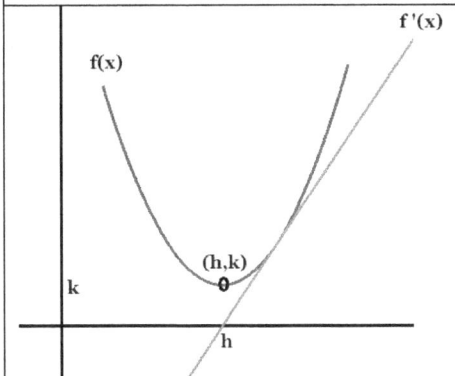

The parabola is continuous and has a positive gradient to the right of the turning point (h, k) which is $x > h$, which is becoming steeper and more steeper, and is **strictly increasing** over the interval $x \in (h, \infty)$ as shown.

Strictly Decreasing continuous functions

A function that is continuous and has a negative gradient, $f'(x) < 0$ that is decreasing in size as x is increasing in a defined interval or sub-domain, can be said to be a strictly decreasing function in that interval or sub-domain.

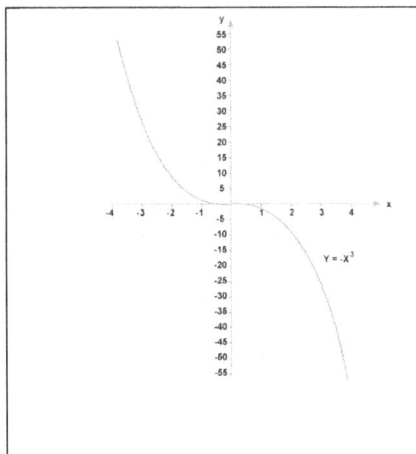

The function to the left is continuous and has a negative gradient, $f'(x) < 0$ everywhere except the origin, which looks like a stationary point.

The gradient of the function is getting more and more negative, **strictly decreasing** over the interval $x > 0$.

For the interval $x < 0$ the gradient is still negative, however as the values of x get larger in this interval, $f'(x)$ approaches less negative gradient values eventually reaching $f'(x) = 0$ which is the stationary point. This function is **not** strictly decreasing where $x < 0$.

Act 3, Scene 1. The drama of identifying key features of functions.

Using technology you can graph any function y=f(x).
We can also graph the derivative function f'(x) and compare it with f(x).
We can use f'(x) to describe f(x).

We need to find the stationary points, local minimums, maximums and points of inflection, and intervals where the function is strictly increasing or decreasing.

Amazing

Algebrains

Key features of functions.
$f(x)$ **related to the derivative function** $f'(x)$

- Continuity
- Smoothness
- Rate of change over x
- Local minimums, maximums
- Stationary points of inflection
- Endpoints due to restricted domains.

We need to find f'(x) where possible and analyse it.

The hybrid function $f(x)$ with many sub-intervals, features sharp corners, local minimum, an asymptote discontinuity and an endpoint.

$$f(x) = \begin{cases} 2x+2, & x \le 0 \\ (x-2)^2 - 2, & 0 \le x \le 4 \\ \dfrac{1}{(x-5)^2} + 1, & 4 < x \le 6 \end{cases}$$

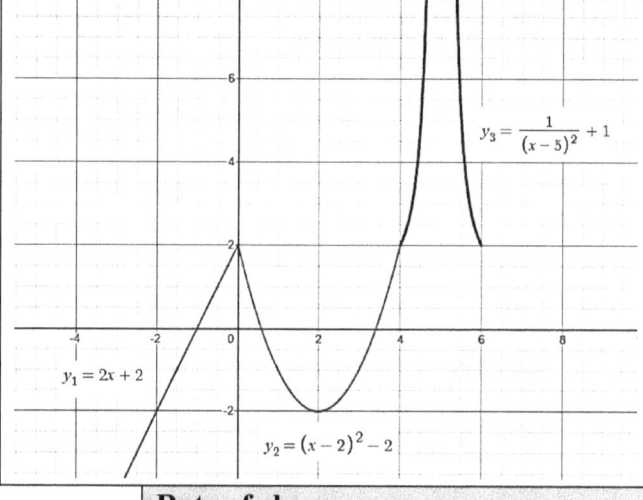

Continuity/ Smoothness/ Endpoints.	**Stationary points**	**Rate of change over x**
Places where the derivative doesn't exist.	Places where the derivative is equal to zero. $f'(x) = 0$	Intervals where the rate of change described by the derivative are strictly increasing or strictly decreasing.

134

Act 3, Scene 2. Scene 1. The drama of identifying key features of functions.

The derivative function.

$$f'(x) = \begin{cases} 2 & ,\{x < 0\} \\ 2x-4 & ,\{0 < x < 4\} \\ \dfrac{-2}{(x-5)^3} & \cup\{5 < x < 6\} \\ & ,\{4 < x < 5\} \end{cases}$$

For this function there is an asymptote at x=5 and an endpoint at (6,2).

The gradient for this function is constant for x<0 and varies elsewhere.

Continuity/ Smoothness/ Endpoints.

$f'(x)$ **does not exist at:**

- $x = 0$ and $x = 4$ due to lack of smoothness when joining the pieces of the function $f(x)$
- $x = 5$ due to the discontinuity at the asymptote
- $x = 6$ due to the lack of smoothness at the endpoint

Rate of change over x in the domain.

- Strictly increasing intervals $x \in (2,4)$ and when

$$\frac{-2}{(x-5)^3} > 4 \text{ Solving } \frac{2}{(x-5)^3} < -4, \text{ that is } x > \sqrt[3]{\frac{-1}{2}} + 5 \text{ up to}$$

the asymptote of $x = 5$

- There are no strictly decreasing intervals for this function.

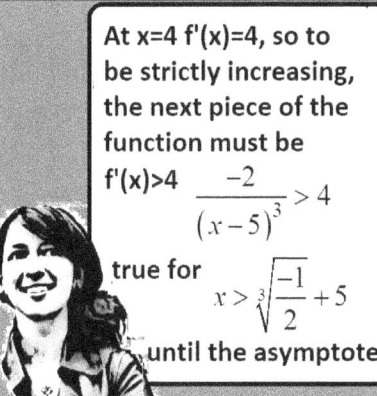

At x=4 f'(x)=4, so to be strictly increasing, the next piece of the function must be f'(x)>4 $\dfrac{-2}{(x-5)^3} > 4$ true for $x > \sqrt[3]{\dfrac{-1}{2}} + 5$ until the asymptote.

Stationary points, only one section is possible for $f'(x) = 0$

$f'(x) = 2x - 4$

$2x - 4 = 0 \therefore x = 2$

$f(2) = -2$

Stationary point $(2, -2)$

Checking the gradient around this point

$f'(1.8) < 0 \quad f'(2.2) > 0$

local minimum.

Amazing

Algebrains

$$f(x) = \begin{cases} 2x+2, & x \le 0 \\ (x-2)^2 - 2, & 0 \le x \le 4 \\ \dfrac{1}{(x-5)^2} + 1, & 4 < x \le 6 \end{cases}$$

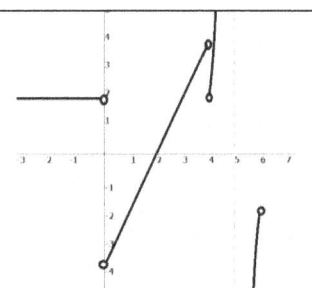

Careful analysis of the function f(x) and its derivative f'(x) is needed to find the key features relating to the rate of change of the function f(x).

$$f'(x) = \begin{cases} 2 & ,\{x < 0\} \\ 2x-4 & ,\{0 < x < 4\} \\ \dfrac{-2}{(x-5)^3} & \cup\{5 < x < 6\} \\ & ,\{4 < x < 5\} \end{cases}$$

Application: Rates of change, the derivative with respect to time.

A cube of ice of side length l melts at a rate that is linearly proportional to its surface area S. The volume V is changing over time measured in minutes. Time is represented by the variable t.

Expressed mathematically $\dfrac{dV}{dt} = kS$, where k is a constant.

Write S in terms of length l, write V in terms of length l. Using the chain rule calculate the rate of change of the length l, that is $\dfrac{dl}{dt}$ with respect to time.

Surface area $S = 6l^2$ and volume $V = l^3$ By the chain rule $\dfrac{dV}{dt} = \dfrac{dV}{dl} \times \dfrac{dl}{dt} = 3l^2 \dfrac{dl}{dt}$ We are given $\dfrac{dV}{dt} = kS = k\left(6l^2\right)$	Equating these $\dfrac{dV}{dt}$ expressions gives $3l^2 \dfrac{dl}{dt} = 6kl^2$ which simplifies to $\dfrac{dl}{dt} = 2k$

What is the sign of the constant k? How is the volume V of the ice cube changing over time? How is the length l changing over time?

The volume of the ice cube V, and the side length l is decreasing as the cube melts. Therefore the sign of the constant k must be negative.

How long in terms of constant k will it take for the ice cube to melt completely?

Initially the ice cube has the side length l of 2cm, therefore at time $t = 0, l = 2$ The side length is changing uniformly at the rate of $\dfrac{dl}{dt} = 2k$ Since the derivative is a multiple of a constant k this means the gradient is constant. Constant gradients belong to linear functions such as $y = mx + c$ where $\dfrac{dy}{dx} = m$.	Let t be the time of observing the ice cube. The side length of the cube is a constantly decreasing, which means that $l(t) = 2 + 2kt$ since $\dfrac{dl}{dt} = 2k$ Solving for $l = 0$ gives $2 + 2kt = 0$ $1 + kt = 0 \Rightarrow kt = -1 \Rightarrow t = \dfrac{-1}{k}$ so time to melt in minutes is $t = \dfrac{-1}{k}$, a positive value since k is negative as the ice cube is shrinking.

It's important to recognize that derivatives are a mathematical model for measuring change, and can be used to model real world systems.

| **Business,** derivatives are used to model share prices, imports, exports, changes in currency values. | **Biology,** derivatives are used to model population growth, ecosystems and the spread of diseases. | **Physics,** rates and changes in motion and energy over time are measured with derivatives. | **Chemistry,** uses derivatives to monitor the variations in concentrations and rates of reactions. | **Social Analysis** uses derivatives to model fluctuations in trends and voting patterns. |

Newton's method of approximation of solutions for $f(x)=0$

For simple linear and quadratic functions $f(x)$ we have formulas for finding solutions for $f(x)=0$.

$f(x) = mx + c$	$x = -{c}/{m}$
$f(x) = ax^2 + bx + c$	$x = \dfrac{-b \pm \sqrt{b^2 - 4ac}}{2a}$

In many real life problems that can't be represented by linear or quadratic functions it is often difficult to find solutions for $f(x)=0$. If the function $f(x)$ is differentiable then **Newton's method** can be used for approximating the solutions for $f(x)=0$, it works by repeatedly taking the derivative at different values of x, using the gradient $f'(x)$ to guide the computations for approximating a solution for $f(x)=0$.

Newton's method starts with an estimate of the solution, say x_0, if $f(x_0)\neq 0$ and if the derivative at the point x_0 is not equal to zero, $f'(x_0)\neq 0$, it uses the tangent that will intersect with the x-axis at some point say $(x_1,0)$ to make a better estimate of the solution.

If $f(x_1)\neq 0$ and $f'(x_1)\neq 0$, we can repeat the process as the tangent to x_1 will intersect with x-axis at some point closer to the actual solution, say $(x_2,0)$, if $f(x_2)\neq 0$ and $f'(x_2)\neq 0$, repeat the process again and so on until we find the tangent intersects the x-axis at $(\alpha,0)$ and $f(\alpha)\approx 0$ is to the accuracy that we require.

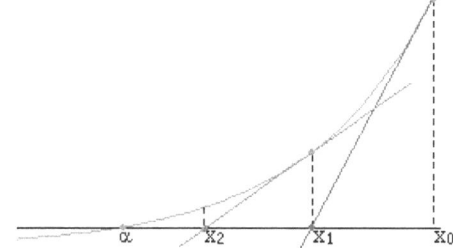

Newton's method is estimating the solution by repeatedly calculating a new approximation x_1 using the previous approximation x_0 according to the formula:

$$x_1 = x_0 - \frac{f(x_0)}{f'(x_0)}$$

Consider from first principals how this works.
A basic linear equation given the gradient m through point (x_0,y_0) is defined by $y - y_0 = m(x - x_0)$
A tangent line at x_0 has $y = f(x_0)$ and the gradient given by the derivative is $m=f'(x_0)$ and the point we want to find is where this line intercepts the x-axis with will be $(x_1,0)$, so the linear equation becomes:
$f(x_0) = f'(x_0)(x_1 - x_0)$ making x_1 the subject gives the formula above.

Newton's method in pseudocode

input x0, tolerance, maxIterations, f, fprime
i ← 0
while |f(x0)| > tolerance and (i < maxIterations)
    y0 ← f(x0)
    yp ← fprime(x0)
    x1 ← x0 – y0/yp
    i ← i + 1
    x0 ← x1
end while
if (i < maximumIterations) **then**
    print x0
else
    print *solution not found after maxIterations*
end if

Example 5.3 Application of Newtons Method

Using Newtons method to approximate a solution for $f(x)=0$ where $f(x)=x^3 - 3x^2 + 2x - 1$ and $f'(x)=3x^2 - 6x + 2$

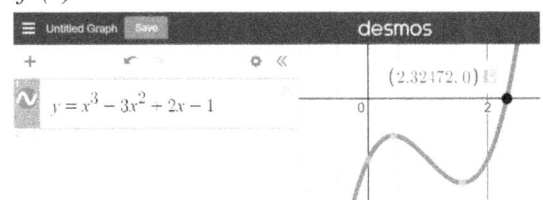

$f(2) = -0.96$ is below the x-axis, $f(3) = 5.03$ is above, solution in interval $[2,3]$ guess $x_0=2.5$

$$x_{i+1} = x_i - \frac{f(x_0)}{f'(x_0)} \rightarrow x_1 = 2.5 - \frac{0.875}{5.75}$$

$$x_1 = 2.3478 \rightarrow x_2 = 2.3478 - \frac{0.1006}{4.4513}$$

$$x_3 = 2.3252 \rightarrow x_4 = 2.3252 - \frac{0.0021}{4.2685}$$

$$x_4 = 2.3247 \text{ accuracy accepted}$$

Example 5.4: Calculus in business and financial modelling.

The fictitious Sunny Solar Company makes portable solar panels for camping, where $M(x)$ is the cost to manufacture x solar panels and $R(x)$ is the revenue earned by selling x solar panels. Each solar panel is sold for $100.00.

$$M(x) = \frac{7}{40000}x^4 - \frac{35}{3000}x^3 + \frac{21}{200}x^2 + 39.3x$$

$$R(x) = 100x, \quad where \ x \geq 0$$

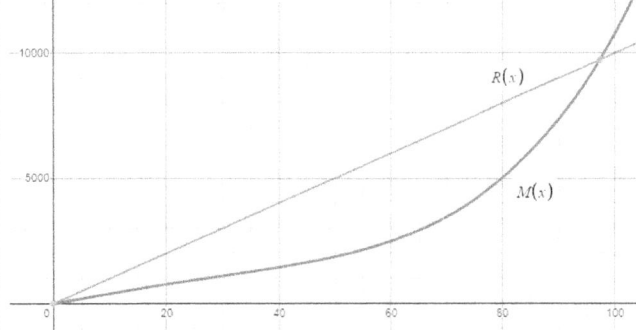

Business/Economics terminology:

- *Marginal Cost: $MC(x) = M'(x)$ is a derivative function representing the change in cost production per unit as the number of solar panels manufactured increases.*
- *Marginal Revenue: $MR(x) = R'(x)$ is a derivative function representing the change in revenue per unit as the number of units increases.*
- *Marginal Profit: $P(x)$ is the difference between Marginal Revenue and Marginal Cost. $P(x) = MR(x) - MC(x)$.*

The Sunny Solar Company has the following Marginal costs, Marginal Revenue and Marginal Profit.

$$MC(x) = 0.0007x^3 - 0.035x^2 + 0.21x + 39.3$$

$$MR(x) = 100$$

$$P(x) = MR(x) - MC(x), \ where \ x \geq 0$$

Graphing these three functions shows the relationship between the functions.

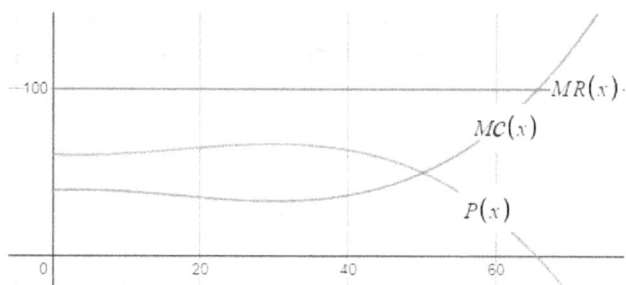

$P(x) = MR(x) - MC(x)$ evaluates to
$$P(x) = -0.0007x^3 + 0.035x^2 - 0.21x + 60.7$$
$$P'(x) = -0.0021x^2 + 0.07x - 0.21$$
$$where \ x \geq 0$$

Under mainstream economic theory, a company will maximize its overall profits when Marginal Revenue equals Marginal Cost, or when Marginal Profit is exactly zero $P(x) = 0$.

Using **Newton's method**
$f(60) = 22.9$ is above the x-axis, $f(20) = -22.6$ is below, solution in interval [60,70] guess $x_0 = 65$

$$x_{i+1} = x_i - \frac{f(x_0)}{f'(x_0)} \rightarrow x_1 = 65 - \frac{2.6875}{-4.5325}$$

$$x_1 = 65.5929 \rightarrow x_2 = 65.5929 - \frac{0.0361}{4.6534}$$

$$x_3 = 65.5851$$
accuracy accepted and an approximation for this value of x so that $P(65.585) \approx 0$ is found.

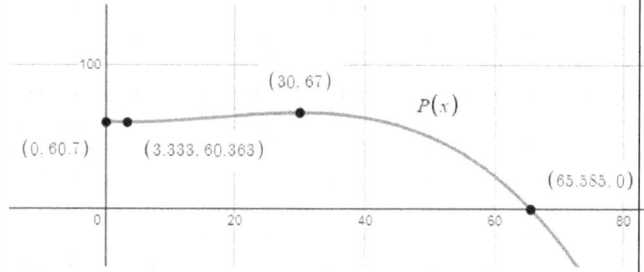

Production should be increased to approximately 65 to 66 solar panel units per run for a maximum profit. Notice that this number of solar panels has the largest difference in ordinates between $R(x)$ and $M(x)$ as observed on the graphs of these functions.

DERIVING DERIVATIVES SUMMARY:

Derivatives are a measure of change of a smooth and continuous function.

Average rate of change between two points is given by: $\dfrac{rise}{run} = \dfrac{f(x_1) - f(x_0)}{x_1 - x_0}$

Instantaneous rate of change at $(x_0, f(x_0))$ is

$f'(x_0) = \dfrac{dy}{dx}\Big|_{x=x_0}$

Tangents to graphs at the point $(x_0, f(x_0))$ is given by $(y - y_0) = m(x - x_0)$ where the gradient

$m = f'(x_0) = \dfrac{dy}{dx}\Big|_{x=x_0}$

Stationary points occur when $f'(x) = \dfrac{dy}{dx} = 0$

The derivative can be used to find stationary points and the nature of stationary points

Nature of Stationary point $f'(x_0) = 0$	Just before $f'(x_0^-)$	$f'(x_0) = 0$	Just after $f'(x_0^+)$
Minimum	$\searheadarrow$	—	/
Maximum	/	—	\
Point of Inflection	/	—	/
Point of Inflection	\	—	\

Rules for finding the derivative

Power Rule $y = uv$

$\dfrac{dy}{dx} = u\dfrac{dv}{dx} + v\dfrac{du}{dx}$

$f(x) = g(x)h(x)$

$f'(x) = g(x)h'(x) + g'(x)h(x)$

Quotient Rule $y = \dfrac{u}{v}$

$\dfrac{dy}{dx} = \dfrac{v\dfrac{du}{dx} - u\dfrac{dv}{dx}}{v^2}$

$f(x) = \dfrac{g(x)}{h(x)}$

$f'(x) = \dfrac{h(x)g'(x) - h'(x)g(x)}{[h(x)]^2}$

Chain Rule $y = u(x)$

$\dfrac{dy}{dx} = \dfrac{dy}{du} \times \dfrac{du}{dx}$

$f(x) = f(g(x))$

$f'(x) = f'(g(x))g'(x)$

Differentiation by recognition

$\dfrac{d}{dx}(x^n) = nx^{n-1}$

$\dfrac{d}{dx}(e^{ax}) = ae^{ax}$

$\dfrac{d}{dx}(\sin(ax)) = a\cos(ax)$

$\dfrac{d}{dx}((ax+b)^n) = an(ax+b)^{n-1}$

$\dfrac{d}{dx}(\log_e(ax)) = \dfrac{a}{x}$

$\dfrac{d}{dx}(\cos(ax)) = -a\sin(ax)$

$\dfrac{d}{dx}(\tan(ax)) = \dfrac{a}{\cos^2(ax)} = a\sec^2(ax)$

A derivative exists at the point $(x, f(x))$ if the function is **continuous** and **smooth** at that point.

Functions have **strictly decreasing** intervals if $f'(x) = \dfrac{dy}{dx}$ is becoming **more negative** in the interval.

Functions have **strictly increasing** intervals if $f'(x) = \dfrac{dy}{dx}$ is becoming **more positive** in the interval.

Newton's method estimating the solution for $f(x)=0$ with the formula: $x_{i+1} = x_i - \dfrac{f(x_0)}{f'(x_0)}$ where initial guess is x_0

DERIVING DERIVATIVES: Check your Understanding.

5.1 Gradient functions in the real world.

Skateboard ramp profiles.

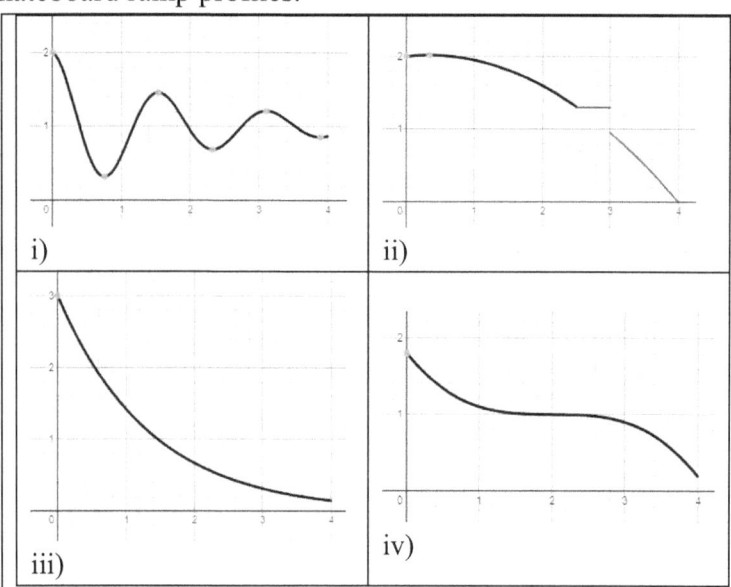

i)

ii)

iii)

iv)

Can you identify the graphs of the function describing several skateboard ramp functions of height $h_1(x), h_2(x), h_3(x), h_4(x)$ as a function of horizontal distance, where $x \in [0, 4]$

a) Match the graphs with the functions.

b) Without finding the actual derivative or gradient function, sketch the gradient function for each skateboard ramp $h_1(x), h_2(x), h_3(x), h_4(x)$.

$h_1(x) = -0.1(x-2)^3 + 1$

$h_2(x) = 3e^{-0.75x}$

$h_3(x) = \begin{cases} -0.15x^2 + 0.1x + 2, & x \in \{[0, 2.5] \cup [3, 4]\} \\ 1.3, & x \in (2.5, 3) \end{cases}$

c) Find the derivatives of each skateboard ramp function $h_1(x), h_2(x), h_3(x), h_4(x)$ using calculus.

$h_4(x) = e^{-0.5x} \cos(4x) + 1$

d) Confirm a), b) and c) with the aid of technology.

e) Find the coordinates of any stationary points that exist on the different skateboard ramp profile functions $h_1(x), h_2(x), h_3(x), h_4(x)$.

f) Which skateboard ramps profile have strictly decreasing gradients in the specified domain?

g) Complete the table below with the instantaneous gradient to 4 decimal places for each skateboard profile at the given values of x.

Function	$\dfrac{dh}{dx}$ at $x=1$	$\dfrac{dh}{dx}$ at $x=2$	$\dfrac{dh}{dx}$ at $x=3$
$h_1(x)$			
$h_2(x)$			
$h_3(x)$			
$h_4(x)$			

2. Derivatives in the real world, measuring change.

a) Bow Wow Wow Company is a theoretical company that makes dog apparel.

The Revenue function of Bow Wow Wow Co is given by $R(x)$ where R is the revenue and x is the number of parcels sold, where each parcel contains 1000 dog jackets.

$R(x) = 100x - 0.25x^2$ where $x \geq 0$.

Marginal Profit: *$P(x)$ is the difference between Marginal Revenue and Marginal Cost.* $P(x) = R'(x) - C'(x)$

This year the Marginal Profit is given by:
$P(x) = -0.9x^3 - 4.25x^2 + 100x + 5$

The change in the rate of revenue per parcel produced is called the Marginal Revenue.

i. Find the Marginal revenue, which is the derivative of revenue.

ii. Find the maximum revenue for parcels sold.

iii. What is the rate of production resulting in maximum revenue?

iv. What is the maximum revenue and how many parcels need to be sold to achieve it?

v. Using Newton's method find when the Marginal Profit is equal to 0 in parcel units.

b) Effective antidote

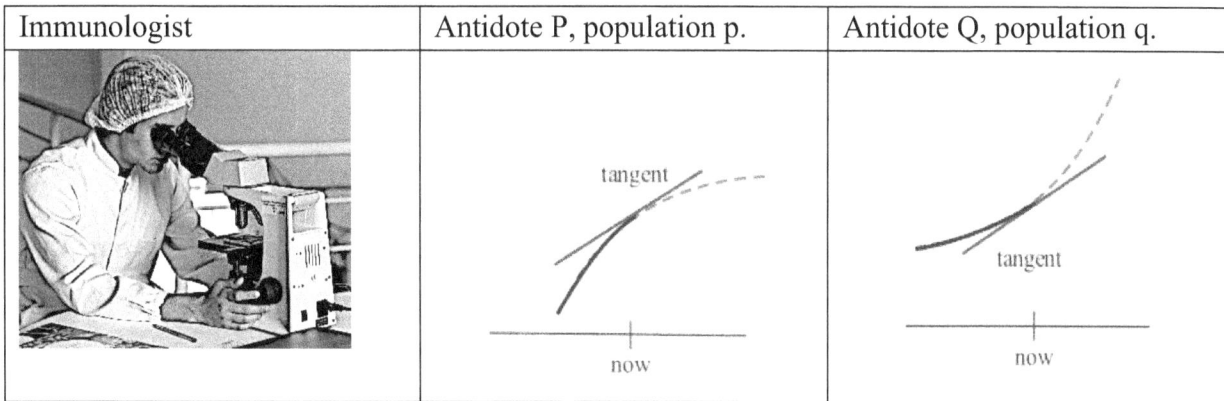

Suppose an epidemic has started in the country of Elbonia, in population p, and population q. Immunologists have developed two antidotes P, and Q, which are separately administered to the two separate affected populations p and q respectively, at the same time. After the antidote is given, the number of people x infected by the disease is monitored against time t, and an interval of the monitoring of the two populations is shown in the diagram above.

a) Identify which rate of spread is increasing, strictly increasing, decreasing, strictly decreasing, from the diagrams shown.

b) What does the tangent line represent when time is "now"?

c) Which antidote appears to be working "now"?

d) By monitoring the continued spread at timely intervals can the immunologists decide which antidote is more effectively fighting the spread of the disease at the time of "now"?

5.2 Derivatives in the real world, measuring change (continued).

c) Fashion trends

A new fashion trend that takes off in the city of Dubbo, which has a population of 100000 people, is modelled by the equation:

$$N(t) = \frac{100000}{99e^{-0.5t} + 1}$$

Where N is the number of people who have adopted the trend and t is the time in weeks.

i. Sketch the graph of $N(t)$ for the first 50 weeks, indicating endpoints and any asymptotes where they exist.

ii. Find the rate of change of the fashion trend over time.

iii. In which intervals of time is the function $N(t)$ strictly increasing?

iv. What does that mean about how the fashion trend is being adopted by the people of Dubbo? How is the rate of the trend changing over time?

v. Do all the people of Dubbo adopt the fashion? Explain.

5.3 Setting up rates of change equations and finding stationary points.

a) Ice cream .

Ice cream starts leaking from the bottom of the ice-cream cone and falls onto a floor forming a circular puddle.

The radius of the puddle increases at a rate of 0.1 cm/s.

How fast is the area of the puddle increasing when the radius is 0.8 cm?

b) Spherical scoop of ice-cream of radius r.

Find the minimum volume of a right circular ice cream cone which can completely contain (circumscribe) a spherical scoop of ice-cream of radius r.

This situation is shown in the diagram.

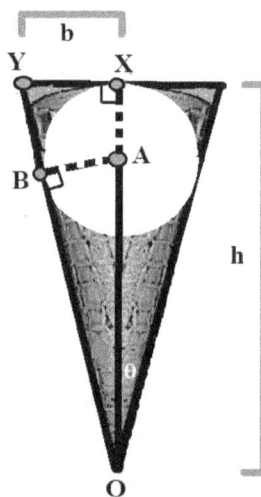

DERIVING DERIVATIVES SOLUTIONS.
5.1 Gradient functions in the real world. (SOLUTIONS)

a) Skateboard ramp profiles matched with functions.

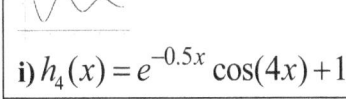	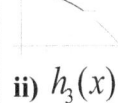
i) $h_4(x) = e^{-0.5x}\cos(4x) + 1$	**ii)** $h_3(x)$

iii) $h_2(x) = 3e^{-0.75x}$	**iv)** $h_1(x) = -0.1(x-2)^3 + 1$

b) Without finding the actual derivative or gradient function, sketch the gradient function for each skateboard ramp $h_1(x), h_2(x), h_3(x), h_4(x)$.

c) Derivatives of $h_1(x), h_2(x), h_3(x), h_4(x)$

$h_1'(x) = -0.3(x-2)^2, \ x \in (0,4)$

$h_2'(x) = -2.25e^{-0.75x}, \ x \in (0,4)$

$h_3'(x) = \begin{cases} -0.3x + 0.1, & x \in \{(0, 2.5) \cup (3,4)\} \\ 0, & x \in (2.5, 3) \end{cases}$

$h_4'(x) = e^{-0.5x}\left(-0.5\cos(4x) - 4\sin(4x)\right)$

e) finding the stationary points $h_1(x), h_2(x), h_3(x), h_4(x)$.

$h_1'(x) = -0.3(x-2)^2, \ x \in (0,4)$ $h_1'(x) = 0 \Rightarrow -0.3(x-2)^2 = 0$ $x = 2, h_1(2) = 1 \ SP(2,1)$	$h_2'(x) = -2.25e^{-0.75x}, \ x \in (0,4)$ $h_2'(x) = 0, \ no\ solution$
$h_3'(x) = \begin{cases} -0.3x + 0.1, & x \in \{(0, 2.5) \cup (3,4)\} \\ 0, & x \in (2.5, 3) \end{cases}$ $-0.3x + 0.1 = 0$ $x = \dfrac{1}{3}, h_3(\dfrac{1}{3}) = \dfrac{365}{180} \ SP\left(\dfrac{1}{3}, \dfrac{73}{36}\right)$	$h_4'(x) = e^{-0.5x}\left(-0.5\cos(4x) - 4\sin(4x)\right)$ $h_4'(x) = 0 \Rightarrow e^{-0.5x}\left(-0.5\cos(4x) - 4\sin(4x)\right) = 0$ $\left(-0.5\cos(4x) - 4\sin(4x)\right) = 0$ $-0.125 = \tan(4x) \ \ x \approx \{0.75, 1.54, 2.33, 3.11\}$ $SP(0.75, 0.32), (1.54, 1.46),$ $(2.33, 0.69), (3.11, 1.21)$

f) $h_1(x), h_3(x), h_4(x)$ have strictly decreasing intervals in the domain [0,4]

g) Complete the table with instantaneous gradients to 4 decimal places for each skateboard profile $h_1(x), h_2(x), h_3(x), h_4(x)$.

Function	$\dfrac{dh}{dx}$ at $x=1$	$\dfrac{dh}{dx}$ at $x=2$	$\dfrac{dh}{dx}$ at $x=3$
$h_1(x)$	-0.3	0	-0.3
$h_2(x)$	-1.0628	-0.5020	-0.2371
$h_3(x)$	-0.2	-0.5	0
$h_4(x)$	0.6035	0.9465	1.1883

2. Derivatives in the real world, measuring change. (SOLUTIONS)

a) Bow Wow Wow Jackets

$$R(x) = 100x - 0.25x^2$$
$$P(x) = -0.9x^3 - 4.25x^2 + 100x + 5$$

i. Find the Marginal Revenue which is the derivative of the revenue function. Differentiating the function we get $R'(x) = 100 - 0.5x$

ii. To find the maximum revenue we need to let the first derivative equal zero and solve for x the number of parcels sold.
$$R'(x) = 0$$
$$100 - 0.5x = 0$$
$$x = 200$$

iii. Substituting $x = 200$ back into the revenue equation $R(x)$ will find the revenue for Bow-Wow Wow, for 200 parcels sold.
$$R(200) = 100(200) - 0.25(200)^2$$
$$= 10000$$

iv. This means that the rate of production resulting in maximum revenue occurs when the number of parcels sold is two hundred resulting in total revenue of $10,000.

v. Using Newton's method finde Marginal Profit is equal to 0
$P(8) = 72.2$ above x-axis, $f(9) = -95.35$ is below, solution in interval $[8,9]$ guess $x_0 = 8.5$

$$x_{i+1} = x_i - \frac{f(x_0)}{f'(x_0)} \rightarrow x_1 = 8.5 - \frac{-4.775}{-167.325}$$
$$x_1 = 8.471 \rightarrow x_2 = 8.471 - \frac{0.055}{-165.695}$$
accuracy accepted $P(8.471) \approx 0$.

b) Effective antidote

Immunologist	Antidote P, population p.	Antidote Q, population q.
	tangent ---- now	tangent now

i. Antidote Q shows an interval where $\frac{dx}{dt}$ is strictly increasing, Antidote P shows an interval where $\frac{dx}{dt}$ is strictly decreasing.

ii. The tangent represents the instantaneous rate of infection $\frac{dx}{dt}$ of the number of people x getting the disease at the time t of "now"?

iii. At the time of $t = "now"$, Antidote P seems to be more effective in reducing the rate of infection in population p, as time around the time of "now" in the interval shown. It looks like the rate of infection is out of control in population q.

iv. There is no information about the long run outcomes as the rate of infection is not shown for a longer time interval.

c) Fashion trends

$$N(t) = \frac{100000}{99e^{-0.5t}+1}$$

i.   Graph below

ii.   Find the rate of change of the fashion trend over time.

$$\frac{dN}{dt} = \frac{-100000(-49.5e^{-0.5t})}{\left(99e^{-0.5t}+1\right)^2}$$

$$= \frac{4950000e^{-0.5t}}{\left(99e^{-0.5t}+1\right)^2}$$

iii. By observing the derivative function $N'(t)$ we see that the rate is always positive, but is only strictly increasing up to approximately 9.19 weeks after the trend has started, thereafter the adoption of the trend slows down.

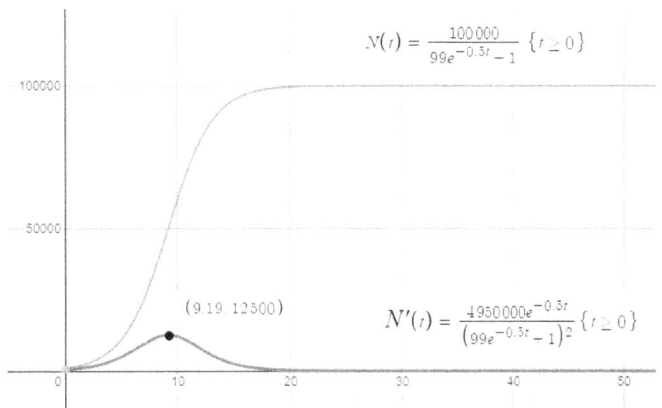

iv.   The rate of adopting the trend is slowing down over time, as the gradient becomes less positive after around 9 weeks.

v.   It is unlikely that all the people of Dubbo will adopt the fashion trend, as some parts of the population will have no interest in the trend.

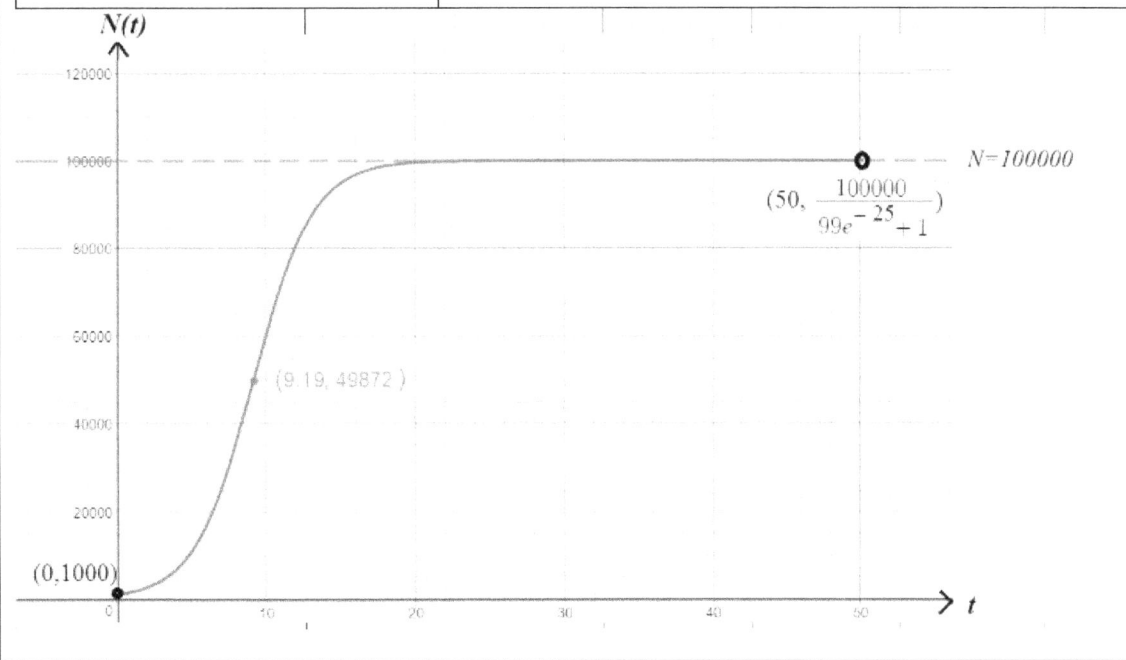

a) Ice cream starts leaking from the bottom of the ice-cream cone.

We can relate the derivatives using the chain rule.

$$\frac{dA}{dr} = 2\pi r \text{ cm}^2/\text{cm}$$

$$\frac{dr}{dt} = 0.1 \text{ cm/s}$$

$$A = \pi r^2$$

$$\frac{dA}{dt} = \frac{dA}{dr} \times \frac{dr}{dt}$$

$$= (2\pi r)(0.1)$$

$$= 0.2\pi r$$

$$\frac{dA}{dt} = 0.2\pi r \text{ cm}^2/\text{s}$$

$$\left.\frac{dA}{dt}\right|_{r=0.8} = 0.2\pi(0.8)$$

$$= 0.16\pi \text{ cm}^2/s$$

b) Ice cream cone with a spherical scoop of ice-cream of radius r.

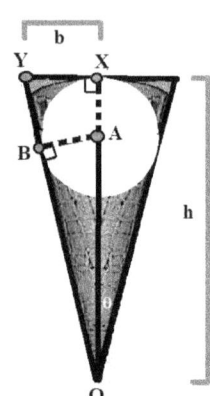

The sphere of ice cream is inside the cone, just touching the sides of the cone, with the top of the sphere level with the top of the cone.

Volume of cone: $V_C = \frac{1}{3}\pi b^2 h$, where b is the radius of the base.

Volume of scoop: $V_S = \frac{4}{3}\pi r^3$, where r is the radius of the scoop.

Using geometry and trigonometry we can relate the quantities of height, radius of the sphere and angle of the cone. We need to define b in terms of h and r.

From an origin O we can take the cross-section, and label several important locations for this problem, A, B, X, Y. ▲OXY forms a right angled triangle at vertex X, which is a similar triangle to the triangle formed by ▲OBA at vertex B.

$$\frac{OX}{OB} = \frac{XY}{BA} = \frac{OY}{OA}$$

$$OX = h, \; OA = h - r$$

$$XY = b, \; AB = r$$

$$OY = \sqrt{h^2 + b^2}$$

$$OB = \sqrt{(h-r)^2 - r^2}$$

$$OB = \sqrt{h^2 - 2hr + r^2 - r^2}$$

$$OB = \sqrt{h^2 - 2hr}$$

$$\therefore \frac{b}{h} = \frac{r}{\sqrt{h^2 - 2hr}}$$

$$b = \frac{rh}{\sqrt{h^2 - 2hr}}$$

5.3 Setting up rates of change equations and finding stationary points.(SOLUTIONS)

c) Find the minimum volume of a right circular ice cream cone which can completely contain (circumscribe) a spherical scoop of ice-cream of radius r.

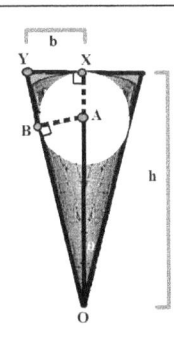

$$\therefore \frac{b}{h} = \frac{r}{\sqrt{h^2 - 2hr}}$$

$$b = \frac{rh}{\sqrt{h^2 - 2hr}}$$

$$V_C = \frac{1}{3}\pi b^2 h$$

$$V_C(h) = \frac{1}{3}\pi \left(\frac{rh}{\sqrt{h^2 - 2hr}} \right)^2 h$$

$$V_C(h) = \frac{1}{3}\pi \left(\frac{r^2 h^2}{h^2 - 2hr} \right) h$$

$$V_C(h) = \frac{1}{3}\pi \left(\frac{r^2 h^2}{h - 2r} \right)$$

$$V_C'(h) = \frac{1}{3}\pi \left(\frac{(h - 2r)(2r^2 h) - (r^2 h^2)(1)}{(h - 2r)^2} \right)$$

$$V_C'(h) = \frac{1}{3}\pi \left(\frac{2r^2 h^2 - 4r^3 h - r^2 h^2}{(h - 2r)^2} \right)$$

$$V_C'(h) = \frac{1}{3}\pi \left(\frac{r^2 h^2 - 4r^3 h}{(h - 2r)^2} \right)$$

$$V_C'(h) = 0$$

$$r^2 h^2 - 4r^3 h = 0$$

$$r^2 h^2 = 4r^3 h$$

$$h = 4r$$

$$V_C(4r) = \frac{1}{3}\pi \left(\frac{r(4r)}{\sqrt{(4r)^2 - 2(4r)r}} \right)^2 (4r)$$

$$V_C(4r) = \frac{1}{3}\pi \left(\frac{64r^5}{(16r^2 - 8r^2)} \right)$$

$$V_C(4r) = \frac{1}{3}\pi \left(\frac{64r^5}{8r^2} \right)$$

$$V_C(4r) = \frac{1}{3}\pi \left(8r^3 \right)$$

$$V_C(4r) = \frac{8}{3}\pi r^3$$

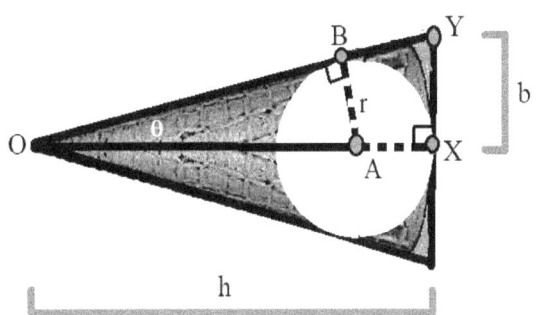

What angle is that θ ?

$$\sin\theta = \frac{r}{3r} = \frac{1}{3}$$

$$\theta \approx 19.47^\circ$$

Possibly the angle on your next ice cream cone.

Too funny?

xkcd   Calvin & Hobbes

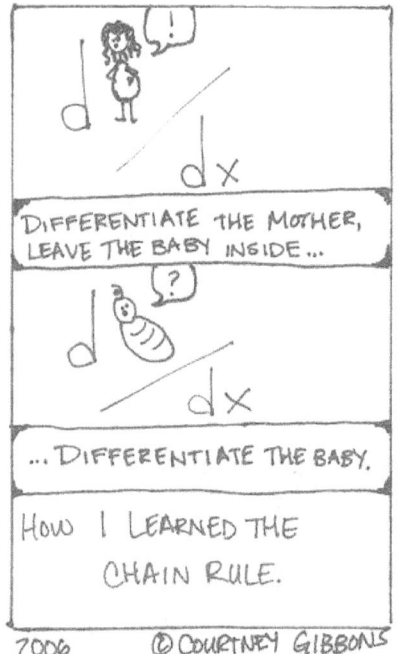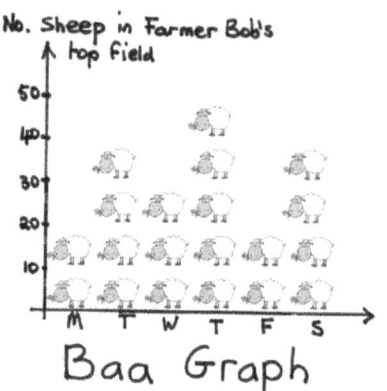

Integrating Integrals.

Integration is mathematical operation that allows quantities such as distance, length, area, volume, money earned and many other quantities to be determined by the addition of infinitesimal data under a function that models how these quantities are changing. Integration in mathematics is often called **Anti-differentiation** as it is the opposite operation to **differentiation**. Differentiation and Integration are the main operations in the area of mathematics called Calculus.

Integration is the connection between the area under a particular function $f(x)$ in a given domain interval and the evaluation of the corresponding antiderivative function $F(x)$ within that same domain interval, for which a cyclic relationship exists $F'(x) = f(x)$ and $\int f(x)\,dx = F(x) + C$.

In a simple example, a bath that is being filled with water, where 6 litres per minute is added to the initially empty bath every minute for 10 minutes. The graph below show s the quantity of water can be calculated by finding the area under the function $f(x) = 6$ where x is the time, in the interval between $x = 0$ and $x = 10$ minutes, where the change in time x is $\Delta x = 1$ minute.	This quantity of water added to the bath can also be calculated by finding the antiderivative of $f(x) = 6$ which could be $F(x) = 6x$ and finding the difference between $F(10)$ and $F(0)$.

In a simple example, a bath that is being filled with water, where 6 litres per minute is added to the initially empty bath every minute for 10 minutes. The graph below show
s the quantity of water can be calculated by finding the area under the function $f(x) = 6$ where x is the time, in the interval between $x = 0$ and $x = 10$ minutes, where the change in time x is $\Delta x = 1$ minute.

$$\text{Water Added} = \sum_{i=1}^{10} f(i)(\Delta x)$$

$$= f(1)(1) + f(2)(1) + f(3)(1) + f(4)(1) + f(5)(1)$$
$$\quad + f(6)(1) + f(7)(1) + f(8)(1) + f(9)(1) + f(10)(1)$$
$$= (6)(1) + (6)(1) + (6)(1) + (6)(1) + (6)(1)$$
$$\quad + (6)(1) + (6)(1) + (6)(1) + (6)(1) + (6)(1)$$
$$= 60 \text{ litres}$$

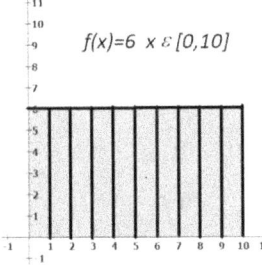

f(x)=6 x ε [0,10]

This quantity of water added to the bath can also be calculated by finding the antiderivative of $f(x) = 6$ which could be $F(x) = 6x$ and finding the difference between $F(10)$ and $F(0)$.

$$F(x) = 6x,\ F(10) = 60,\ F(0) = 0$$
$$F(10) - F(0) = 60 \text{ litres}$$

Which can be written as a *definite integral*.

$$\text{Volume} = \int_0^{10} f(x)\,dx \qquad = \int_0^{10} 6\,dx$$
$$= \left[F(x) \right]_0^{10} \qquad = \left[6x \right]_0^{10}$$
$$= \left[F(10) \right] - \left[F(0) \right] \quad = \left[6(10) \right] - \left[6(0) \right]$$
$$\qquad\qquad\qquad\qquad = 60 \text{ litres}$$

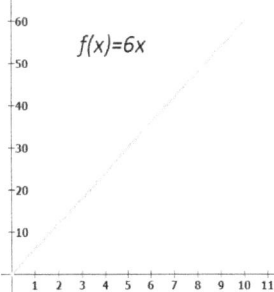

f(x)=6x

NOTE: The gradient of this function is 6.

The area under a function, by integration and estimation using rectangles and trapeziums.

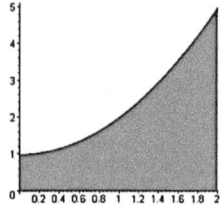

If function $f(x) = x^2 + 1$ is modelling the change in some quantity, showing the increase in $f(x)$ for every increase in x. We can find the exact quantity in a given interval of x by finding the area A under this function bound by the x-axis, using the definite integral.

$$A = \int_0^2 f(x)dx = \int_0^2 (x^2 + 1)dx$$

$$= \left[\frac{x^3}{3} + x\right]_0^2 = \frac{14}{3} \approx 4.3333$$

From first principles the area under $f(x)$ can be estimated by creating rectangles of width Δx with the height informed by $f(x)$ and adding all their areas. The thinner the rectangles, the more accurate the estimate will be.

Depending on which value of $f(x)$ is used to find the height of the rectangles the estimate will either be less than or greater than the actual area in the interval bound by the x-axis.

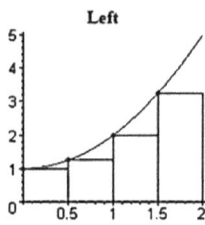

Rectangles of width $\Delta x = 0.5$ on an increasing function such as $f(x) = x^2 + 1$ using the left x value for the height $f(x)$ will underestimate the actual area.

x	0	0.5	1	1.5
$f(x)$	1	1.25	2	3.25

$$Area_{left} = (1)(0.5) + (1.25)(0.5)$$
$$+ (2)(0.5)$$
$$+ (3.25)(0.5)$$
$$= 3.75$$

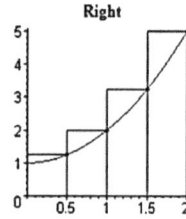

Rectangles of width $\Delta x = 0.5$ for $f(x) = x^2 + 1$ using the right x value for the height f(x) will overestimate the actual area.

x	0.5	1	1.5	2
$f(x)$	1.25	2	3.25	5

$$Area_{right} = (1.25)(0.5) + (2)(0.5)$$
$$+ (3.25)(0.5)$$
$$+ (5)(0.5) = 5.75$$

Conversely on an decreasing function using the left value of x of the rectangles to get the heights $f(x)$ results in an estimate that is more than the actual area.

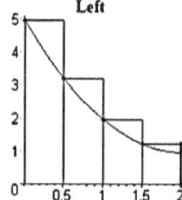

Using the right side of the rectangle, the area estimate is less than the actual area.

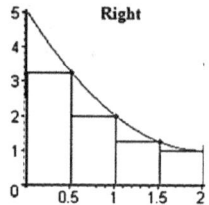

Adding the areas of trapeziums of width Δx in the required interval instead of rectangles will improve the area estimation as the heights are averaged: $\frac{\Delta x}{2}(f(x_{i+1}) + f(x_i))$

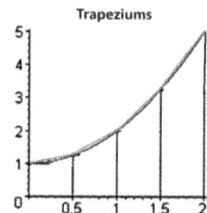

x	0	0.5	1	1.5
$f(x_i)$	1	1.25	2	3.25
$f(x_{i+1})$	1.25	2	3.25	5

$$Area_{trap} = (1 + 1.25)(0.5)$$
$$+ (1.25 + 2)(0.5)$$
$$+ (2 + 3.25)(0.5)$$
$$+ (3.25 + 5)(0.5)$$
$$= 4.75$$

Rewriting the *Area* expression allows the pattern to be seen of the $f(x_i)$ values.

$$Area_{trap} = 0.5(1 + 1.25 + 1.25 + 2$$
$$+ 2 + 3.25 + 3.25$$
$$+ 5) = 4.75$$

This method of finding the area under a function is known as the **trapezoidal rule**.

The **trapezoidal rule** estimates the area under a function between $x=a$ and $x=b$ using n trapeziums of width Δx and is expressed in general form as:

$$\int_a^b f(x)dx \approx \frac{\Delta x}{2}(f(x_0) + 2f(x_1) + 2f(x_2) + 2f(x_3)+\ldots 2f(x_{n-1}) + f(x_n))$$

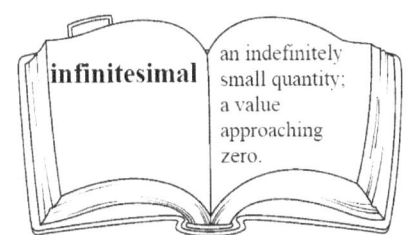

The definition of the word "**infinitesimal**"

Adjective: extremely small. *Synonyms:* tiny, miniscule, very small
Noun: MATHEMATICS an indefinitely small quantity, a value
approaching zero. (*Oxford English Dictionary*)

With the simple idea of the addition of infinitesimals the whole idea of Calculus is formed.

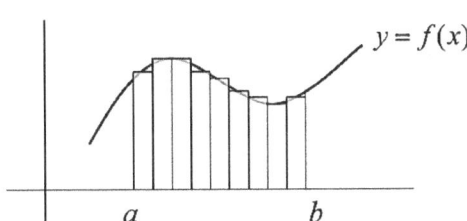

If a function $f(x)$ is continuous, the area under the graph in a given domain of x is given by the addition of infinite infinitesimal rectangular areas whose height is given by $f(x)$ and whose width is given by Δx, a value that is approaching zero.

$$\text{Area} = \lim_{n \to \infty} \sum_{k=0}^{n-1} f(x_k)\Delta x = \int_a^b f(x)dx$$

The elongated S symbol $\int$ used to represent the antiderivative also called the integral, indicates a summation of infinitesimals.

This amazing finding was discovered by the manual calculation of infinitesimals numerically, and the realisation of the connection between the derivative function and the antiderivative function.

The antiderivative $F(x)$ is also called the integral, it is found by integration and algebraically represents an area function between a function $f(x)$ and the *x-axis*, where $F'(x) = f(x)$.

By substitution of values the antiderivative $F(x)$ can be used to find the area bound by the function $f(x)$ and the *x-axis*. If the antiderivative function is restricted to a domain it is called a definite integral.

Properties of definite integrals.

$$\int_a^b f(x)dx = \Big[F(x)\Big]_a^b = F(b) - F(a)$$

Called the definite integral of $f(x)$ with respect to x, from $x = a$ to $x = b$

- $F(x)$ is the antiderivative function $F'(x) = f(x)$

- a is the lower limit of integration

- b is the upper limit of integration

- $f(x)$ is continuous between $x = a$ to $x = b$

The difference between Differentiation and Integration.

Differentiation creates the derivative

$$f'(x) = \frac{d(f(x))}{dx}$$ of the function $f(x)$, which

represents the gradient function, and gives the gradient of the curve at any value of x

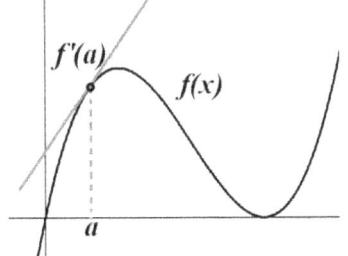

Integration creates the antiderivative

$F(x) = \int f(x)\,dx$ of the function $f(x)$, which can
be used to find the area under the function bound
by the *x-axis*, where $F'(x) = f(x)$.

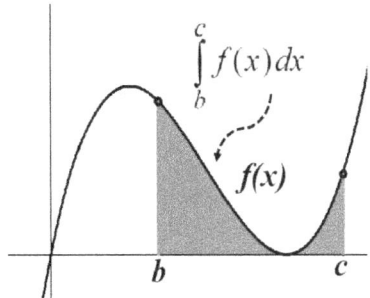

If $A(x)$ is a function that finds the area under a function $f(x)$ bound by the x-axis, then $f(x) = A'(x)$.

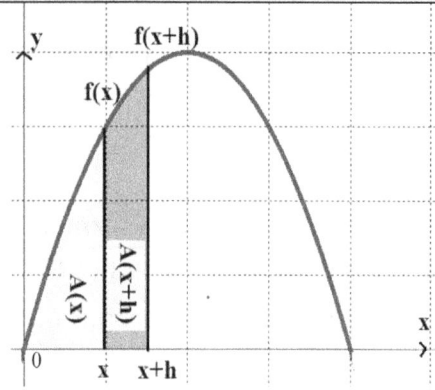

The area of the strip whose height ranges between $f(x)$ and $f(x+h)$ is the difference between $A(x+h) - A(x)$, where the change in x is $\Delta x = h$
This approximation to the area becomes more accurate as $\lim_{h \to 0}$

An infinitesimal approximation of the area between x and $(x+h)$

$$f(x) \times h = \lim_{h \to 0} A(x+h) - A(x)$$

Dividing both sides by h (an infinitesimal width)

$$f(x) = \lim_{h \to 0} \frac{A(x+h) - A(x)}{h}$$

Notice that the right hand side in the above expression looks like differentiation from first principles:

$f(x) = A'(x)$ which implies that $\int f(x)\,dx = A(x)$

$A(x+h) - A(x) = \int_{x}^{x+h} f(x)\,dx$ which is the

Fundamental Theorem of Calculus.

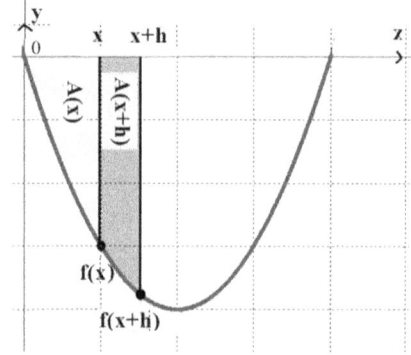

Areas between the function and the x-axis found by integration where the function has negative values in the interval, are **negative areas**.

Since $f(x) \times h = \lim_{h \to 0} A(x+h) - A(x)$ and $f(x) < 0$

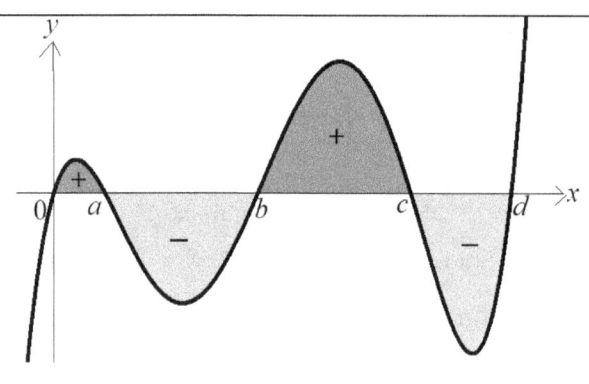

Areas bound by a function $f(x)$ and the x-axis where the function is above the x-axis are positive, areas where the function is below the x-axis are negative.

Finding the x-intercepts for a function $f(x)$ can identify which areas are above and below the x-axis.

Definite Integrals	**Indefinite Integrals**
$\int_{a}^{b} f(x)dx = \left[F(x)\right]_{a}^{b} = F(b) - F(a)$	$\int f(x)\,dx = F(x) + C$
where $F'(x) = f(x)$	where C is a constant and $F'(x) = f(x)$
Integration is restricted to the interval of x, from $x = a$ to $x = b$	Integration is not restricted to an interval and represents an infinite area bounded by the x-axis.

The constant of integration.

$\int f(x)\,dx = F(x) + C$, the constant of integration C should always be stated for a general solution.

Every antiderivative of $f(x)$ is of the form $F(x) + C$ where C is a constant and can take any numeric value. $F(x) + C$ represents every possible antiderivative of $f(x)$, since $\dfrac{d\left(F(x)+C\right)}{dx} = f(x)$.

Integration properties.	Example	
Sum and Difference Rule. $\int (f \pm g)\,dx = \int f\,dx \pm \int g\,dx$	$\int \left(x^2 + x\right)dx = \int x^2 dx + \int x\,dx$ $= \dfrac{x^3}{3} + \dfrac{x^2}{2} + C$	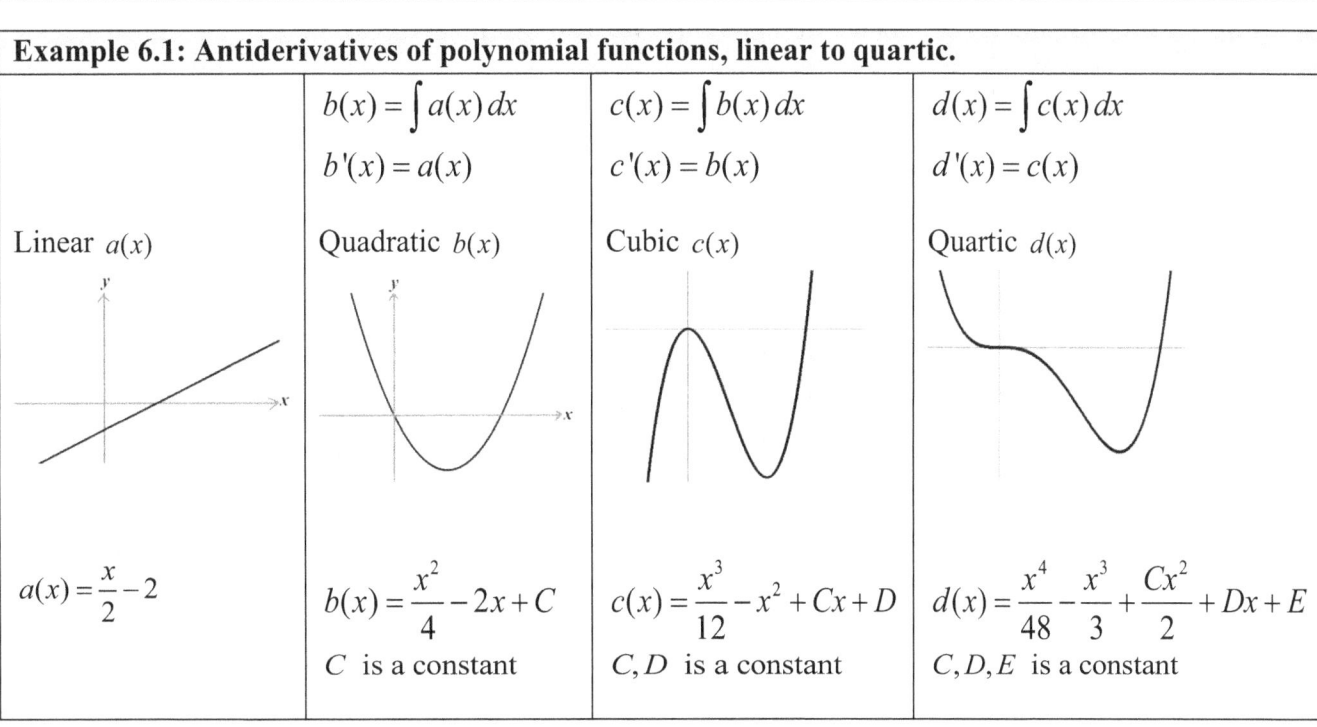
Constant Multiple Rule. $\int kf(x)\,dx = k\int f(x)\,dx$ (k constant)	$\int \left(2x^3 + 4x\right)dx = 2\int \left(x^3 + 2x\right)dx$ $= 2\left(\dfrac{x^4}{4} + x^2\right) + C$ $= \dfrac{x^4}{2} + 2x^2 + C$	

As for differentiation, there are rules that can be applied to antidifferentiate certain functions bypassing methods using the first principles.

Power rule for antiderivatives of x^n	
Case where $n \neq -1$	$\int x^n dx = \dfrac{x^{n+1}}{n+1} + C$ if $n \neq -1$
Case where $n = -1$	$\int x^{-1} dx = \int \dfrac{1}{x}\,dx = \ln|x| + C$

Example 6.1: Antiderivatives of polynomial functions, linear to quartic.				
	$b(x) = \int a(x)\,dx$ $b'(x) = a(x)$	$c(x) = \int b(x)\,dx$ $c'(x) = b(x)$	$d(x) = \int c(x)\,dx$ $d'(x) = c(x)$	
Linear $a(x)$	Quadratic $b(x)$	Cubic $c(x)$	Quartic $d(x)$	
$a(x) = \dfrac{x}{2} - 2$	$b(x) = \dfrac{x^2}{4} - 2x + C$ C is a constant	$c(x) = \dfrac{x^3}{12} - x^2 + Cx + D$ C,D is a constant	$d(x) = \dfrac{x^4}{48} - \dfrac{x^3}{3} + \dfrac{Cx^2}{2} + Dx + E$ C,D,E is a constant	

Calculus formulas using rules to Differentiate and Antidifferentiate

because $\dfrac{d}{dx}\big(F(x)+C\big)=f(x)$ **when** $\displaystyle\int f(x)\,dx = F(x)+C$

$\dfrac{d}{dx}\left(x^n\right)=nx^{n-1}$	$\displaystyle\int x^n dx = \dfrac{x^{n+1}}{n+1}+C \ \ \text{if } n\neq -1$ $\displaystyle\int x^{-1}dx = \int \dfrac{1}{x}dx = \ln|x|+C \ \ \text{if n=1}$
$\dfrac{d}{dx}\left(e^{ax+b}\right)=ae^{ax+b}$	$\displaystyle\int e^{ax+b}dx = e^b\int e^{ax}dx = \dfrac{1}{a}e^{ax+b}+C$
$\dfrac{d}{dx}\big(\log_e(ax+b)\big)=\dfrac{a}{ax+b}$	$\displaystyle\int \dfrac{a}{ax+b}dx = \int \dfrac{1}{x+\frac{b}{a}}\,dx = \log_e\left|x+\dfrac{b}{a}\right|+C$
$\dfrac{d}{dx}\big(\sin(ax+b)\big)=a\cos(ax+b)$	$\displaystyle\int \sin(ax+b)\,dx = -\dfrac{1}{a}\cos(ax+b)+C$
$\dfrac{d}{dx}\big(\cos(ax+b)\big)=-a\sin(ax+b)$	$\displaystyle\int \cos(ax+b)\,dx = \dfrac{1}{a}\sin(ax+b)+C$
$\dfrac{d}{dx}\big(\tan(ax+b)\big)=\dfrac{a}{\big[\cos(ax+b)\big]^2}$	

Example 6.2: Using formulas, properties and rules for Indefinite integrals.

$\displaystyle\int \dfrac{a}{ax+b}dx = \int \dfrac{1}{x+\frac{b}{a}}\,dx = \log_e\left|x+\dfrac{b}{a}\right|+C$	$\displaystyle\int \dfrac{25}{5x+3}dx = 5\int \dfrac{1}{x+\frac{3}{5}}\,dx = 5\log_e\left|x+\dfrac{3}{5}\right|+C$
$\displaystyle\int \sin(ax+b)\,dx = -\dfrac{1}{a}\cos(ax+b)+C$	$\displaystyle\int \sin(2x+\pi)\,dx = -\dfrac{1}{2}\cos(2x+\pi)+C$

Combinations

$$\int \sin(2x+\pi)+3x^2+2\,dx = -\dfrac{1}{2}\cos(2x+\pi)+x^3+2x+C$$

$$\int \dfrac{25}{5x+3}dx + 2\int e^{-7x}\,dx + \int\left(x^2+5x+6\right)dx$$

$$=\int\left[\dfrac{25}{5x+3}+2e^{-7x}+\left(x^2+5x+6\right)\right]dx$$

$$=5\log_e\left|x+\dfrac{3}{5}\right|-\dfrac{2}{7}e^{-7x}+\dfrac{x^3}{3}+\dfrac{5}{2}x^2+6x+C$$

Fibonacci Blue/Flickr

154

Example 6.3: Using formulas, properties and rules for Definite integrals.

$\int e^{ax+b} \, dx = e^b \int e^{ax} \, dx$

$\qquad = \dfrac{1}{a} e^{ax+b} + C$

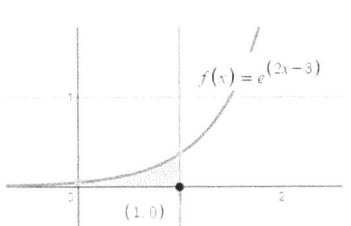

Shaded Area $= \dfrac{1}{2}\left(e^{-1} - e^{-3}\right)$

$\displaystyle\int_0^1 e^{2x-3} \, dx = e^{-3} \int_0^1 e^{2x} \, dx = \left[e^{-3}\left(\dfrac{1}{2}\right)e^{2x} \right]_0^1$

$\qquad = \left[\left(\dfrac{1}{2}\right)e^{2x-3} \right]_0^1$

$\qquad = \left[\left(\dfrac{1}{2}\right)e^{2(1)-3} \right] - \left[\left(\dfrac{1}{2}\right)e^{2(0)-3} \right]$

$\qquad = \left[\left(\dfrac{1}{2}\right)e^{-1} \right] - \left[\left(\dfrac{1}{2}\right)e^{-3} \right]$

$\qquad = \dfrac{1}{2}\left(e^{-1} - e^{-3}\right)$

$\int \cos(ax+b) \, dx = \dfrac{1}{a}\sin(ax+b) + C$

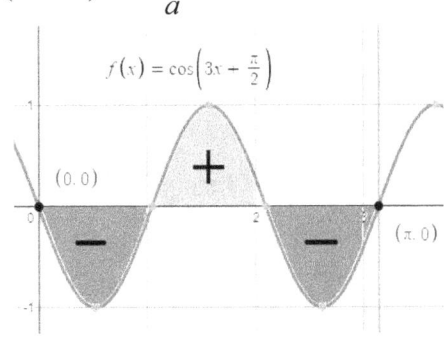

Net Shaded Area $= \dfrac{-2}{3}$

$\displaystyle\int_0^\pi \cos\left(3x + \dfrac{\pi}{2}\right) dx = \left[\dfrac{1}{3}\sin(3x + \dfrac{\pi}{2}) \right]_0^\pi$

$\qquad = \left[\dfrac{1}{3}\sin(3(\pi) + \dfrac{\pi}{2}) \right] - \left[\dfrac{1}{3}\sin(3(0) + \dfrac{\pi}{2}) \right]$

$\qquad = \dfrac{1}{3}\left(\sin\left(\dfrac{7\pi}{2}\right) - \sin\left(\dfrac{\pi}{2}\right) \right)$

$\qquad = \dfrac{1}{3}(-1 - 1)$

$\qquad = \dfrac{-2}{3}$

Definite Integrals may be combined when the terminals are the same.

$\displaystyle\int_{-6}^{-4} \dfrac{25}{5x+3} \, dx + \int_{-6}^{-4} \left(x^2 + 5x + 6\right) dx$

$\qquad = \displaystyle\int_{-1}^{4} \dfrac{25}{5x+3} + \left(x^2 + 5x + 6\right) dx$

$\qquad = \left[5\log_e \left| x + \dfrac{3}{5} \right| + \dfrac{x^3}{3} + \dfrac{5}{2}x^2 + 6x \right]_{-6}^{-4}$

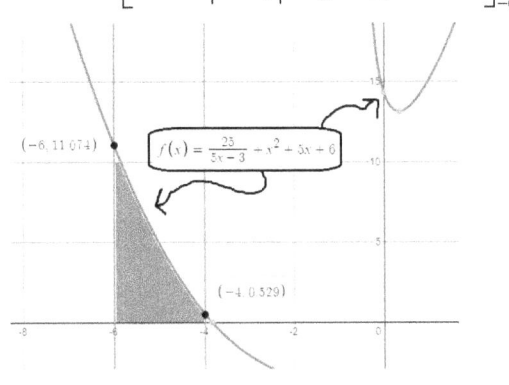

Definite Integrals **cannot be combined** when the terminals are different.

$\displaystyle\int_{-4}^{-2} \dfrac{25}{5x+3} \, dx + \int_{-2}^{0} \left(x^2 + 5x + 6\right) dx$

$\qquad = \left[5\log_e \left| x + \dfrac{3}{5} \right| \right]_{-4}^{-2} + \left[\dfrac{x^3}{3} + \dfrac{5}{2}x^2 + 6x \right]_{-2}^{0}$

Net Shaded Area

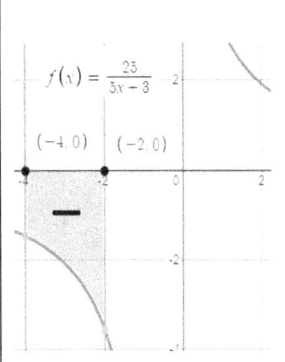

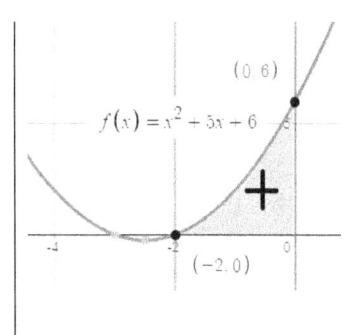

When asked to find the area bounded by the function and the *x-axis*, you will need to find all the *x-intercepts* and work out which sub-intervals are above and below the *x-axis*.

The sum of the bounded areas is the added magnitudes or sizes of the areas, where the sign of each area in a sub-interval is ignored.

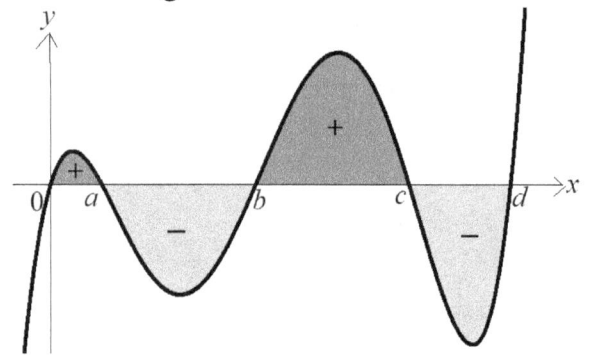

Work out the Positive areas	Work out the Negative areas
$Area_{\{0<x<a\}} = \int_0^a f(x)dx$	$Area_{\{a<x<b\}} = \int_a^b f(x)dx$
$Area_{\{b<x<c\}} = \int_b^c f(x)dx$	$Area_{\{c<x<d\}} = \int_c^d f(x)dx$

Then add the size of the areas, note the modulus sign around the negative areas below the *x-axis*.

$$Area_{\{0<x<a\}} + \left| Area_{\{a<x<b\}} \right| + Area_{\{b<x<c\}} + \left| Area_{\{c<x<d\}} \right|$$

Another important property of definite integrals is the order of the limits of integration to change the sign of the area.

$$\int_a^b f(x)\,dx = F(b) - F(a) \qquad \int_b^a f(x)\,dx = F(a) - F(b) \qquad \therefore \int_a^b f(x)\,dx = -\int_b^a f(x)\,dx$$

You can get the opposite signed area if you swap the terminals of integration, this may be useful in some situations where you need to work out the area bound by the function and the *x-axis*..

NOTE: If the upper and lower limits are the same then the integral is zero. $\int_a^a f(x)\,dx = 0$

Example 6.4: Areas bounded by the function and the x-axis.

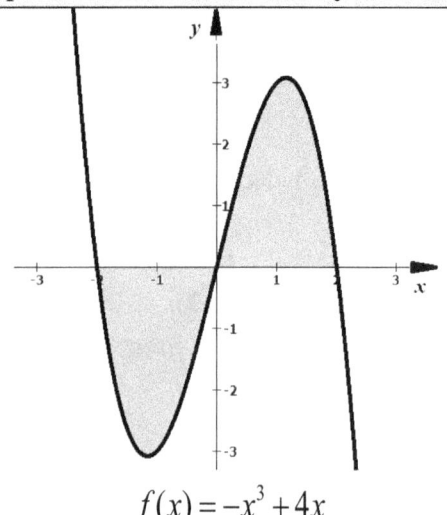

$$f(x) = -x^3 + 4x$$

3 *x*-intercepts exist at $(-2,0), (0,0), (2,0)$

The shaded area can be found by evaluating the integrals with a modulus sign about the negative area.

$$ShadedArea = \left| \int_{-2}^0 f(x)dx \right| + \int_0^2 f(x)dx$$

Which is equivalent to swapping the terminals for the negative area below the x-axis.

$$ShadedArea = \int_0^{-2} f(x)dx + \int_0^2 f(x)dx$$

Example 6.5: Calculus in business and financial modelling.

The Sunny Solar Company is a theoretical company that makes portable solar panels for camping.

The Sunny Solar Company has the following Marginal costs $MC(x)$ Marginal Revenue $MR(x)$ and Marginal Profit $P(x)$ associated with the manufacture of x solar units.

$$MC(x) = 0.0007x^3 - 0.035x^2 + 0.21x + 39.3$$

$$MR(x) = 100, \text{ where } x \geq 0$$

$$P(x) = MR(x) - MC(x)$$

These functions represent the change in cost for every solar unit manufactured.

Graphing these three functions shows the relationship between the functions.

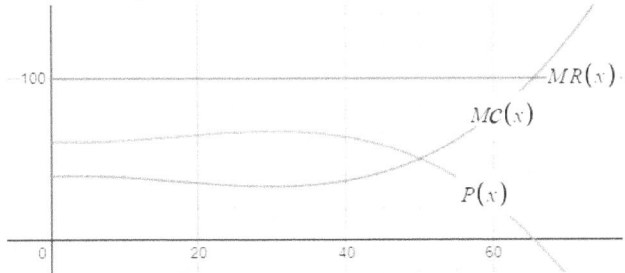

The cost to manufacture x solar panels is $M(x)$

$$M(x) = \int_0^x MC(x)\,dx$$

$$M(x) = \frac{7}{40000}x^4 - \frac{35}{3000}x^3 + \frac{21}{200}x^2 + 39.3x + C$$

Since $M(0) = 0, C = 0$

$R(x)$ is the revenue earned by selling x solar panels. Each solar panel is sold for $100.00 as shown by the Marginal Revenue function .

$$MR(x) = 100, \text{ where } x \geq 0,$$

$$R(x) = \int_0^x MR(x)\,dx = 100x + C$$

$C = 0$, because $R(0) = 0$

The Sunny Solar Company has the following Manufacturing costs, and Revenue as a function of x solar units.

$$M(x) = \frac{7}{40000}x^4 - \frac{35}{3000}x^3 + \frac{21}{200}x^2 + 39.3x$$

$$R(x) = 100x$$

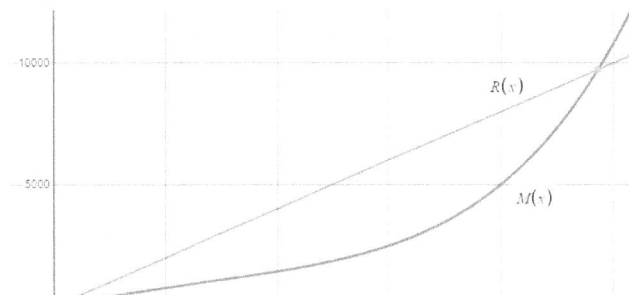

The largest profit can be made when there is the largest difference in ordinates (y values) between $R(x)$ and $M(x)$ as observed on the graphs of these functions.

Business/Economics terminology:

- ***Marginal Cost:*** *$MC(x) = M'(x)$ is a derivative function representing the change in cost production per unit as the number of solar panels manufactured increases.*

- ***Marginal Revenue:*** *$MR(x) = R'(x)$ is a derivative function representing the change in revenue per unit as the number of units increases.*

- ***Marginal Profit:*** *$P(x)$ is the difference between Marginal Revenue and Marginal Cost. $P(x) = MR(x) - MC(x)$.*

157

Application 6.1: Analysis of the 100 metre sprint race, where area=distance.

Usain Bolt ran the 100 metres sprint in a world record performance in Berlin in 2009 in 9.58 s. At each second an approximation of his velocity in metres/second is recorded in the table shown. Usain's velocity is approximated by a function that fits these points in the first 7 seconds: $U(t) = 12.2 - 12.2e^{-0.725t}$.

Amos is a medium level sprinter, and his best performance for the 100 metres sprint is 11.4 s. He reaches a maximum velocity of 9.9 m/s at the 6 second mark and sustains this until the end of the race.
$A(t) = 9.9 - 9.9e^{-0.77t}$

Bob is a recreational sprinter who can reach a maximum velocity of 8.7 m/s after 7 seconds, maintained until the end of the race. $B(t) = 8.7 - 8.7e^{-0.77t}$

Time (s)	Usain m/s	Amos m/s	Bob m/s
0	0	0	0
1	6.3	5	4.7
2	9.4	7.5	6.8
3	11	8.7	7.8
4	11.6	9.4	8.3
5	12	9.7	8.5
6	12.2	9.9	8.6
7	12.2	10	8.7

$U(t) = \dfrac{dx_u}{dt}$, $A(t) = \dfrac{dx_a}{dt}$, $B(t) = \dfrac{dx_b}{dt}$

Velocity v in metres/second is the change in position x metres divided by time t seconds: $v(t) = \dfrac{dx}{dt}$. The area under the function $v(t)$ is the quantity of

$\left(\dfrac{\text{metres}}{\text{second}} \times \text{second} \right) = \text{metres}$.

A definite integral of the velocity with respect to time will give the distance travelled for each runner in a given time interval.

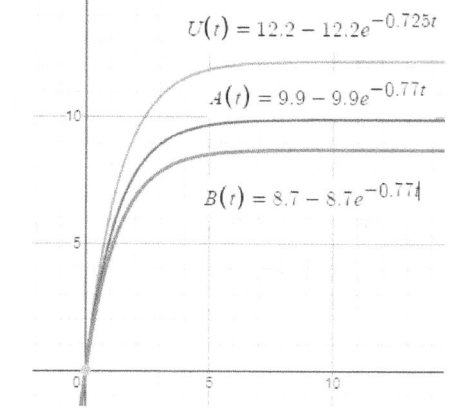

$\displaystyle\int_0^{9.8} U(t)dt = \int_0^{9.8} \left(12.2 - 12.2e^{-0.725t}\right)dt$

$= 100.065 \approx 100\,m$

$\displaystyle\int_0^{11.4} A(t)dt = \int_0^{11.4} \left(9.9 - 9.9e^{-0.77t}\right)dt$

$= 100.005 \approx 100\,m$

$\displaystyle\int_0^{k} B(t)dt = \int_0^{k} \left(8.7 - 8.7e^{-0.77t}\right)dt$

$\qquad k \approx 12.8s$

$\approx 100\,m$

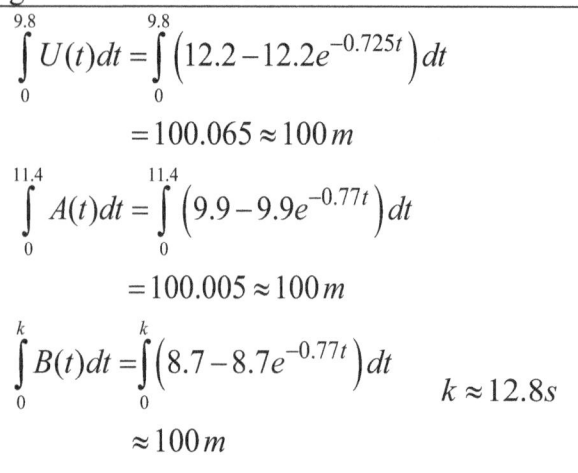

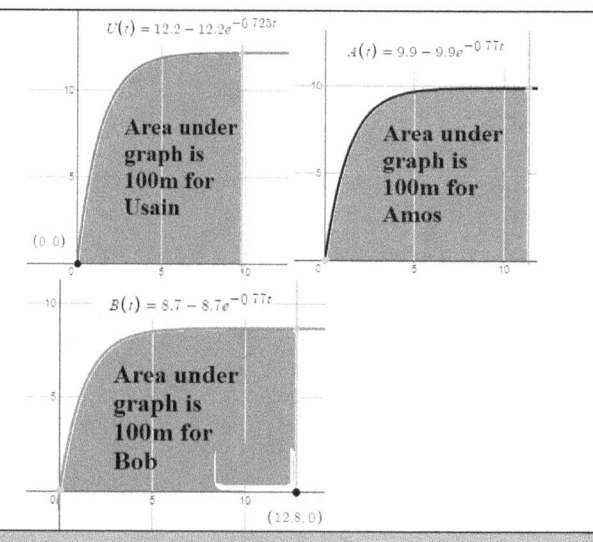

Areas between intersecting functions $f(x)$ and $g(x)$

In some cases the functions being integrated in an given domain or interval of x will cross each other. For these cases the points of intersection will need to be calculated and the order of subtraction will depend on which function has the largest value in any sub-interval.

We need to find all the intersections and then determine which function has the highest ordinates in any subintervals. $f(x) = g(x)$ at $x = \{a, b, c, d\}$

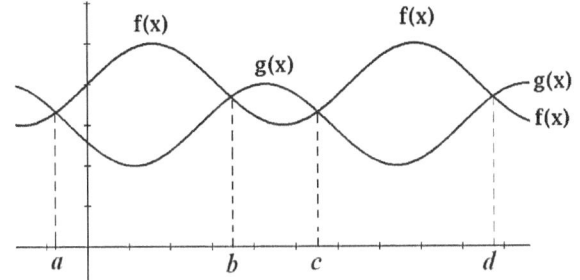

$f(a) = g(a),\ f(b) = g(b),\ f(c) = g(c), f(d) = g(d)$

The area under $f(x)$ bound by the x-axis between

$x = a$ and $x = d$ is given by $Area_1 = \int\limits_a^d f(x)\,dx$

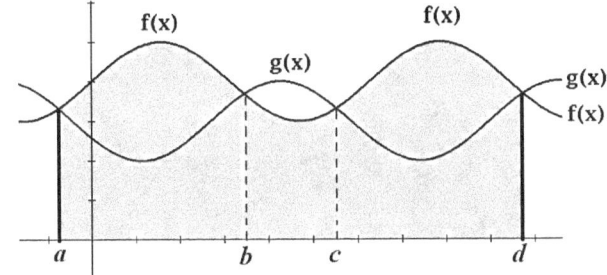

The area under $g(x)$ bound by the x-axis between

$x = a$ and $x = d$ is given by $Area_2 = \int\limits_a^d g(x)\,dx$

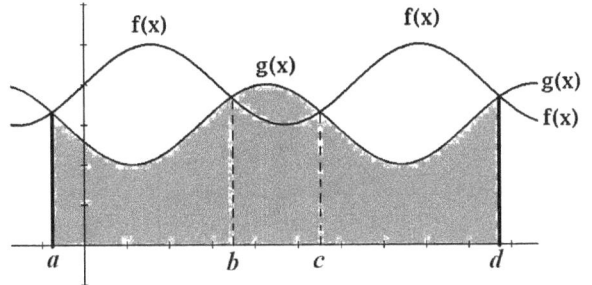

In this example

$f(x) > g(x)$ when $\{a < x < b\}$ or $\{c < x < d\}$

$g(x) > f(x)$, when $\{b < x < c\}$

Subtract the area of the function with the lower ordinates (y values) in each of the intervals from the area of the function with the higher ordinates.

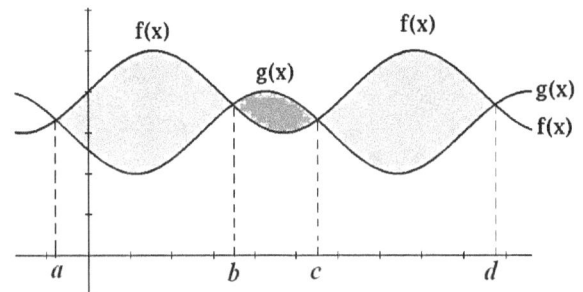

$$Area_{\{a<x<b\}} = \int\limits_a^b f(x)\,dx - \int\limits_a^b g(x)\,dx$$

$$Area_{\{b<x<c\}} = \int\limits_b^c g(x)\,dx - \int\limits_b^c f(x)\,dx$$

$$Area_{\{c<x<d\}} = \int\limits_c^d f(x)\,dx - \int\limits_c^d g(x)\,dx$$

Then add the areas.

$$Area_{\{a<x<b\}} + Area_{\{b<x<c\}} + Area_{\{c<x<d\}}$$

Area enclosed between $f(x)$ and $g(x)$ in the interval $\{a \le x \le d\}$ is given by:

$$\int\limits_a^b \left[f(x)-g(x)\right]dx + \int\limits_b^c \left[g(x)-f(x)\right]dx + \int\limits_c^d \left[f(x)-g(x)\right]dx$$

NOTE: when the integration terminals are the same the definite integrals may be combined.

$$\int\limits_a^b f(x)\,dx - \int\limits_a^b g(x)\,dx = \int\limits_a^b \left[f(x)-g(x)\right]dx$$

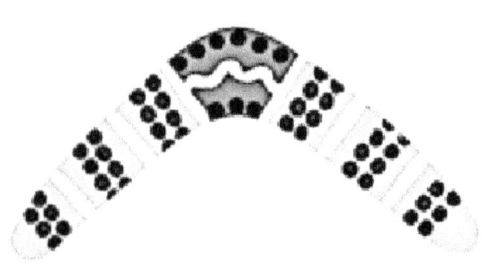

Approximate points are: $(0,0), (15,15), (30,0)$ where the x and y units are in centimetres (cm).

To find the area of a boomerang, find the function describing the outline of the top of the boomerang, then find a function describing the outline of the bottom of the boomerang.

The units for $y = f(x)$ and $y = g(x)$ are centimetres. The horizontal axis x is a scale of centimetres.

The fitted curve for the top line is $f(x)$.

The fitted curve for the bottom line is $g(x)$.

We can find the area between the curves using integration.

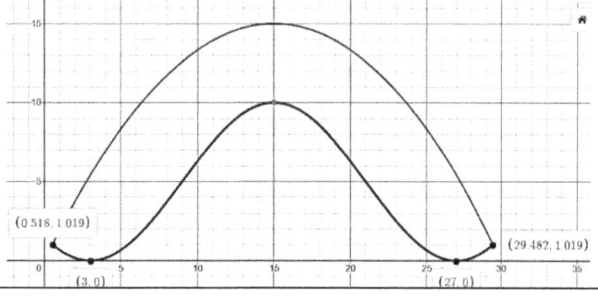

$f(x) = -\dfrac{1}{15}x^2 + 2x$ curve height in cm

$g(x) = -5\cos\left(\dfrac{\pi}{12}(x-3)\right) + 5$, curve height in cm

The area between the functions is in cm^2

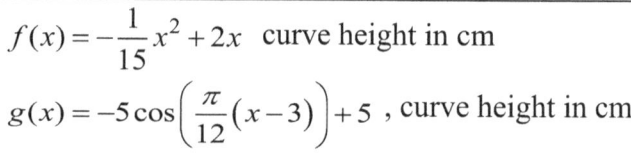

Find the intersection points for $f(x) = g(x)$ to determine the area under $g(x)$ to subtract from the parabola $f(x)$ to make the boomerang shaped outline.

These functions intersect approximately at the coordinates of $(0.518, 1.019)$ and $(29.482, 1.019)$

The definite integral for finding the area of the boomerang.

$$Area = \int_{0.518}^{29.482} f(x)\,dx - \int_{0.518}^{29.482} g(x)\,dx$$

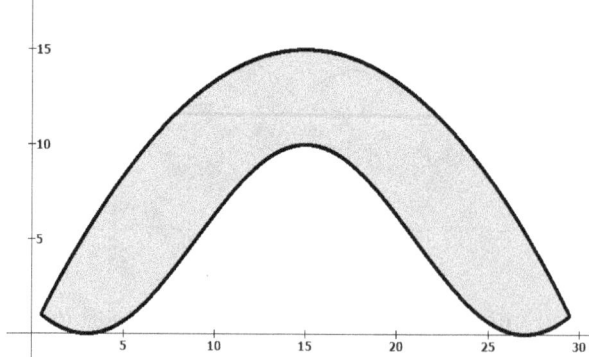

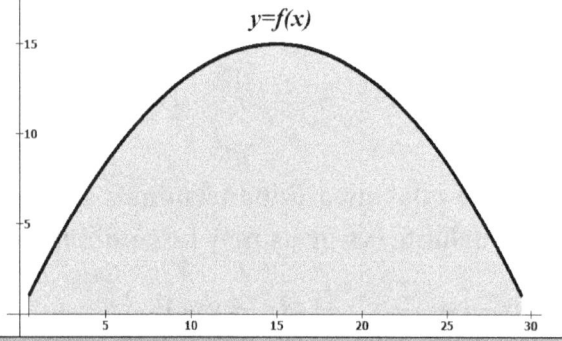

$y=f(x)$

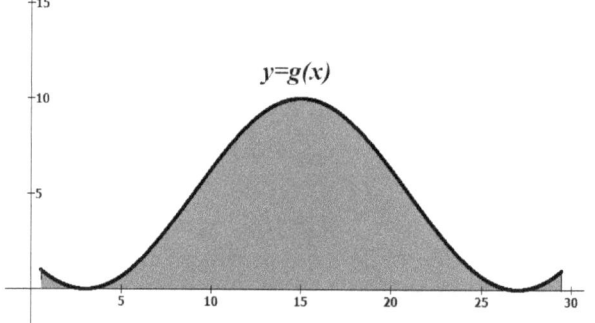

$y=g(x)$

THE FUNDAMENTAL THEOREM OF CALCULUS

$$F'(x) = f(x) \Leftrightarrow \int f(x)dx = F(x) + C \Leftrightarrow \int_a^b f(x)dx = F(b) - F(a)$$

The area under a function represents a quantity that is related to the function and its independent variable on the horizontal axis. The quantity represented by the area under a function is a product of the function's $f(x)$ units and the x units on the horizontal axis, a common example is when $f(x)$ is a quantity per time unit or quantity/time, and x is a time scale, then the area under the function $f(x)$ is a quantity, which is. $\left(\dfrac{\text{quantity}}{\text{time}} \times \text{time} \right) = \text{quantity}$, in a summation of infinitesimal areas

of height $f(x)$ by width Δx. $\text{Area} = \lim\limits_{n \to \infty} \sum\limits_{k=0}^{n-1} f(x_k)\Delta x = \int_a^b f(x)dx$

- If for example $y = f(x)$ represents velocity in metres/second, and x represents time in seconds, the area under the function $y = f(x)$ is the quantity of $\left(\dfrac{\text{metres}}{\text{second}} \times \text{second} \right) = \text{metres}$.

- If for example $y = g(x)$ represents a leaking rate of liquid in litres/days, and x represents time in days, the area under the function $y = g(x)$ is the quantity of $\left(\dfrac{\text{litres}}{\text{days}} \times \text{days} \right) = \text{litres}$.

Example 6.6: A demonstration of the Fundamental Theorem of calculus.

The mass (M) in grams of a radio-active isotope is decaying each year (t) according to the rule
$m(t) = \dfrac{-4}{t^2}$ where $M'(t) = m(t)$ and $t \geq 1$

$\dfrac{dM}{dt} = M'(t) = \dfrac{-4}{t^2}$, the area under the function

$M'(t)$ is the quantity of $\left(\dfrac{\text{grams}}{\text{years}} \times \text{years} \right) = \text{grams}$.

If the mass M=10 grams, when t=1, we can determine that mass at any time $M(t)$ by integration and evaluation of the constant of integration. $\int M'(t)\,dt = M(t) + C$

$$M(t) = \int \dfrac{-4}{t^2}dt$$
$$= -4\int t^{-2}dt$$
$$= 4t^{-1} + C$$

$M(1) = 4(1)^{-1} + C = 10$
$\therefore C = 6$
$M(t) = 4t^{-1} + 6$

The mass after 100 years is given by substitution.
$$M(100) = 4(100)^{-1} + 6$$
$$= 6.04 \, grams$$

The amount of quantity in grams of the decay of M between 2 and 7 years is the shaded area between the graph and the t-axis.

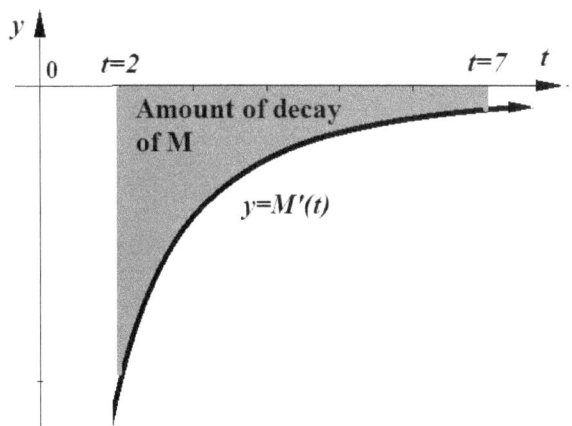

Since the matter is decaying the change is negative, and the area is below the *t-axis*.

The decay in mass between t=2 years and t=7 years is given by $M(7) - M(2) = \dfrac{46}{7} - 8 = -\dfrac{10}{7}$

This is equal to $\int_2^7 M'(t)dt = M(7) - M(2)$

The Average value of a function.

Any average can be used for comparison lower than average will give a smaller quantity, higher than average will give a larger quantity. For a function describing a **variable rate** of change such as

$f(x) = \dfrac{dy}{dx}$, $f(t) = \dfrac{dx}{dt}$, $f(x) = F'(x)$ or $f(t) = F'(t)$, the average value of the function determines a

constant rate value in the interval that produces the same quantity of area, distance, volume, height or other quantity represented by the area under the curve of a function.

For a continuous function $f(x)$ in the domain $x \in [a,b]$, the average value f_{ave} of $f(x)$ is given by the

evaluation of the definite integral $f_{ave} = \dfrac{1}{(b-a)} \displaystyle\int_a^b f(x)\,dx$. The area under $f(x)$ from $x = a$ to $x = b$ is

equal to the area made by a rectangle of dimensions $(b-a) \times f_{ave}$.

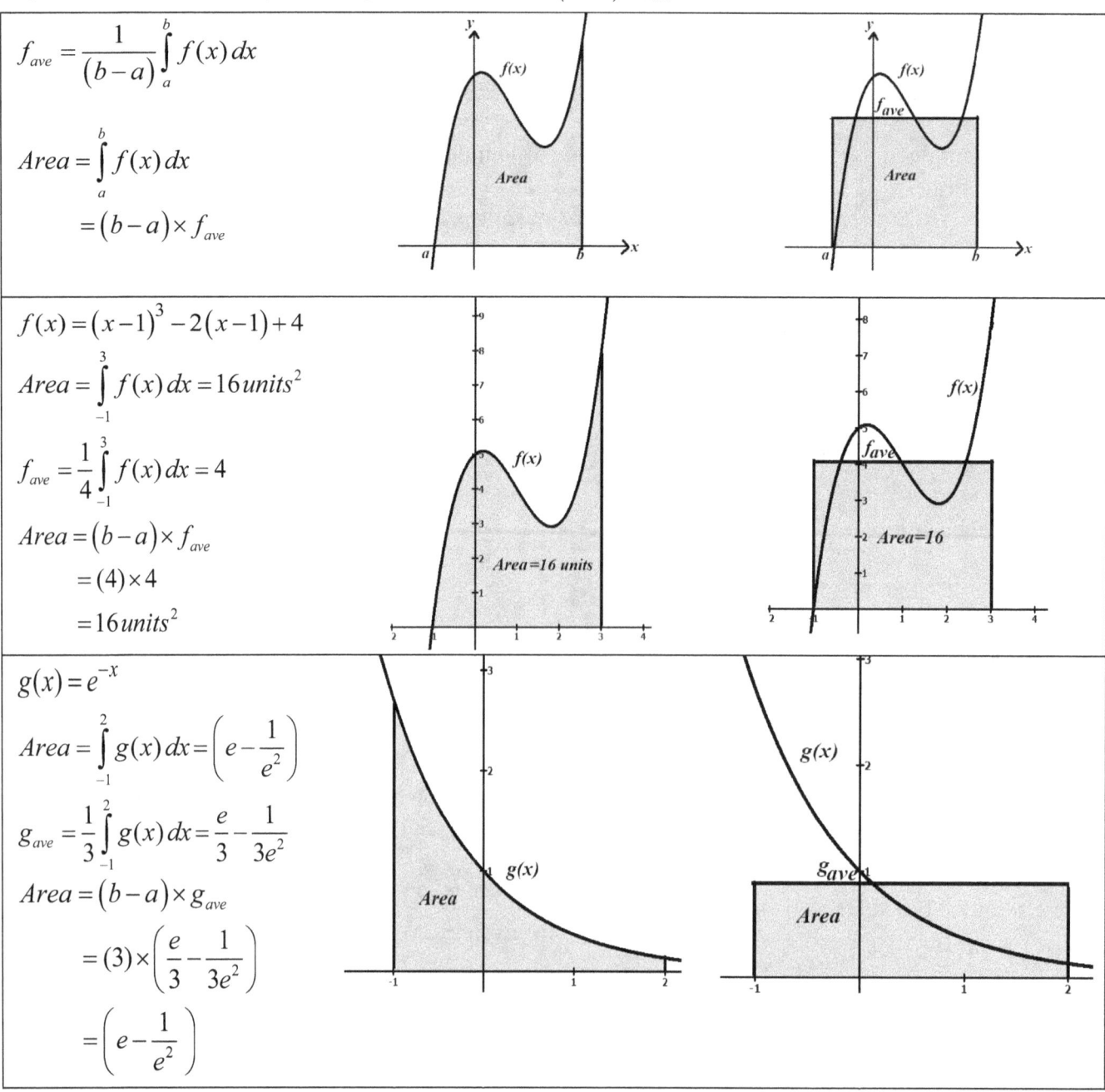

$$f_{ave} = \frac{1}{(b-a)} \int_a^b f(x)\,dx$$

$$Area = \int_a^b f(x)\,dx$$

$$= (b-a) \times f_{ave}$$

$$f(x) = (x-1)^3 - 2(x-1) + 4$$

$$Area = \int_{-1}^3 f(x)\,dx = 16\,units^2$$

$$f_{ave} = \frac{1}{4} \int_{-1}^3 f(x)\,dx = 4$$

$$Area = (b-a) \times f_{ave}$$

$$= (4) \times 4$$

$$= 16\,units^2$$

$$g(x) = e^{-x}$$

$$Area = \int_{-1}^2 g(x)\,dx = \left(e - \frac{1}{e^2} \right)$$

$$g_{ave} = \frac{1}{3} \int_{-1}^2 g(x)\,dx = \frac{e}{3} - \frac{1}{3e^2}$$

$$Area = (b-a) \times g_{ave}$$

$$= (3) \times \left(\frac{e}{3} - \frac{1}{3e^2} \right)$$

$$= \left(e - \frac{1}{e^2} \right)$$

162

INTEGRATING INTEGRALS SUMMARY:

Differentiation creates the derivative $f'(x) = \dfrac{d(f(x))}{dx}$

Integration creates the antiderivative $F(x) = \int f(x)\,dx$

where

Indefinite Integrals

$\int f(x)\,dx = F(x) + C$

Definite Integrals

$\int_a^b f(x)\,dx = \left[F(x)\right]_a^b = F(b) - F(a)$

Area under $f(x)$ in a given domain of x is given by the addition of infinite infinitesimal areas:

$\text{Area} = \lim_{n\to\infty} \sum_{k=0}^{n-1} f\left(x_k\right)\Delta x = \int_a^b f(x)\,dx$

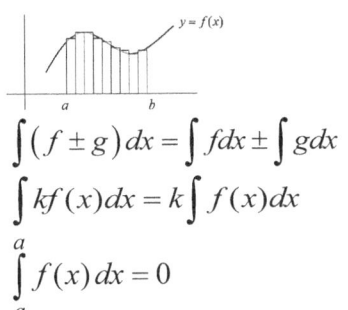

$\int (f \pm g)\,dx = \int f\,dx \pm \int g\,dx$

$\int kf(x)\,dx = k\int f(x)\,dx$

$\int_a^a f(x)\,dx = 0$

The fundamental theorem of Calculus

$F'(x) = f(x) \Leftrightarrow \int f(x)\,dx = F(x) + C \Leftrightarrow \int_a^b f(x)\,dx = F(b) - F(a)$

The quantity represented by the area under a function is a product of the function's $f(x)$ units and the x units.

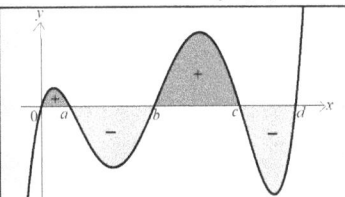 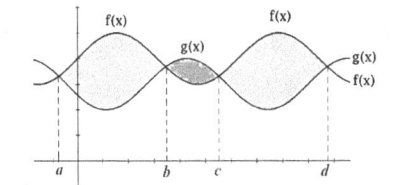

Areas bound by $f(x)$ above the *x-axis* are positive.
Areas below the *x-axis* are negative.

$\int_a^b f(x)\,dx = -\int_b^a f(x)\,dx$

$Area_{\{a<x<b\}} = \int_a^b f(x)\,dx - \int_a^b g(x)\,dx$

$Area_{\{b<x<c\}} = \int_b^c g(x)\,dx - \int_b^c f(x)\,dx$

$Area_{\{c<x<d\}} = \int_c^d f(x)\,dx - \int_c^d g(x)\,dx$

If terminals are the same definite integrals can be combined

$\int_a^b f(x)\,dx - \int_a^b g(x)\,dx = \int_a^b \left[f(x) - g(x)\right]\,dx$

The Average value of a function in an interval

$f_{ave} = \dfrac{1}{(b-a)} \int_a^b f(x)\,dx$.

Rules for derivatives and antiderivatives are provided in end of year exams.	
$\dfrac{d}{dx}(x^n) = nx^{n-1}$	$\int x^n\,dx = \dfrac{x^{n+1}}{n+1} + C \;$ if $n \neq -1$ $\int x^{-1}\,dx = \int \dfrac{1}{x}\,dx = \ln|x| + C$ if n=1
$\dfrac{d}{dx}(e^{ax+b}) = ae^{ax+b}$	$\int e^{ax+b}\,dx = e^b \int e^{ax}\,dx = \dfrac{1}{a}e^{ax+b} + C$
$\dfrac{d}{dx}(log_e(ax+b)) = \dfrac{a}{ax+b}$	$\int \dfrac{a}{ax+b}\,dx = \int \dfrac{1}{x+\frac{b}{a}}\,dx = log_e\left|x + \dfrac{b}{a}\right| + C$
$\dfrac{d}{dx}(sin(ax+b)) = a\,cos(ax+b)$	$\int sin(ax+b)\,dx = -\dfrac{1}{a}cos(ax+b) + C$
$\dfrac{d}{dx}(cos(ax+b)) = -a\,sin(ax+b)$	$\int cos(ax+b)\,dx = \dfrac{1}{a}sin(ax+b) + C$
$\dfrac{d}{dx}(tan(ax+b)) = \dfrac{a}{[cos(ax+b)]^2}$	

The **trapezoidal rule** for estimating the area under a function is expressed in general form as:

$$\int_a^b f(x)\,dx \approx \dfrac{\Delta x}{2}\left(f(x_0) + 2f(x_1) + 2f(x_2) + 2f(x_3) + \ldots 2f(x_{n-1}) + f(x_n)\right)$$

INTEGRATING INTEGRALS: Check your Understanding.

6.1 Area under functions and between.

a)

For the curve shown, determine approximately the sum of rectangles bound by the function and the x-axis for the interval over the interval from $x = 0$ to $x = 5$ using $\Delta x = 0.5$ using the height of the left value of the function in the sub-interval.

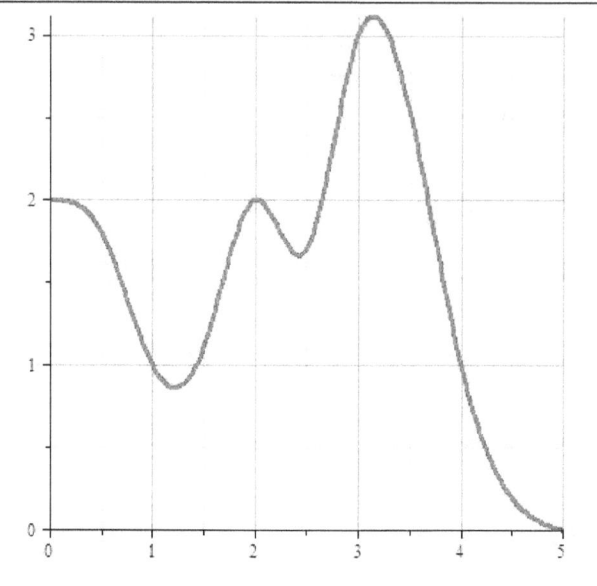

b) Determine the sum of the areas of rectangles bound by the function $f(x)$ and the x-axis for

$f(x) = x^2 - 1$ over the interval from $x = -4$ to $x = 1$ using $n = 5$ sub-intervals and evaluation of $f(x)$ at the **right end** of the sub-intervals.

Complete the table with the area of each rectangle, and state the sum of the areas.

i	x_i	Δx	$f(x_i)$	$\Delta x \times f(x_i)$
1				
2				
3				
4				
5				

c) Determine the sum of the areas of rectangles bound by the function $f(x)$ and the x-axis for

$f(x) = x^2 - 1$ over the interval from $x = -4$ to $x = 1$ using $n = 5$ sub-intervals and evaluation of $f(x)$ at the **left end** of the sub-intervals.

Complete the table with the area of each rectangle, and state the sum of the areas.

i	x_i	Δx	$f(x_i)$	$\Delta x \times f(x_i)$
1				
2				
3				
4				
5				

d) Find the net area bound by $f(x) = x^2 - 1$ and the x-axis over the interval from $x = -4$ to $x = 1$ using the definite integral. Sketch the function and the area.

e) Write down (but do not evaluate) the definite integral(s) for the bounded shaded areas shown.

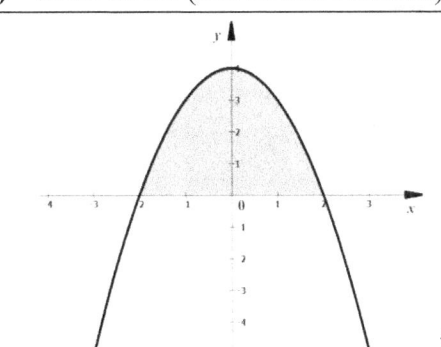

$$f(x) = 4 - x^2$$

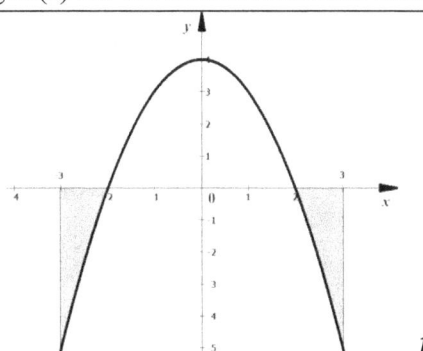

$$f(x) = 4 - x^2$$

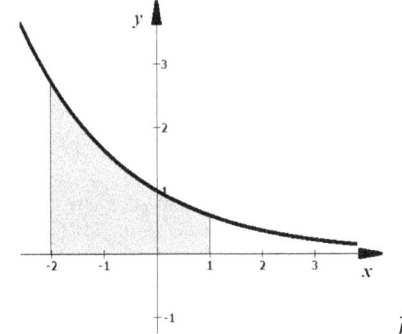

$$f(x) = e^{-0.5x}$$

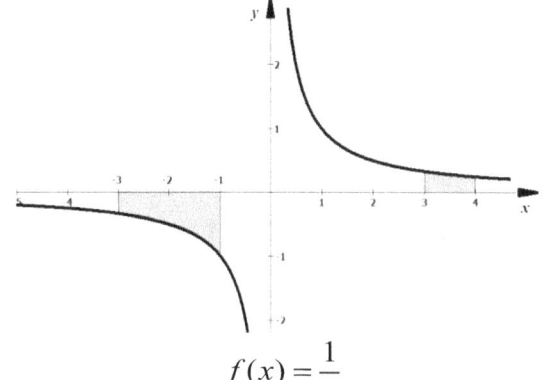

$$f(x) = \frac{1}{x}$$

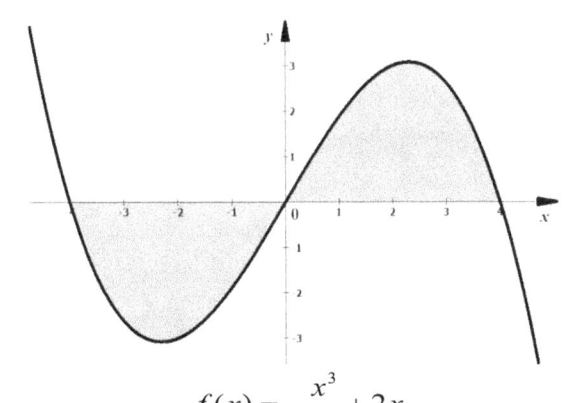

$$f(x) = -\frac{x^3}{8} + 2x$$

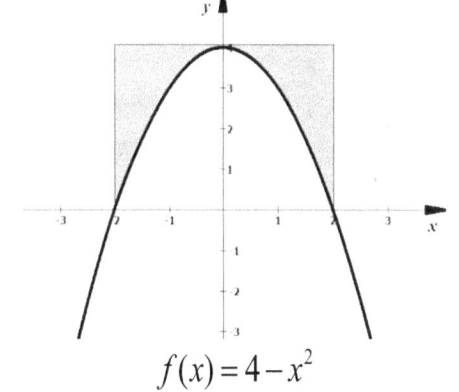

$$f(x) = 4 - x^2$$

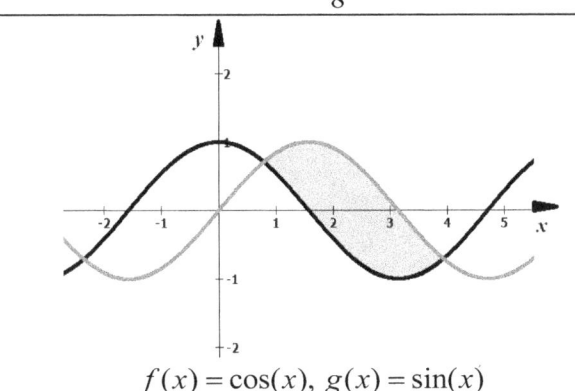

$$f(x) = \cos(x),\ g(x) = \sin(x)$$

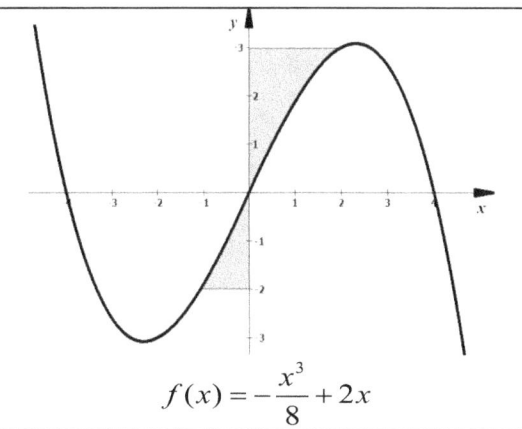

$$f(x) = -\frac{x^3}{8} + 2x$$

6.2 Quantities by integration.

a) Two swimmers, Bo and Jo are swimming to win at the Parkdale Pool.

For part of the race their speed of swimming in metres per minute is given by the following functions, where t is the time in minutes.

Bo: $b(t) = 0.01(t-5)^3 + 4, \quad t \in [1,7]$

Jo: $j(t) = 0.05t + 1.5, \quad t \in [1,7]$

   i.   Sketch the functions and shade the difference in distance for the two swimmers in the interval $t \in [1,7]$.

   ii.   Set up and solve the definite integrals to find the difference in distance for the two swimmers in the interval $t \in [1,7]$.

b)

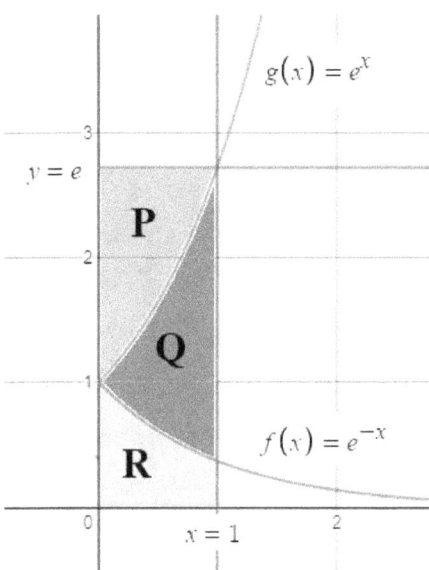

Find the areas P,Q and R , bound by the *x-axis* and the *y-axis* and the areas formed by the equations:

$f(x) = e^{-x}, x \ge 0$

$g(x) = e^{x}, x \ge 0$

$y = e, y = 0$

$x = 1, x = 0$

Show that $P + Q + R = e$

c) An empty cylinder is being filled with water at a time varying rate of $0.1t^2 \ cm^3/\sec$

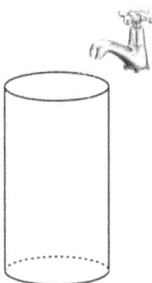

   i.   What is the function that represents this rate of filling?

   ii.   Find how much water is in the cylinder after 5 seconds.

  iii.   Find the average value of the function in this interval

  iv.   Using the average value to find volume of water after 5 seconds

6.2 Quantities by integration.

d)

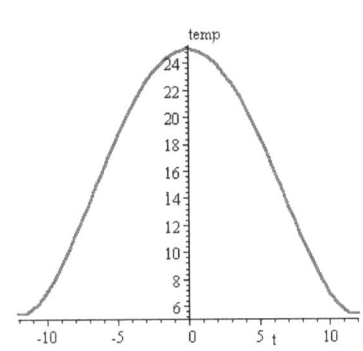

The temperature T (in °C) recorded during a day followed the curve shown.

$T = 0.001t^4 - 0.280t^2 + 25$ where t is the number of hours from noon $(-12 \leq t \leq 12)$

What was the average temperature during the day?

e)

Using the trapezoidal rule formula find the area under the function
$$f(x) = e^x \sin(x)$$
between x=0 and x=1 at intervals of $\Delta x = 0.2$ given it passes through the following points:

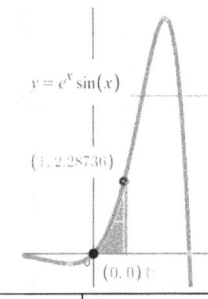

x	0	0.2	0.4	0.6	0.8	1.0
$f(x)$	0	0.2427	0.5809	1.0288	1.5965	2.2874

f)

The following algorithm accepts as inputs f is the f(x), lowest x value x=a, highest x value x=b, and number of trapeziums n in the interval. Complete the algorithm defined in pseudocode below for computing the area using the trapezoidal rule formula:

```
input f,a,b,n
sum ← 0
a ← lowest x-value
b ← highest x-value
n ←
dx ← (b-a)/
left ← a
right ← a + dx
for i     1 to n
    strip ← 0.5(f(left) + f(right)) ×
    sum ← sum +
    left ← left + dx
    right ← right + dx

print
```

167

6.3 6.3. Application of Integration to real world problems.

a)

Factory A

Two factories are manufacturing product xyz, the following functions represent the rate of money earned in $1000's per 100 units produced, represented by x.

Factory A: $f(x) = 0.3x + 1, x > 0$

Factory B: $g(x) = e^{0.25x} - 0.5x, x > 0$

Factory B

i.   Sketch the two functions and find any intercepts and end points for 1000 units by each factory.

ii.   Explain the meaning of the intercepts and endpoints relative the quantities represented.

iii. Find the difference in money earned for each factory for the production of 1000 units, and compare the results.

b)

The rate of outflow from a wine vat over a 24-hour period is given by the equation $\dfrac{dW}{dt} = 30 + \sin\left(\dfrac{\pi t}{6}\right)$, where W is the volume of wine in the vat in litres and t is the number of hours after 3am on Monday.

i.   Find $\dfrac{dW}{dt}$ to 3 decimal places when t = 3.5 and interpret this result

ii.   Sketch the graph of $\dfrac{dW}{dt} = W'(t)$ versus t for $t \in [0, 24]$. Find the times when the wine flow has been at the minimum rate to 4 decimal places.

iii. Find the times in format HH:MM:SS (hours minutes seconds) for which $W'(t) \approx 30.9$ *Litres / hour* and rising.

iv.   Interpret the integral $\displaystyle\int_0^6 W'(t)\, dt$.

v.   Find the volume to the nearest litre of wine that has been removed between 3 pm Monday and 2 am Tuesday

INTEGRATING INTEGRALS SOLUTIONS.
6.1 Area under functions and between. (SOLUTIONS)

a) $\Delta x = 0.5$ using the height of the left value of the function in the sub-interval.

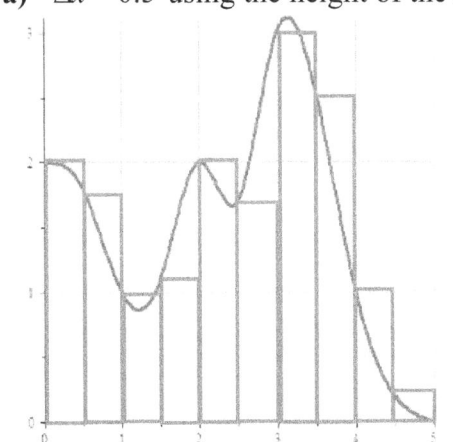

1	2×0.5
2	1.75×0.5
3	1×0.5
4	1.1×0.5
5	2×0.5
6	1.7×0.5
7	3×0.5
8	2.5×0.5
9	1×0.5
10	0.25×0.5

Area=8.15 units^2

b) right end of the sub-intervals.

1	-3	1	8	8
2	-2	1	3	3
3	-1	1	0	0
4	0	1	-1	-1
5	1	1	0	0

Total net area = 10

c) left end of the sub-intervals.

1	-4	1	15	15
2	-3	1	8	8
3	-2	1	3	3
4	-1	1	0	0
5	0	1	-1	-1

Total net area = 25

d)

$$Area = \int_{-4}^{1} f(x)\,dx$$

$$= \int_{-4}^{1} (x^2 - 1)\,dx$$

$$= \left| \frac{x^3}{3} - x \right|_{-4}^{1}$$

$$= \left(\frac{(1)^3}{3} - (1) \right) - \left(\frac{(-4)^3}{3} - (-4) \right)$$

$$= \frac{50}{3}$$

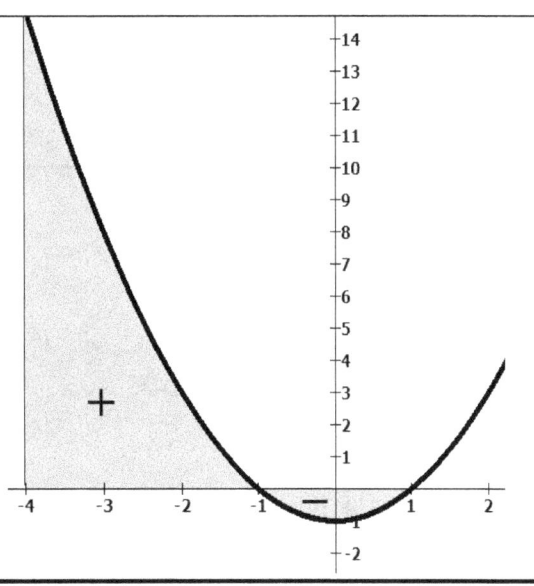

e) Note the swapping of definite integral terminals to change the sign of the area.

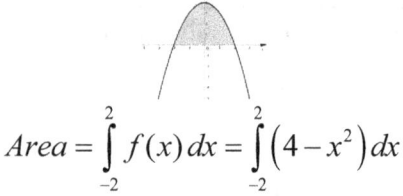

$$Area = \int_{-2}^{2} f(x)\,dx = \int_{-2}^{2} \left(4 - x^2\right)\,dx$$

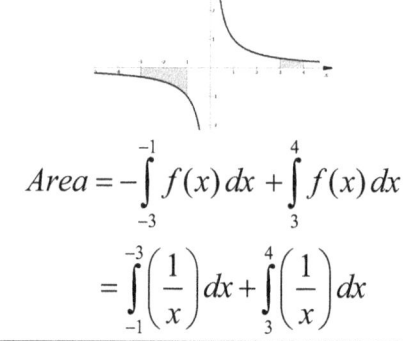

$$Area = -\int_{-3}^{-2} f(x)\,dx + -\int_{2}^{3} f(x)\,dx$$

$$= \int_{-2}^{-3} \left(4 - x^2\right)\,dx + \int_{3}^{2} \left(4 - x^2\right)\,dx$$

swapping of definite integral terminals

$$Area = \int_{-2}^{1} f(x)\,dx = \int_{-2}^{1} \left(e^{-0.5x}\right)\,dx$$

$$Area = -\int_{-3}^{-1} f(x)\,dx + \int_{3}^{4} f(x)\,dx$$

$$= \int_{-1}^{-3} \left(\frac{1}{x}\right)\,dx + \int_{3}^{4} \left(\frac{1}{x}\right)\,dx$$

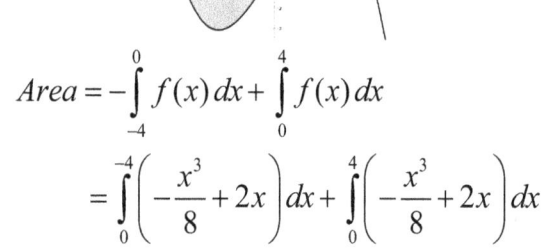

$$Area = -\int_{-4}^{0} f(x)\,dx + \int_{0}^{4} f(x)\,dx$$

$$= \int_{0}^{-4} \left(-\frac{x^3}{8} + 2x\right)\,dx + \int_{0}^{4} \left(-\frac{x^3}{8} + 2x\right)\,dx$$

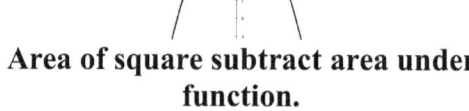

Area of square subtract area under function.

$$Area = \left(4 \times 4\right) - \int_{-2}^{2} f(x)\,dx$$

$$= 16 - \int_{-2}^{2} \left(4 - x^2\right)\,dx$$

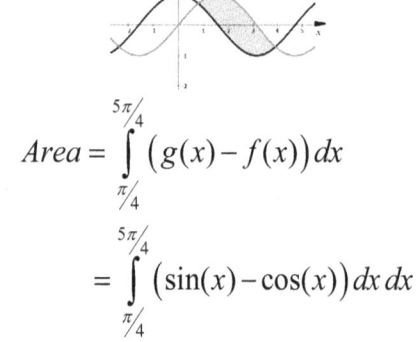

$$Area = \int_{\pi/4}^{5\pi/4} \left(g(x) - f(x)\right)\,dx$$

$$= \int_{\pi/4}^{5\pi/4} \left(\sin(x) - \cos(x)\right)\,dx\,dx$$

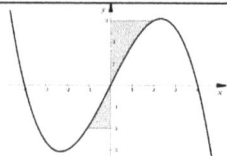

Area of rectangles subtract areas under functions.

$$Area = \left[\left(1 \times 2\right) - -\int_{-1}^{0} f(x)\,dx\right] + \left[\left(3 \times 2\right) - \int_{0}^{2} f(x)\,dx\right]$$

$$= \left[2 - \int_{-1}^{0} \left(-\frac{x^3}{8} + 2x\right)\,dx\right] + \left[6 - \int_{0}^{2} \left(-\frac{x^3}{8} + 2x\right)\,dx\right]$$

6.2 Quantities by integration. (SOLUTIONS)

a) Two swimmers Bo and Jo are swimming to win at the Parkdale Pool.

$$\text{Bo: } b(t) = 0.01(t-5)^3 + 4, \quad t \in [1,7]$$

$$\text{Jo: } j(t) = 0.05t + 1.5, \quad t \in [1,7]$$

i. interval $t \in [1,7]$.

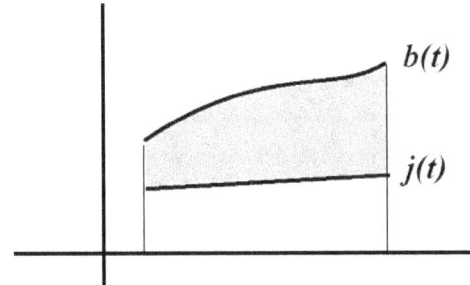 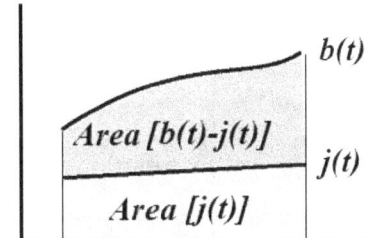

ii. in the interval $t \in [1,7]$.

$$\int_1^7 b(t)\,dt - \int_1^7 j(t)\,dt = \int_1^7 \big(b(t) - j(t)\big)\,dt$$

$$= \int_1^7 \Big(\big(0.01(t-5)^3 + 4\big) - \big(0.05t + 1.5\big)\Big)\,dt$$

$$= 13.2\, metres$$

Bo is 13.2 metres ahead of Jo.

b)

$$R = \int_0^1 f(x)\,dx \qquad\qquad Q = \int_0^1 g(x)\,dx - \int_0^1 f(x)\,dx \qquad\qquad P = \int_0^1 e\,dx - \int_0^1 g(x)\,dx$$

$$= \int_0^1 e^{-x}\,dx \qquad\qquad = \int_0^1 e^x\,dx - R \qquad\qquad = \int_0^1 e\,dx - \int_0^1 e^x\,dx$$

$$= \big[-e^{-x}\big]_0^1 \qquad\qquad = \big[e^x\big]_0^1 - 1 + \frac{1}{e} \qquad\qquad = \big[xe\big]_0^1 - \big[e^x\big]_0^1$$

$$= -e^{-(1)} + e^{-(0)} \qquad\qquad = e - 1 - 1 + \frac{1}{e} \qquad\qquad = e - 0 - e + 1$$

$$= 1 - \frac{1}{e} \qquad\qquad\qquad = e - 2 + \frac{1}{e} \qquad\qquad\qquad = 1$$

$$P + Q + R = (1) + \left(e - 2 + \frac{1}{e}\right) + \left(1 - \frac{1}{e}\right) = e \quad \text{shown}$$

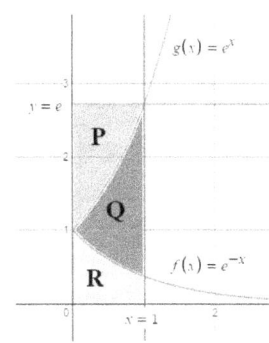

c)

i. The change in volume V over time t is given by $\dfrac{dV}{dt} = 0.1t^2$

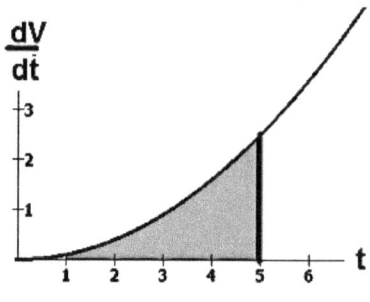

ii. Volume

$$V = \int_0^5 0.1t^2\, dt = \left[\frac{t^3}{30}\right]_0^5$$

$$= \frac{(5)^3}{30} - \frac{(0)^3}{30}$$

$$= \frac{125}{30}\, cm^3 = \frac{25}{6}\, cm^3$$

iii. average value of the function

$$f_{ave} = \frac{1}{(5-0)} \int_0^5 0.1t^2\, dt$$

$$= \frac{125}{150}\, cm^3 = \frac{5}{6}\, cm^3$$

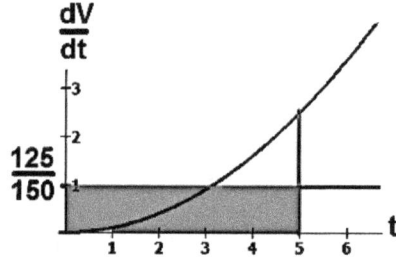

iv. Notice that $V = \dfrac{25}{6} = f_{ave} \times 5$

d)

We consider the graph of the situation and estimate that the average should be around 14 to 16 degrees.

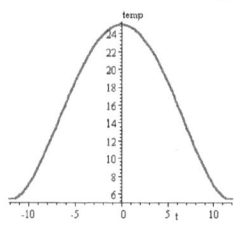

The average value of the function T will give the same area value over the time $(-12 \leq t \leq 12)$

$$f_{ave} = \frac{1}{(12 - -12)} \int_{-12}^{12} \left(0.001t^4 - 0.280t^2 + 25\right) dt$$

$$= \frac{1}{24}\left[0.0002t^5 - 0.093t^3 + 25t\right]_{-12}^{12}$$

$$\approx 15.71^\circ C$$

e)

Using the trapezoidal rule area under the function: $f(x) = e^x \sin(x)$ between x=0 and x=1 at intervals of $\Delta x = 0.2$ given it passes through the following points:

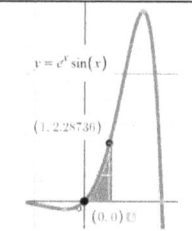

x	0	0.2	0.4	0.6	0.8	1.0
$f(x)$	0	0.2427	0.5809	1.0288	1.5965	2.2874

$$\int_0^1 f(x)dx \approx \frac{0.2}{2}(0 + 2(0.2427) + 2(0.5809) + 2(1.0288) + 2(1.5965) + 2.2874)$$

6.2 Quantities by Integration. (SOLUTIONS)

f)

Complete the algorithm defined in pseudocode below for computing the trapezoidal rule formula:

input f,a,b,n

sum ← 0

a ← lowest x-value

b ← highest x-value

n ← number of trapeziums

dx ← (b-a)/n

left ← a

right ← a + dx

for i **from** 1 **to** n

    strip ← 0.5(f(left) + f(right)) × dx

    sum ← sum + strip

    left ← left + dx

    right ← right + dx

end for

print sum

6.3 Applications of Integration (SOLUTIONS)

a)

Factory A: $f(x) = 0.3x + 1, x > 0$

Factory B: $g(x) = e^{0.25x} - 0.5x, x > 0$

Factory A

Factory B

i.	Intercepts (8.01,3.40), Endpoint factory A (10,4) Endpoint factory B (10,7.2)
ii.	The instantaneous rate of earnings of per production by each factory are equal at approximately $3400 per 800 units which is the intercept.
iii.	At 10 weeks factory A the instantaneous rate of earnings/production is $4000 per 1000 units. At 10 weeks factory B the instantaneous rate of earnings/production is $7200 per 1000 units.

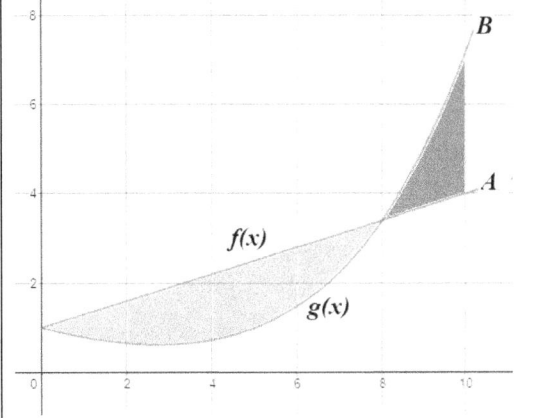

b)

i $W'(t) \approx 30.966 \; Litres \, / \, hour$ represents the outflow of wine from the vat at exactly $t=3.5$ hours which is 6:30 am Monday

ii Using technology, find the endpoints

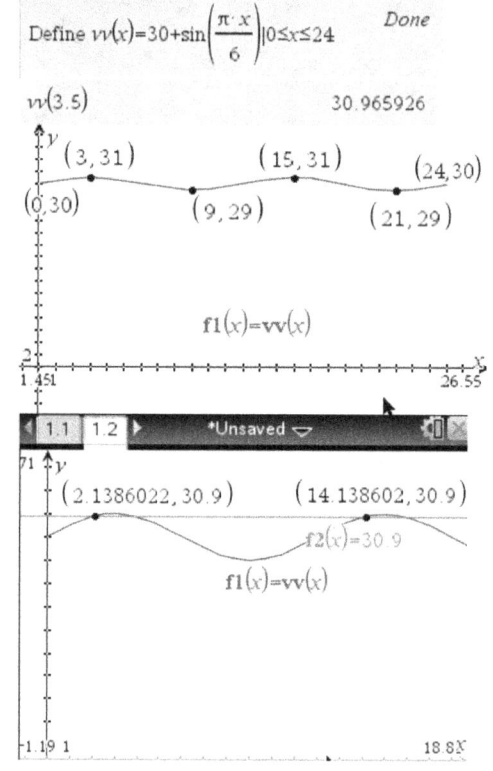

iii First time happens at $t_1 \approx 2.1386$ hours after 3am Monday which is 5:07:43 am Monday

Second time happens at $t_2 \approx 14.1386$ hours after 3am Monday which is 17:08:19 pm Monday

$(2.1286022°) \blacktriangleright DMS \qquad 2°7'42.96792''$

$(14.138602°) \blacktriangleright DMS \qquad 14°8'18.9672''$

iv this integral represents the quantity of wine in litres that has flowed out of the vat from 3am to 9am on Monday

v From 3pm Monday $t=12$ until 2am Tuesday $t=23$

$$Volume = \int_{12}^{23} W'(t) \, dt$$
$$= 330 \; litres$$

$\int_{12}^{23} vv(x) \, dx \qquad \dfrac{3 \cdot (110 \cdot \pi - \sqrt{3} + 2)}{\pi}$

$\int_{12}^{23} vv(x) \, dx \qquad 330.25587$

Discrete Probability

What are the chances of an event happening? The aim of probability is to examine this question using basic counting and mathematical operations involving finding combinations and consideration of sets of information and events. To be able to work out the probability of certain events or outcomes we need to group together relevant information belonging to the all the possible outcomes in the scope of the problem we wish to study. **Discrete probability is concerned with outcomes that have countable results represented by whole numbers.**

In theoretical probability all the discrete outcomes in the event space, which is all the possible outcomes are equally likely. So we need to be able to count the number of events we're looking for and count all the possible events that can happen and work out the ratio.

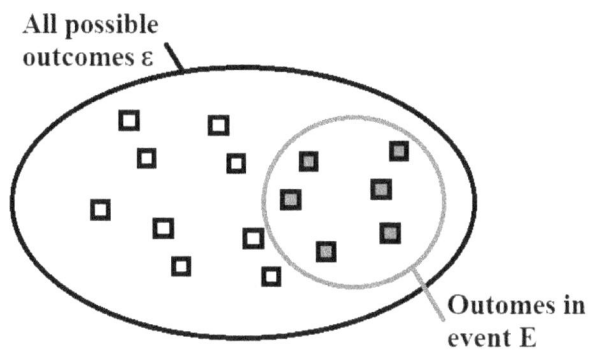

The symbol $n(E)$ represents the number of E outcomes, and $n(\varepsilon)$ is the total number of outcomes that are possible for the entire scope of the trial or experiment.

$$\Pr(E) = \frac{n(E)}{n(\varepsilon)}$$

The total event space is ε, and the size of event space $n(\varepsilon)$.

The probability of an outcome is a ratio of the number of outcomes we're looking for divided by the total number of possible outcomes.

The probability of outcome E,

$$\Pr(E) = \frac{\text{Number of E outcomes}}{\text{Total number of outcomes}}$$

The symbol ε is often used to represent all the outcomes that are possible for the problem we are studying. It is sometimes referred to as the universal set.

This means that probability of any event E is a fraction between 0 and 1, $0 \le \Pr(E) \le 1$.

If event E is impossible then $\Pr(E) = 0$.

If event E is certain to happen then $\Pr(E) = 1$.

The probability of *any* outcome happening in this event space is $\Pr(\varepsilon) = \dfrac{n(\varepsilon)}{n(\varepsilon)} = 1$

$$P(E) + P(E') = 1$$

Where $(E' = not\ E)$, E' is the symbol representing the complement of outcome E.

Example 7.1: Counting favourable events and possible outcomes for Discrete Random Variables.

Consider the word "Crocodile", it has 9 letters, some of which are repeats. We want to select a random letter from this word where $\varepsilon=\{C,R,O,C,O,D,I,L,E\}$. The size of this event space is $n(\varepsilon)=9$. The number of "L" s in this word is $n(L)=1$. The number of vowels in this word is $n(vowels)=4$

$$Pr(L)=\frac{n(L)}{n(\varepsilon)}=\frac{1}{9}, Pr(vowels)=\frac{n(vowels)}{n(\varepsilon)}=\frac{4}{9}$$

Using Tree diagrams to count for probability

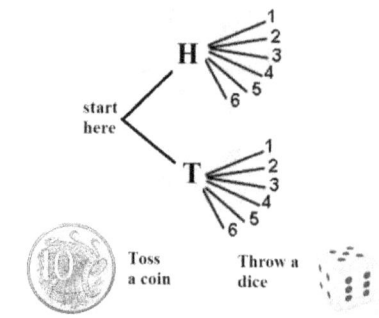

$\varepsilon = \{H1,H2,H3,H4,H5,H6,$
$\quad T1,T2,T3,T4,T5,T6\}$ $n(\varepsilon)=12,$

$$Pr(H3)=\frac{n(H3)}{n(\varepsilon)}=\frac{1}{2}$$

Using Venn diagrams to count for probability.

If we throw a six sided dice $\varepsilon=\{1,2,3,4,5,6\}$, $n(\varepsilon)=6$. We can represent event E as the even numbers and the complement E' will be "not even" numbers commonly called the odd numbers. $n(E) = n(E')=3$. Theoretically the probability of not getting an even number after a roll of this dice

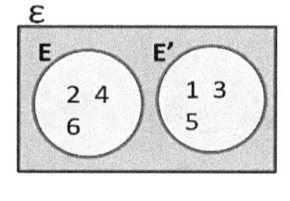

is $Pr(E')=\frac{n(E')}{n(\varepsilon)}=\frac{3}{6}$

Using tables to count for probability.

The sum of two rolls of the dice. The total possible outcomes can be represented in a table where $n(\varepsilon)=36$.

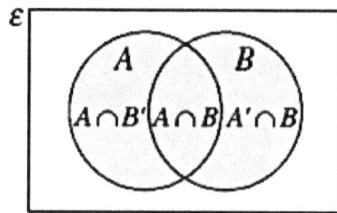

$Pr(2)=\frac{1}{36}$

$Pr(6)=\frac{5}{36}$

$Pr(10)=\frac{3}{36}$

	1	2	3	4	5	6
1	2	3	4	5	6	7
2	3	4	5	6	7	8
3	4	5	6	7	8	9
4	5	6	7	8	9	10
5	6	7	8	9	10	11
6	7	8	9	10	11	12

Using Venn diagrams to count for probability.

In a group of 47 students, 5 students study Latin only, 30 students study French only, 2 students study both Latin and French, and 10 students study neither French or Latin. If we represent each student as a possible outcome then $n(\varepsilon)=47$.

$$Pr(French)=\frac{n(French)}{n(\varepsilon)}=\frac{32}{47}$$

$$Pr(Latin\,and\,French)=\frac{n(Latin\,and\,French)}{n(\varepsilon)}=\frac{2}{47}$$

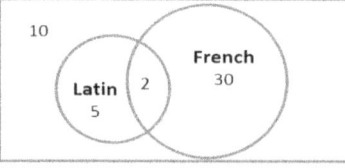

Using Venn diagrams and tables to represent outcomes and events and to count for probability.

	B	B'
A	$A\cap B$	$A\cap B'$
A'	$A'\cap B$	$A'\cap B'$

Quick Probability Rule Revision	Example
Addition rule (non-mutually exclusive events) The probability of two events A or B occurring: $\Pr(A\,or\,B) = \Pr(A) + \Pr(B) - \Pr(A\,and\,B)$ Or expressed using set notation $\Pr(A \cup B) = \Pr(A) + \Pr(B) - \Pr(A \cap B)$ 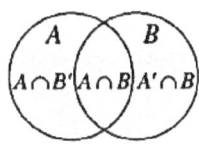 This rule works to stop counting $\Pr(A \cap B)$ twice since it is included in both A and B.	The probability Mr. X catches the train is 0.40, the bus is 0.30, both the train and the bus is 0.20. What is the probability that Mr. X catches the train or the bus or both? <ul><li>Let event A = Mr. X catches train</li><li>Let event B = Mr. X catches bus</li></ul> $\Pr(A\,or\,B) = \Pr(A) + \Pr(B) - \Pr(A\,and\,B)$ $= 0.40 + 0.30 - 0.20$ $= 0.5$
Addition rule (mutually exclusive events) If A and B are **mutually exclusive** events then: $\Pr(A \cap B) = 0$ giving $\Pr(A \cup B) = \Pr(A) + \Pr(B)$ 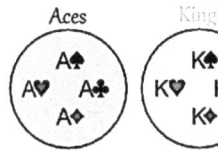 Consider the disjoint sets of Aces and Kings from a regular deck of 52 cards. They have nothing in common.	Pr(Ace or King)=Pr(Ace) + Pr(King) $\Pr(Ace \cup King) = \Pr(Ace) + \Pr(King)$ $= \dfrac{4}{52} + \dfrac{4}{52}$ $= \dfrac{2}{13}$
Conditional Probability is the probability that event A occurs given event B has already occurred: $\Pr(A/B) = \dfrac{\Pr(A \cap B)}{\Pr(B)}$	Given a red card has been selected from the full deck, what is the probability that an Ace has been selected. $\Pr(\text{Red}) = \dfrac{26}{52} = \dfrac{1}{2}$, $\Pr(\text{Ace}) = \dfrac{4}{52} = \dfrac{1}{13}$ $\Pr(Ace/\text{Red}) = \dfrac{\Pr(Ace \cap \text{Red})}{\Pr(\text{Red})} = \dfrac{\frac{2}{26}}{\frac{1}{2}} = \dfrac{4}{26} = \dfrac{2}{13}$
Multiplication Rule (dependent events occurring jointly or consecutively) $\Pr(A \cap B) = \Pr(A) \times \Pr(B)$	The probability that an Ace is randomly selected first and **removed** from a deck of cards, followed by a randomly selected King as second selection. $\Pr(Ace1 \cap King2) = \Pr(Ace1) \times \Pr(King2)$ 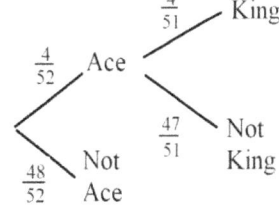 $= \dfrac{4}{52} \times \dfrac{4}{51}$ $= \dfrac{4}{663}$
If the events A and B are **independent** then the probability that event A occurs given B has already occurred or vice-versa is given by: $\Pr(A/B) = \Pr(A)$ or $\Pr(B/A) = \Pr(B)$.	The probability that a King is randomly selected first and **returned to the** deck of cards, followed by a randomly selected Ace as second selection. $\Pr(Ace2/King1) = \Pr(Ace)$

Discrete Probability Distributions and their representations.

Tabular form, the n possible values that the Discrete Random variable X can take in the event space are placed in the first row $x_1, x_2, ..., x_{n-1}, x_n$ the corresponding probabilities $p_1, p_2, ..., p_{n-1}, p_n$ for these values are placed in below each value in the second row. The sum of the probabilities is equal to 1.

x	x_1	x_2	x_3	x_4
$\Pr(X = x_i)$	p_1	p_2	p_3	p_4

When selecting 3 cards randomly from a full deck of 52 cards, showing the probabilities of 0,1,2,3 ♣ clubs being selected in any order.
Let C be the event a club card is selected, where C'=not C.

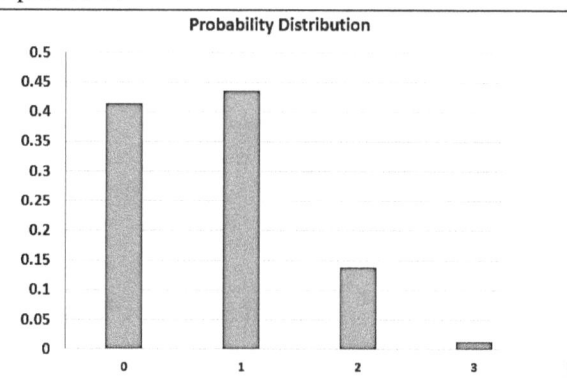

x	0	1	2	3
$\Pr(X = x)$	$\dfrac{39}{52} \times \dfrac{38}{51} \times \dfrac{37}{50}$	$3\left(\dfrac{13}{52} \times \dfrac{39}{51} \times \dfrac{38}{50}\right)$	$3\left(\dfrac{13}{52} \times \dfrac{12}{51} \times \dfrac{39}{50}\right)$	$\dfrac{13}{52} \times \dfrac{12}{51} \times \dfrac{11}{50}$
	$= \dfrac{54834}{132600}$	$= \dfrac{57798}{132600}$	$= \dfrac{18252}{132600}$	$= \dfrac{1716}{132600}$
	≈ 0.4135	≈ 0.4359	≈ 0.1376	≈ 0.0129

Graphical form the n possible values that the Discrete Random variable X can take in the event space are placed in the horizontal axis with tick marks of $x_1, x_2, ..., x_{n-1}, x_n$ the corresponding probabilities $p_1, p_2, ..., p_{n-1}, p_n$ for these values are represented as bars with heights representing the corresponding probabilities. The sum of the probabilities is equal to 1.

When selecting 3 cards randomly from a full deck of 52 cards, showing the probabilities of 0,1,2,3 ♣ clubs being selected in any order.

The bar graph shows the relative probability distribution for this experiment for each possible outcome.

Probability Distribution

178

Binomial Distributions of Discrete Random variables $X \sim Bi(n,p)$

A binomial distribution is one where the discrete random variable X can **only have one of two distinct outcomes**. Examples of this are getting an even number on a dice roll, getting to school on time or a particular football team winning a game.

Properties of binomial distributions which are represented symbolically with $X \sim Bi(n,p)$

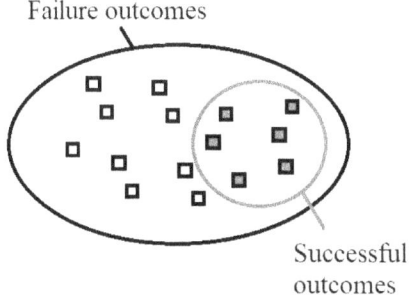

Failure outcomes

Successful outcomes

1. Each experiment consist of n identical trials
2. Each trial has only two outcomes, **success** or **failure**
3. $\Pr(success) = p$ and $\Pr(failure) = 1 - p$
4. The trials are independent
5. The discrete random variable of interest X, is the number of success observed for n trials.
6. Probabilities sum to 1, $\Pr(success) + \Pr(failure) = 1$
7. The trials conducted are known as Bernoulli sequences.

Example 7.2 A binomial distribution, where a die is rolled and a 'success' is noted as obtaining a square number {1,4}, and failure is rolling a non-square number {2,3,5,6}.

Success is the probability of getting a square number when a die is rolled,

$$p = \frac{1}{3}$$

Failure is the probability of not getting a square number when a die is rolled

$$(1 - p) = \frac{2}{3}$$

S	Bernoulli Sequences	Probability
0	FFFFF	$1\left(\frac{2}{3}\right)\left(\frac{2}{3}\right)\left(\frac{2}{3}\right)\left(\frac{2}{3}\right)\left(\frac{2}{3}\right)$
1	SFFFF, FSFFF, FFSFF, FFFSF,FFFFS	$5\left(\frac{1}{3}\right)\left(\frac{2}{3}\right)\left(\frac{2}{3}\right)\left(\frac{2}{3}\right)\left(\frac{2}{3}\right)$
2	SSFFF, SFSFF, SFFSF, SFFFS, FSSFF, FSFSF, FSFFS, FFSSF, FFSFS, FFFSS	$10\left(\frac{1}{3}\right)\left(\frac{1}{3}\right)\left(\frac{2}{3}\right)\left(\frac{2}{3}\right)\left(\frac{2}{3}\right)$
3	SSSFF, SSFSF, SSFFS, SFSSF, SFSFS, SFFSS, FSSSF, FSSFS, FSFSS, FFSSS	$10\left(\frac{1}{3}\right)\left(\frac{1}{3}\right)\left(\frac{1}{3}\right)\left(\frac{2}{3}\right)\left(\frac{2}{3}\right)$
4	FSSSS, SFSSS, SSFSS, SSSFS,SSSSF	$5\left(\frac{1}{3}\right)\left(\frac{1}{3}\right)\left(\frac{1}{3}\right)\left(\frac{1}{3}\right)\left(\frac{2}{3}\right)$
5	SSSSS	$1\left(\frac{1}{3}\right)\left(\frac{1}{3}\right)\left(\frac{1}{3}\right)\left(\frac{1}{3}\right)\left(\frac{1}{3}\right)$

If the process is repeated 5 times the sequence of success and failure results are known as Bernoulli Sequences.

The random variable X is the number of successes that can occur in the trial. The 6 possible values that the Discrete Random variable X can take in the event space of the Binomial distribution are placed in the first row of a probability distribution table with the corresponding probabilities to 4 decimal places. The sum of the probabilities is equal to 1.

x	0	1	2	3	4	5
$\Pr(X = x)$	0.1317	0.3292	0.3292	0.1646	0.0412	0.0041

Graphical representation of the Binomial Probability Distribution. The sum of the probabilities is equal to 1

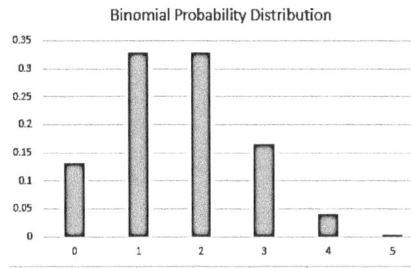

Binomial Probability Distribution

Act 1, Scene 1. The drama of counting Bernoulli sequences and Binomial combinations.

The Bernoulli Sequences are listing all the **possible combinations** for the trial, for two outcomes, success or failure.

If we have many trials it can become time consuming to list all the possible combinations.

The number of combinations can be worked out with the formula

$$C^n_x = \binom{n}{x} = \frac{n!}{x!(n-x)!}$$

And because there are only two outcomes in each trial we can also use Pascal's triangle to count the combinations.

Renaissance

Figures.

In 4 binomial trials, 2 successes be arranged in 6 Bernoulli sequences:
SSFF, SFSF, SFFS FSSF, FSFS,FFSS

Calculation using the combinations formula, $n = 4, x = 2$

$$\binom{n}{x} = \frac{n!}{x!(n-x)!}$$

$$\binom{4}{2} = \frac{4!}{2!(4-2)!} = \frac{4\times3\times2\times1}{(2\times1)(2\times1)} = 6$$

The combinations formula can be used to calculate the probabilities for the Binomial Distribution.

x	0	1	2	3	4
$\Pr(X=x)$	$\binom{4}{0}p^0q^4$	$\binom{4}{1}p^1q^3$	$\binom{4}{2}p^2q^2$	$\binom{4}{3}p^3q^1$	$\binom{4}{4}p^4q^1$
	$(1)p^0q^4$	$(4)p^1q^3$	$(6)p^2q^2$	$(4)p^3q^1$	$(1)p^4q^1$

Where p is the probability of success,

and $q = (1-p)$ is the probability of failure.

The combination coefficients of the probabilities for this trial are:
(1),(4),(6),(4),(1) which correspond to the $(4+1)^{\text{th}}$ row of Pascal's triangle.

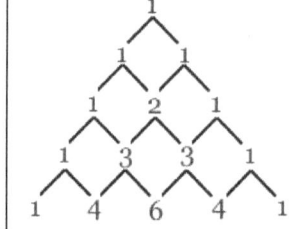

Act 1, Scene 2. The drama of counting Bernoulli sequences and Binomial combinations.

A binomial distribution can be defined for a random variable **X** with two outcomes only (success/failure). Pascal's triangle is a triangular array of the binomial coefficients for counting the combinations of the two outcomes.

The probability of x successes from n trials where p= probability of success, can also be calculated using combinations and the the values of x, n and p

$$P(X = x) = \binom{n}{x} p^x (1 - p)^{n-x}$$

Renaissance

Figures.

Sequences, $n = 4, x = \{0, 1, 2, 3, 4\}$		Combinations
0	FFFF	$1 = \binom{4}{0}$
1	SFFF, FSFF, FFSF, FFFS	$4 = \binom{4}{1}$
2	SSFF, SFSF, SFFS, FFSS, FSFS, FSSF	$6 = \binom{4}{2}$
3	SSSF, SSFS, SFSS, FSSS	$4 = \binom{4}{3}$
4	SSSS	$1 = \binom{4}{4}$

Sequences $n = 5, x = \{0, 1, 2, 3, 4, 5\}$		Combinations
0	FFFFF	$1 = \binom{5}{0}$
1	SFFFF, FSFFF, FFSFF, FFFSF, FFFFS	$5 = \binom{5}{1}$
2	SSFFF, SFSFF, SFFSF, SFFFS, FSSFF, FSFSF, FSFFS, FFSSF, FFSFS, FFFSS	$10 = \binom{5}{2}$
3	SSSFF, SSFSF, SSFFS, SFSSF, SFSFS, SFFSS, FSSSF, FSSFS, FSFSS, FFSSS	$10 = \binom{5}{3}$
4	SSSSF, SSSFS, SSFSS, SFSSS, FSSSS	$5 = \binom{5}{4}$
5	SSSSS	$1 = \binom{5}{5}$

Technology is very handy for working out the calculations.

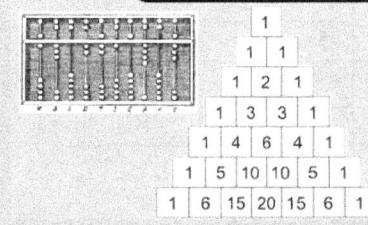

Analysis of a Discrete Probability Distribution.

When X is the countable discrete outcome of a trial, we can calculate the discrete probability distribution of that variable. The probability of each outcome is between 0 and 1, $0 \le \Pr(X = x) \le 1$.

The sum of all the probabilities for all possible outcomes is 1, $\sum_x \Pr(X = x) = 1$.

Measures of centre, Mean.

The **Mean** is referred to as the expected value of the trial: $E(X) = \mu = \sum_x x \times \Pr(X = x)$.

For **linear combinations** of the random variable X:
$E(aX + b) = aE(X) + b$.

For a **binomial distribution** this calculation simplifies to $E(X) = np$.

Example of a discrete probability distribution.

x	1	3	6
$\Pr(X = x)$	0.3	0.3	0.4

$E(X) = (1 \times 0.3) + (3 \times 0.3) + (6 \times 0.4) = 3.6$

Linear Combination $5X + 1$
$E(5X + 1) = 5E(X) + 1 = 5(3.6) + 1 = 19$

Measures of spread around the mean, Variance and standard deviation.

Variance $Var(X) = \sigma^2$

$Var(X) = E\left[(X - \mu)^2\right] = E(X^2) - \left[E(X)\right]^2$

Standard Deviation $\sigma = \sqrt{Var(X)}$

For **linear combinations** of the random variable X:

$Var(aX + b) = a^2 Var(X)$.

For a **binomial distributions** this simplifies to

- Variance $\sigma^2 = np(1 - p)$
- Standard deviation $\sigma = \sqrt{np(1 - p)}$

$E(X^2) = (1^2 \times 0.3) + (3^2 \times 0.3) + (6^2 \times 0.4) = 17.4$

$\left[E(X)\right]^2 = [3.6]^2 = 12.96$

$Var(X) = E(X^2) - \left[E(X)\right]^2 = 17.4 - 12.96$
$\qquad = 4.44$

$\sigma = \sqrt{Var(X)} = \sqrt{4.44} \approx 2.11$

Linear Combination $5X + 1$

$Var(5X + 1) = 5^2 Var(X) = 25(4.44) = 111$

Standard Deviation

$sD(5X + 1) = \sqrt{Var(5X + 1)} \approx 10.54$

Example 7.3: Application of the expected value for a discrete probability distribution.

A gambling game is based on tossing three coins simultaneously. If three heads or three tails are obtained, the player wins $25. Otherwise the player loses $5. It costs each person $5 dollars to play.

x	0	1	2	3
Gain $	20	-5	-5	20
Pr(X=x)	$\frac{1}{8}$	$\frac{3}{8}$	$\frac{3}{8}$	$\frac{1}{8}$

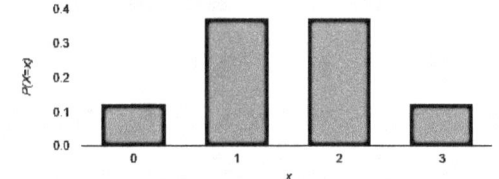

$E(X) = \left(0 \times \frac{1}{8}\right) + \left(1 \times \frac{3}{8}\right) + \left(2 \times \frac{3}{8}\right) + \left(3 \times \frac{1}{8}\right)$
$\qquad = 1.5$

$E(Gain) = \left(20 \times \frac{1}{8}\right) + \left(-5 \times \frac{3}{8}\right) + \left(-5 \times \frac{3}{8}\right) + \left(20 \times \frac{1}{8}\right)$
$\qquad = 1.25$

Expected gain per game is $1.25.

$\sigma^2 = Var(X) = 0.75$

The probability distribution is symmetrical.

The probability of x successes in n trials for a Binomial Distribution is given by: $P(X = x) = \binom{n}{x} p^x (1-p)^{n-x}$

On a good day Miss Archer can hit the bullseye on an archery target 15 times out of 20. Using this statistic $p = 0.75, (1-p) = q = 0.25$, we work out the probability distribution for 10 shots at the target.

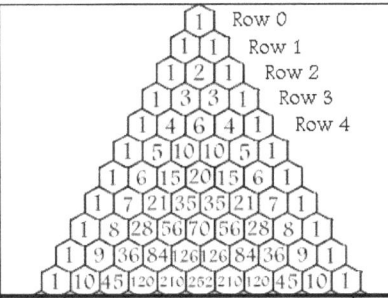

The coefficients in the 10th row of Pascal's triangle, correspond the to the required combinations to calculate the probabilities of success and failure in 10 trials.

1,   10,   45,   120,   210,   252,   210,   120,   45,   10,   1

$\binom{10}{0}, \binom{10}{1}, \binom{10}{2}, \binom{10}{3}, \binom{10}{4}, \binom{10}{5}, \binom{10}{6}, \binom{10}{7}, \binom{10}{8}, \binom{10}{9}, \binom{10}{10}$

x	Pr($X = x$)	$\approx$ Pr($X = x$)
0	$\binom{10}{0}(0.75)^0 (0.25)^{10}$	0.00000
1	$\binom{10}{1}(0.75)^1 (0.25)^9$	0.00003
2	$\binom{10}{2}(0.75)^2 (0.25)^8$	0.00039
3	$\binom{10}{3}(0.75)^3 (0.25)^7$	0.00309
4	$\binom{10}{4}(0.75)^4 (0.25)^6$	0.01622
5	$\binom{10}{5}(0.75)^5 (0.25)^5$	0.05840
6	$\binom{10}{6}(0.75)^6 (0.25)^4$	0.14600
7	$\binom{10}{7}(0.75)^7 (0.25)^3$	0.25028
8	$\binom{10}{8}(0.75)^8 (0.25)^2$	0.28157
9	$\binom{10}{9}(0.75)^9 (0.25)^1$	0.18771
10	$\binom{10}{10}(0.75)^{10} (0.25)^0$	0.05631

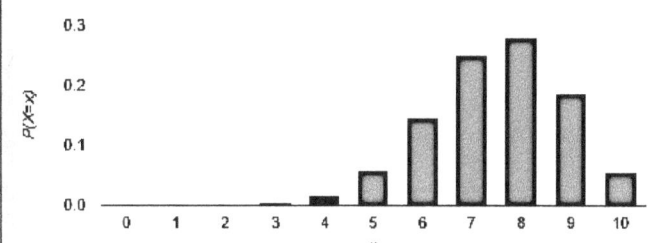

We can see from the probability distribution graph the distribution is not symmetrical.

- Expected value $E(X) = \mu = np = 7.5$,

- Variance $\sigma^2 = np(1-p) = 1.875$

- Standard deviation $\sigma = \sqrt{np(1-p)} = 1.369$

The probability that Miss Archer hits fewer than 3 bullseye targets is very low.

$$\text{Pr}(X < 3) = \text{Pr}(X = 0) + \text{Pr}(X = 1) + \text{Pr}(X = 2)$$
$$\approx 0.00000 + 0.00003 + 0.00039$$
$$\approx 0.00042$$

The lack of symmetry in the distribution can be highlighted by comparing the probabilities of the lower values of x with the higher values of x.

The probability that Miss Archer hits more than 7 bullseye targets is higher.

$$\text{Pr}(X > 7) = \text{Pr}(X = 8) + \text{Pr}(X = 9) + \text{Pr}(X = 10)$$
$$\approx 0.28157 + 0.18771 + 0.05631$$
$$\approx 0.52559$$

The effect of changing *n* and *p* on binomial distribution $X \sim Bi(n, p)$ on the bar graphs representing the probability distribution, can be fun to explore. We need to think about the general shape and whether there is any skewedness or lack of symmetry in the shape of the distribution.

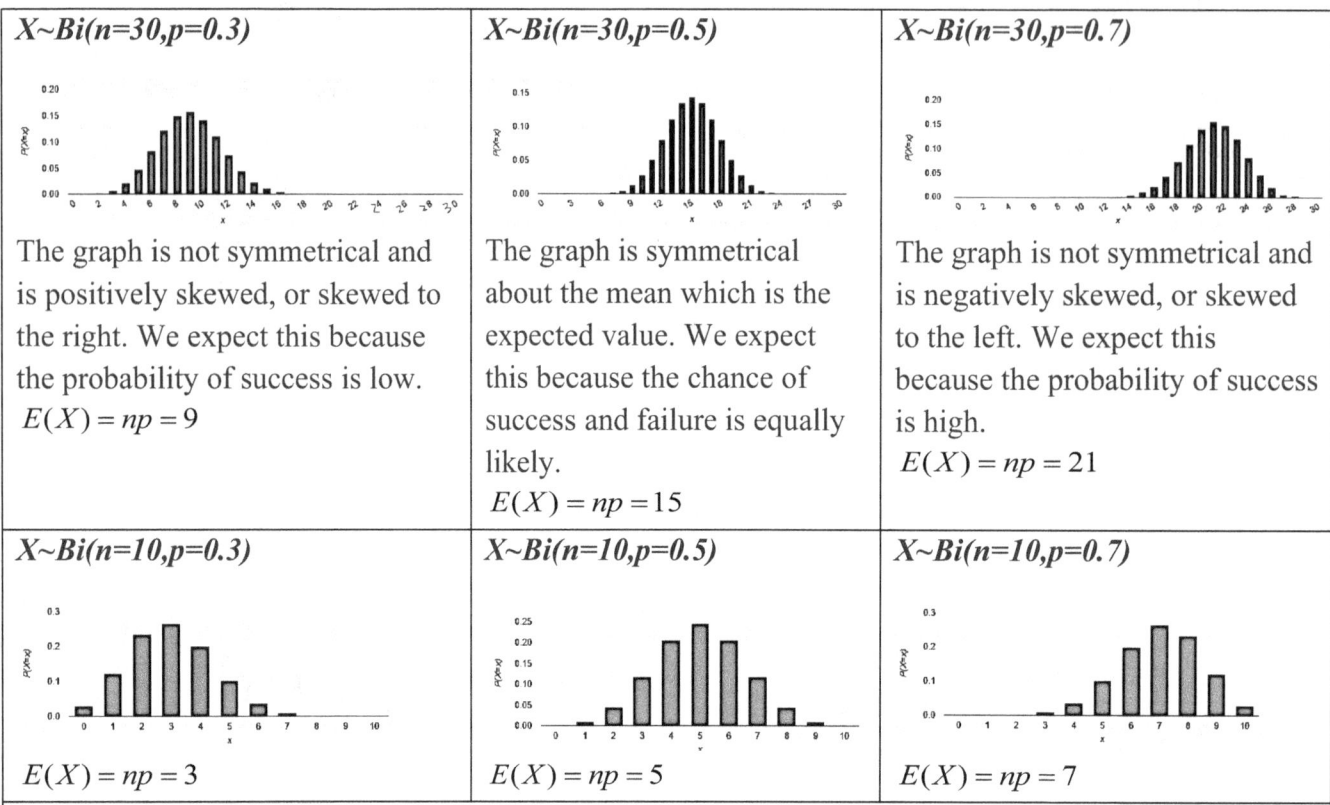

X~Bi(n=30,p=0.3)	**X~Bi(n=30,p=0.5)**	**X~Bi(n=30,p=0.7)**
The graph is not symmetrical and is positively skewed, or skewed to the right. We expect this because the probability of success is low. $E(X) = np = 9$	The graph is symmetrical about the mean which is the expected value. We expect this because the chance of success and failure is equally likely. $E(X) = np = 15$	The graph is not symmetrical and is negatively skewed, or skewed to the left. We expect this because the probability of success is high. $E(X) = np = 21$
X~Bi(n=10,p=0.3)	**X~Bi(n=10,p=0.5)**	**X~Bi(n=10,p=0.7)**
$E(X) = np = 3$	$E(X) = np = 5$	$E(X) = np = 7$

Reducing the number of trials *n* while not changing *p* results in fewer bars with higher probabilities, but the shape of the distribution remains very similar. As compared in the bar graphs above for *n=30* and *n=10*.

X~Bi(n=10,p=0.15)	Changing the *p* value, while *n* remains the same.	**X~Bi(n=10,p=0.85)**
Chance of success is made lower causing a greater skew to the right. $E(X) = np = 1.5$		Chance of success is made higher causing a greater skew to the left. $E(X) = np = 8.5$

A comparison of Binomial distributions for different values of *p* for *n=5*.

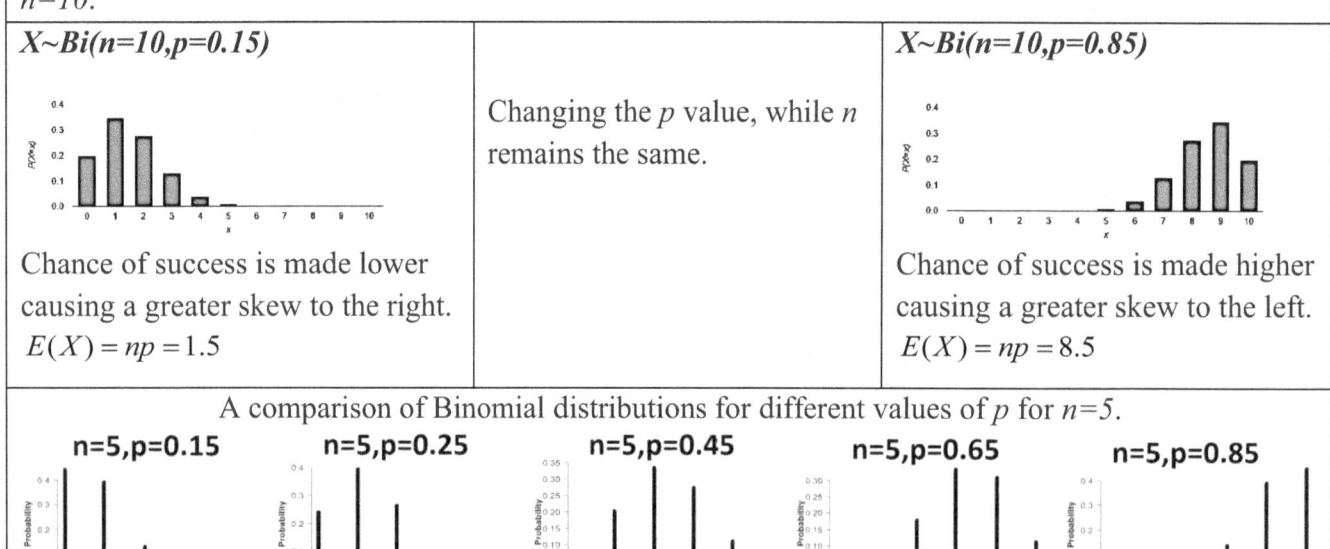

These Graphs created using https://homepage.divms.uiowa.edu/~mbognar/applets/bin.html

https://shiny.rit.albany.edu/stat/binomial/

Two events such as A and B are said to be **independent** if when event A has occurred it does not affect the probability that the event B will occur. If the events A and B are **independent** then the probability that event A occurs given B has already occurred or vice-versa is given by: $\Pr(A/B) = \Pr(A)$ or $\Pr(B/A) = \Pr(B)$

When one event occurs and does affect the probability that the other event will occur, then the two events are said to be **dependent**.

Conditional Probability Pr(A/B)

For dependent events, **Conditional Probability** is the probability that event A occurs given event B has already occurred:

<table>
<tr>
<td>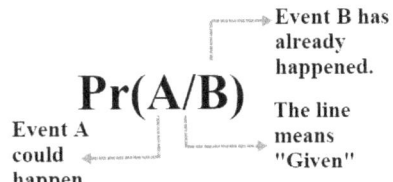</td>
<td>When we know that B has already occurred, every outcome that is outside B should be discarded. So, our sample space is reduced to the set B.</td>
<td>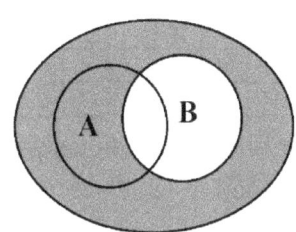

Discard events outside B.</td>
</tr>
<tr>
<td>$\Pr(B/B) = \dfrac{\Pr(B \cap B)}{\Pr(B)}$

$= \dfrac{\Pr(B)}{\Pr(B)}$

$= 1$</td>
<td>$\Pr(A/B) = \dfrac{\Pr(A \cap B)}{\Pr(B)}$

The only way that event A can happen is when the event belongs to the set $A \cap B$.</td>
<td>If event B has already happened where $0 < n(B) < n(\varepsilon)$ then the event space will be reduced for further outcomes such as event A.</td>
</tr>
</table>

Special cases of conditional probability, some thinking is required before applying the rule.

<table>
<tr>
<td>Events A and event B are mutually exclusive so $n(A \cap B) = 0$</td>
<td>Event B is a subset of event A, so when B happens, part of A also happens. $\Pr(A \cap B) = \Pr(B)$</td>
<td>Event A is a subset of event B, so when B happens, part of B, that is equal to A also happens. $\Pr(A \cap B) = \Pr(A)$</td>
</tr>
<tr>
<td>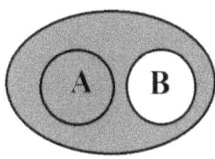</td>
<td>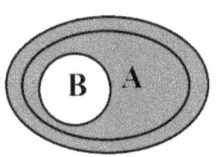</td>
<td>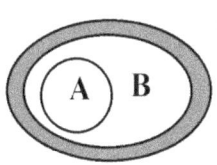</td>
</tr>
<tr>
<td>$\Pr(A/B) = \dfrac{\Pr(A \cap B)}{\Pr(B)}$

$= \dfrac{0}{\Pr(B)}$

$= 0$</td>
<td>$\Pr(A/B) = \dfrac{\Pr(A \cap B)}{\Pr(B)}$

$= \dfrac{\Pr(B)}{\Pr(B)}$

$= 1$</td>
<td>$\Pr(A/B) = \dfrac{\Pr(A \cap B)}{\Pr(B)}$

$= \dfrac{\Pr(A)}{\Pr(B)}$</td>
</tr>
</table>

Example 7.4 Conditional probability. In the country of Greece the probability of having one of the major blood types, is shown in the table below. (source Wikipedia)

Blood Type	O+	A+	B+	AB+	O-	A-	B-	AB-
Pr(X=Blood type)	0.37	0.33	0.11	0.04	0.07	0.05	0.02	0.01

The compatibility table shows which blood donors can donate blood to which recipients.

Q:What is the probability that a randomly chosen person can donate blood for a person with blood type A+?

A recipient with blood type A+, can receive a blood donation from a donor with blood type:
$O+, O-, A+, A-$.

$$\Pr(A\pm \text{ or } O\pm) = \Pr(A\pm \cup O\pm)$$
$$= \Pr(A+) + \Pr(A-) + \Pr(O+) + \Pr(O-)$$
$$= 0.33 + 0.05 + 0.37 + 0.07$$
$$= 0.82$$

Q:Given a blood donor has blood type B, what is the probability that a randomly selected person can receive blood from the donor?

The donor could have blood type B+ or blood type B-. This has not been specified, although the given condition has reduced the donor event space.

$$\Pr(B+/B) = \frac{\Pr(B+\cap B)}{\Pr(B)} \quad \Pr(B-/B) = \frac{\Pr(B-\cap B)}{\Pr(B)}$$

$$= \frac{0.11}{0.13} \qquad\qquad = \frac{0.02}{0.13}$$

$$\approx 0.8462 \qquad\qquad \approx 0.1538$$

If a person with blood type B+ is the donor , they can only donate to people with blood type B+ or AB+.

$$\Pr(B+ \text{ or } AB+) = \Pr(B+ \cup AB+)$$
$$= \Pr(B+) + \Pr(AB+)$$
$$= 0.11 + 0.04$$
$$= 0.15$$

If a person with blood type B- is the donor , they can only donate to people with blood type B-,or B+ or AB- or AB+.

$$\Pr((B\pm \text{ or } AB+) = \Pr(B- \text{ or } B+ \text{ or } AB- \text{ or } AB+)$$
$$= \Pr(B-) + \Pr(B+) + \Pr(AB-) + \Pr(AB+)$$
$$= 0.02 + 0.11 + 0.05 + 0.04$$
$$= 0.22$$

The random selection of the recipient makes these events independent.

Let X be the event of a random receiver of blood selected.

The probability that a randomly selected person can receive blood from a donor with blood type B is:

$$\Pr(X/B) = \Pr(B+/B) \times \Pr(B+ \text{ or } AB+)$$
$$+ \Pr(B-/B) \times \Pr(B\pm \text{ or } AB+)$$
$$\approx (0.8462)(0.15) + (0.1538)(0.22)$$
$$\approx 0.1608$$

The quality control at a factory has found that in a batch of 100 products there are 5 units which are defective. If we randomly pick three units from this batch without replacement. What is the probability that none of them are defective?

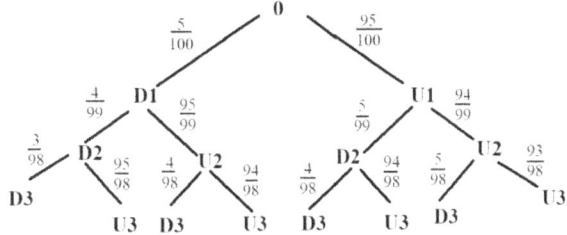

These events are dependent, since each outcome affects subsequent outcomes due to non-replacement. Let us define U_i as the event that the ith chosen unit is not defective, for i=1,2,3. Let us define D_i as the event that the ith chosen unit is defective, for i=1,2,3. We are interested in $\Pr(U_1 \cap U_2 \cap U_3)$.

This is an example of conditional probability when we select without replacement, we are finding the probability of dependent events. $\Pr(U_2 \cap U_1) = \Pr(U_2 / U_1)\Pr(U_1)$.

$\Pr(U_1) = \dfrac{95}{100}$ then $\Pr(U_2 / U_1) = \dfrac{94}{99}$ then $\Pr(U_3 / U_2 \, and \, U_1) = \dfrac{93}{98}$ $\Pr(U_1 \cap U_2 \cap U_3) = \Pr(U_1) \times \Pr(U_2 / U_1) \times \Pr(U_3 / U_2 \, and \, U_1)$ $= \left(\dfrac{95}{100}\right)\left(\dfrac{94}{99}\right)\left(\dfrac{93}{98}\right)$ ≈ 0.8560	Quality control decides to make this the benchmark for delivering batches to stores. Three units are randomly selected from all batches and if they are all working then this is classed as a successful batch where p=0.856. The factory makes 50 batches for testing: $X \sim Bi(n=50, p=0.856)$

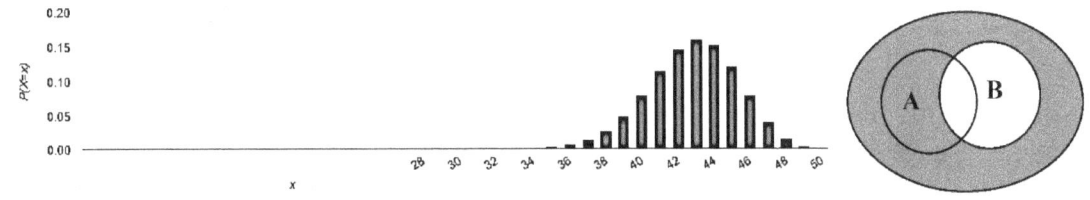

Given that at the least the first 40 batches have been passed and have been sent to stores, find the probability that 5 more batches will also be sent to stores. This is an example of conditional probability. Let B be the event at least 40 batches have passed the quality control, reducing the event space. Let A be the event that up to 45 batches are ok. The events A and B are subsets of the sample space consisting of all batches.

$$\Pr(A / B) = \frac{\Pr(A \cap B)}{\Pr(B)} = \frac{\Pr(40 \le A \le 45)}{\Pr(B \ge 40)}$$

$$\approx \frac{0.8647 - 0.1748}{1 - 0.1748}$$

$$\approx 0.8360$$

Act 2, Scene 1. The drama of finding *n* for a Binomial Distribution.

Sometimes you need to work out the number of trials required in Binomial Distribution to reach a threshold of chance.

If you are given the value of p, you may be able to work it out using some algebra.

Say p=0.1 and you want to have at least a 0.7 probability of success. You can work out the value of n required.

Positive Skewed

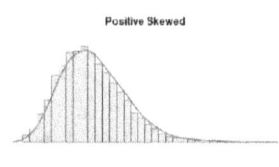

Renaissance

Figures.

The clue is at least how many is required. Think about the combinations at the extremes of the distribution, at Pr(X=0) and at Pr(X=n).

$$P(X \geq 1) = 1 - \Pr(X = 0) = 0.7$$
$$1 - \binom{n}{0} 0.1^0 (1 - 0.1)^n = 0.7$$
$$\binom{n}{0} 0.1^0 (1 - 0.1)^n = 0.3$$
$$(0.9)^n = 0.3$$

$$\log_{0.9}\left(0.9^n\right) = \log_{0.9}\left(0.3\right)$$
$$n = \log_{0.9}\left(0.3\right)$$
$$n \approx 11.4272$$

Act 2, Scene 2. The drama of finding *n* for a Binomial Distribution.

With a value of n that is between 11 and 12. We can check if 11 or 12 trials are needed for probability of more that 0.7

For $X\sim Bi(n=11,p=0.1)$ or
$X\sim Bi(n=12,p=0.1)$
find $\Pr(X>0)=1-\Pr(X=0)$

binomCdf$(11,0.1,1,11)$	0.686189
binomCdf$(12,0.1,1,12)$	0.71757

Renaissance

Figures.

At the extremes of the distribution, the probability calculations are simpler.

If you are given either n or p and asked to find the other...

$\binom{n}{0}p^0\left(1-p\right)^n$ and $\binom{n}{n}p^n\left(1-p\right)^0$

simplifies to become a one variable algebra problem..

$(1)(1)\left(1-p\right)^n$ and $(1)\left(p\right)^n(1)$

$X\sim Bi(n=11,p=0.1)$	$X\sim Bi(n=12,p=0.1)$
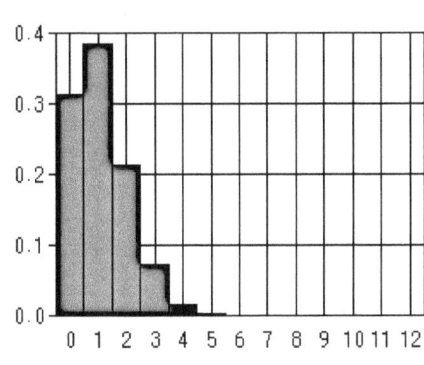	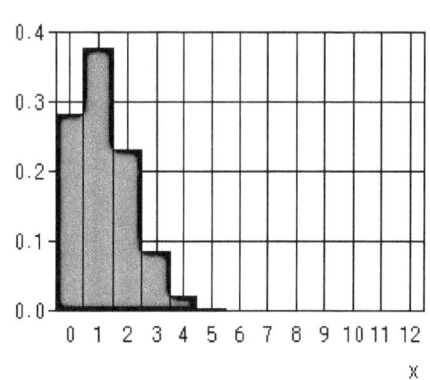

BinomPDF is the name of the functionality used on most technologies for finding the probability for a binomial distribution for a single value, for example, Pr(X=3), while the function **BinomCDF** is a cumulative probability for finding an inequality, for example Pr(X<3). See example 7.5.

Considering Combinations and Possible Outcomes.

An important part of working out the probabilities of outcomes in discrete probability calculations is the consideration of how many favourable outcome combinations can occur as a ratio of the total overall combinations that can occur.

This can get quite complicated to calculate in cases where there is also some conditional probability involved, which can occur when we are told an event has already happened and that event affects the size of the new event space.

The total combinations can get very large and are dependent on the number of items to select from.

It may not be possible to list all the Bernoulli sequences, or use Pascal's triangle to find the combinations.

Technology is helpful to calculate the number of combinations possible.

n	x	Combinations
24	0	1
24	1	24
24	2	276
24	3	2024
24	4	10626
24	5	42504
24	6	134596
24	7	346104
24	8	735471
24	9	1307504
24	10	1961256
24	11	2496144
24	12	2704156
24	13	2496144
24	14	1961256
24	15	1307504
24	16	735471
24	17	346104
24	18	134596
24	19	42504
24	20	10626
24	21	2024
24	22	276
24	23	24
24	24	1
	TOTAL	16777216

Some Binomial Distribution Real Life Examples

Any event that you can think of that can only be a success or a failure can be represented by a binomial distribution.

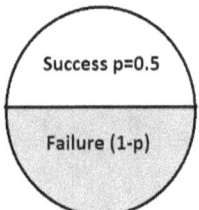 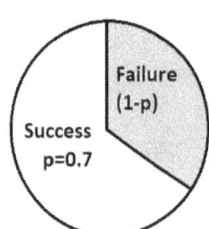  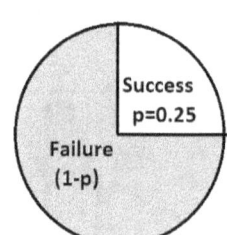

- A new drug is introduced to cure a disease, it works or it doesn't.
- You buy a Lottery ticket, you win or you don't.
- The number of defective or non-defective products in a production run.
- Responses to a Yes/No Survey

In certain conditions the **Normal Distribution** $X \sim N\left(\mu = np, \sigma = np(1-p)\right)$ can be used to approximate a **Binomial Distribution** $X \sim Bi(n, p)$ in cases where n is very large, or p is close to 0.5, or the shape of the Binomial Distribution is symmetrical.

The Binomial Distribution is supported by many technologies, apps and software.

Example 7.5: The Binomial Distribution is supported by many technologies, apps and software.			
Finding probabilities for discrete random variables with Binomial Distributions using technology. $X = Bi(n = 8, p = 0.4)$	**Tinspire CAS**	**Casio Classpad**	**Mathematica**
$Pr(X = 6)$	binomPdf(8,0.4,6) =0.0413	binomialPDF(6,8,0.4) =0.0413	PDF[BinomialDistribution[8,0.4],6] 0.0413
$Pr(5 \leq X \leq 7)$	binomCdf(8,0.4,5,7) =0.1730	binomialCdf(5,7,8,0.4) =0.1730	=CDF[BinomialDistribution[8,0.4],7]- CDF[BinomialDistribution[8,0.4],4] 0.1730
$Pr(X < 5)$	binomCdf(8,0.4,0,4) =0.8263	binomialCdf(0,4,8,0.4) =0.8263	=CDF[BinomialDistribution[8,0.4],4] 0.8263
$Pr(X > 5)$	binomCdf(8,0.4,6,8) =0.0500	binomialCdf(6,8,8,0.4) =0.0500	=CDF[BinomialDistribution[8,0.4],8]- CDF[BinomialDistribution[8,0.4],5] 0.0500
Some more binomial distributions graphed using technology.		Graphs created using applets at: https://homepage.divms.uiowa.edu/~mbognar/applets/bin.html https://keisan.casio.com/exec/system/1180573198 https://shiny.rit.albany.edu/stat/binomial/	

DISCRETE PROBABILITY SUMMARY:

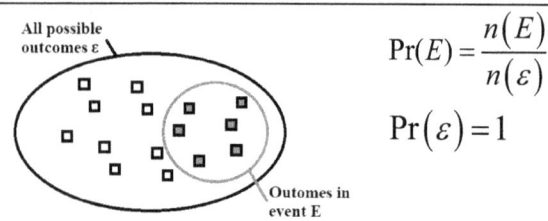

$$Pr(E) = \frac{n(E)}{n(\varepsilon)}$$

$$Pr(\varepsilon) = 1$$

$$0 \le Pr(E) \le 1 \qquad Pr(E') = 1 - Pr(E),$$

Two events A or B occur:

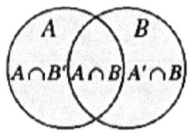

$$Pr(A \cup B) = Pr(A) + Pr(B) - Pr(A \cap B)$$

If A and B are **mutually exclusive** $Pr(A \cap B) = 0$

$$Pr(A \cup B) = Pr(A) + Pr(B)$$

Conditional Probability

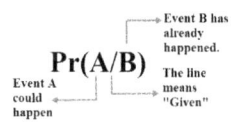 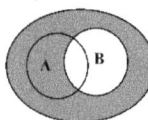

Discard events outside B.

$$Pr(A / B) = \frac{Pr(A \cap B)}{Pr(B)}$$

If events A and B are **independent**

$$Pr(A / B) = Pr(A) \quad \text{or} \quad Pr(B / A) = Pr(B)$$

Each discrete outcome of X has:

$$0 \le Pr(X = x) \le 1, \ \sum_x Pr(X = x) = 1$$

and can be tabulated.

x	1	2	3	...	k
$Pr(X=x)$	p_1	p_2	p_3		p_k

Mean/Expected value:

$$E(X) = \mu = \sum_x x \times Pr(X = x).$$

Variance $Var(X) = \sigma^2$

$$Var(X) = E(X^2) - \left[E(X)\right]^2$$

Standard Deviation $\sigma = \sqrt{Var(X)}$

Linear combinations of random variable X:
$$E(aX + b) = aE(X) + b$$

$$Var(aX + b) = a^2 Var(X)$$

Binomial distributions $X \sim Bi(n,p)$

Random variable X has only two outcomes, success or failure.

$$Pr(success) = p, \ Pr(failure) = 1 - p$$

$$Pr(success) + Pr(failure) = 1$$

Probability of x successes from n trials:

$$P(X = x) = \binom{n}{x} p^x (1-p)^{n-x}$$

Coefficients of binomial expansion can be found using Pascals triangle.

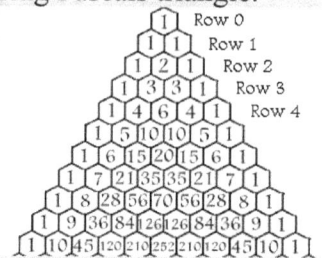

Extreme values of Pascal's triangle

$$\binom{n}{0} = \binom{n}{n} = 1$$

Mean/Expected value:
$$\mu = E(X) = np$$

Variance $Var(X) = \sigma^2$

$$\sigma^2 = np(1-p)$$

Standard deviation $\sigma = \sqrt{np(1-p)}$

Binomial distributions can be tabulated and graphed, and then analysed for symmetry and skewedness.

$E(X) = np = 1.5$	$E(X) = np = 5$	$E(X) = np = 8.5$

The **Normal Distribution**

$$X \sim N\left(\mu = np, \sigma = np(1-p)\right)$$ can approximate a

Binomial Distribution $X \sim Bi(n, p)$ where n is very large, or p is close to 0.5, or the shape of the Binomial distribution is symmetrical.

DISCRETE PROBABILITY: Check your Understanding.

7.1 Sample space and outcomes.

a) Discrete probability uses addition and multiplication to find the total number of possible outcomes that are possible in a trial.

   i.   How many outcomes are possible for tossing 3 fair coins?

   ii.   How many outcomes are possible for tossing 3 fair dice?

   iii.   In how many ways can 8 people swimming in a race finish in first, second and third place?

   iv.   How many odd numbers with distinct digits can be created using all the digits in the set: {4, 6, 7, 1, 3}?

   v.   How many distinct arrangements of the letters of the word IGLOO are there?

7.2 Probability Rules.

a) Roulette is a game of chance with numbered cells, where a ball is rolled one way and the wheel is spun in the opposite direction. Players bet on which cell the ball will land on.

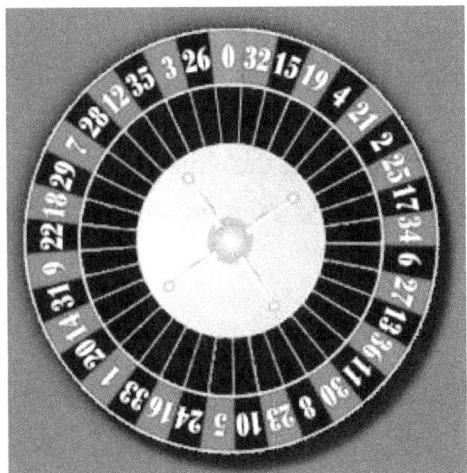

This roulette wheel has 37 cells, where each cell is equally likely for the ball to land on.

- 1 green cell has the label "0", this number is reserved for the "house", and you cannot bet on it.
- 18 red cells and 18 black cells alternating around the wheel.
- You can only bet on the red and black cells.
- The red and black cells are labelled with different numbers 1 to 36 non-sequentially.

   i.   If you bet $1 on a single number, what are your and the house's chances of winning?

   ii.   If you bet $36 dollars on 36 different numbers, what are your and the house's chances of winning?

   iii.   Show the event space for three consecutive single number bets on the roulette wheel.

   iv.   For three consecutive rolls on this roulette wheel what are the chances that the ball lands on a black cell first, followed by a red cell, followed by another red cell?

   v.   Given the first of three consecutive rolls is a black cell, show the possible sample space for the three rolls.

   vi.   Can you explain part v. in terms of conditional probability.

7.2 Probability Rules (Continued)

b) Chloe has a part time job selling new cars. She receives a base salary of $2000 per month and $1000 bonus payment for every new car she sells to a customer.

The company she works for has a table of probabilities of the number of new car sales per month by a salesperson.

This quantity is represented by the variable X, and is used to calculate the bonus based on new cars, x, being sold per month.

x	1	2	3	4	5	6	7	8
$Pr(X=x)$	0.22	0.26	0.18	0.14	0.08	0.06	0.04	0.02

i.   What would be the expected salary Chloe would receive each month?

ii. What is the standard deviation of each month's salary?

c) A family has two adults and two children. The gender and birth order of the children is within the universal set of $\varepsilon = \{GG, GB, BG, BB\}$, where G represents a girl, and B represents a boy, and we assume that each outcome is equally likely for this family.

i.   What is the probability that both children are boys given that the first born child is a boy?

ii.   We asked the mother if she had at least one son and she "Yes, I do.". With this extra information, what is the probability that both children are boys?

iii. If we randomly choose a child from this family and find out he's a boy. What is the probability that both children are boys?

7.3 Binomial Probability Distributions.

a) Should you use the binomial distribution? In each of the following situations, is it reasonable to use a binomial distribution for the random variable X? Give reasons for your answer in each case.

    i.   In a random sample of students in a fitness study, X is the mean daily exercise time of the sample.

    ii.   A manufacturer of shirts picks a random sample of 20 shirts from the production of shirts each day for a detailed inspection. X is the number of shirts with a defect.

    iii.   A nutrition study chooses a group of year 10 students. They are asked whether or not they usually eat at least five servings of fruits or vegetables per day. X is the number who say that they do.

    iv. X is the number of days during the school year when it is raining during Physical education classes.

b) A certain brand of burglar alarm has an acceptable probability of failure of 0.01. In a factory batches of 20 alarms are regularly tested. If there is 3 or more defective alarms, the machinery on the production line needs to be recalibrated.

    i.   Find the probability that 0,1,2,3,4,5 burglar alarms fail in any factory test.

    ii.   Create a graph of the probability distribution, explain the properties.

    iii.   What is the probability that the machinery needs to be recalibrated?

c) An exam has 20 multiple choice questions, where each questions has 5 options A, B, C, D, E. A student guesses the answer for each question.

    i.   What is the expected number of correct answers using the random guessing method?

    ii.   What is the standard deviation of correct answers using random guessing?

    iii.   Find the probability at least 50% of the questions are answered correctly.

    iv.   Find the probability at least 50% are answered correctly given at least 6 answers have been answered correctly.

d) Eight people board an empty bus. We are interested in the people whose birthday is in May.

    i.   Show the probability distribution of the chance of the people on the bus having a birthday in May.

    ii.   Find the probability that at least 3 people on the bus have a birthday in May.

    iii.   Find the probability that exactly 4 people on the bus have a birthday in May, given there is at least one person with a birthday in May.

    iv.   What is the least number of people that need to board the bus so that at least one person has a birthday in May is greater than 80% or 90% or 99%?

7.3 Binomial Probability Distributions (Continued)

e) The *Itchy and Scratchery* lottery company claims that 20% of all tickets sold win a prize.

How many *Itchy and Scratchery* tickets should you buy to have at least 50% chance of winning a prize?

f) Match the graphs with the following binomial distributions.

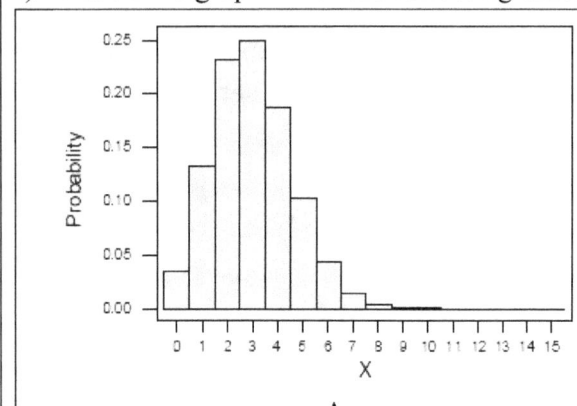

A

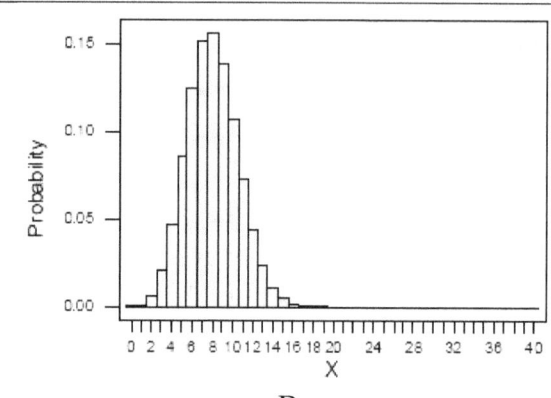

B

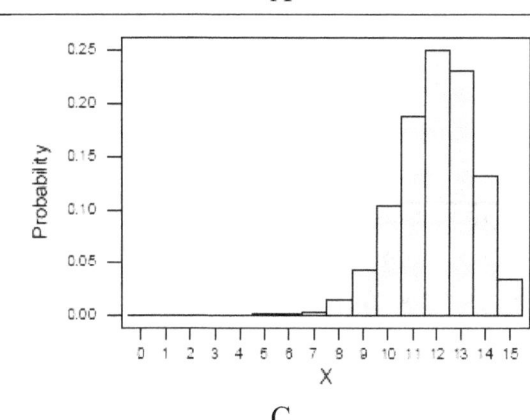

C

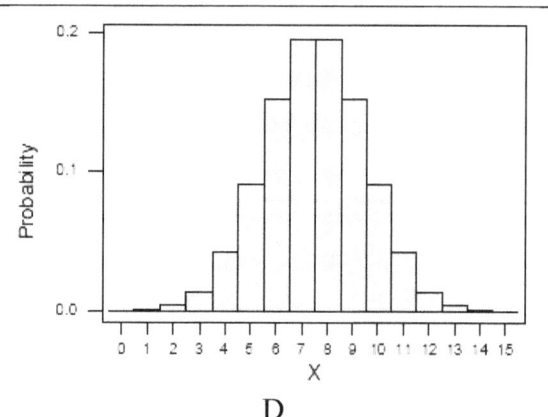

D

Match the graphs with the following binomial distributions.

    i. $X = Bi(n = 15, p = 0.5)$

    ii. $X = Bi(n = 40, p = 0.5)$

    iii. $X = Bi(n = 15, p = 0.8)$

    iv. $X = Bi(n = 15, p = 0.6)$

    v. $X = Bi(n = 40, p = 0.2)$

    vi. $X = Bi(n = 15, p = 0.2)$

DISCRETE PROBABILITY SOLUTIONS.
7.1 Sample space and outcomes (SOLUTIONS)

a)

i. $2^3 = 8$ ways, since each coin thrown has two possibilities

ii. $6^3 = 216$ ways, since each die thrown has 6 possibilities

iii. $8 \times 7 \times 6 = 336$ ways

iv. 72 ways.

Odd numbers with 5 digits from this set must end with digits 1,3,7. One way to work this out is to work backwards using the restriction to work out the samples space.

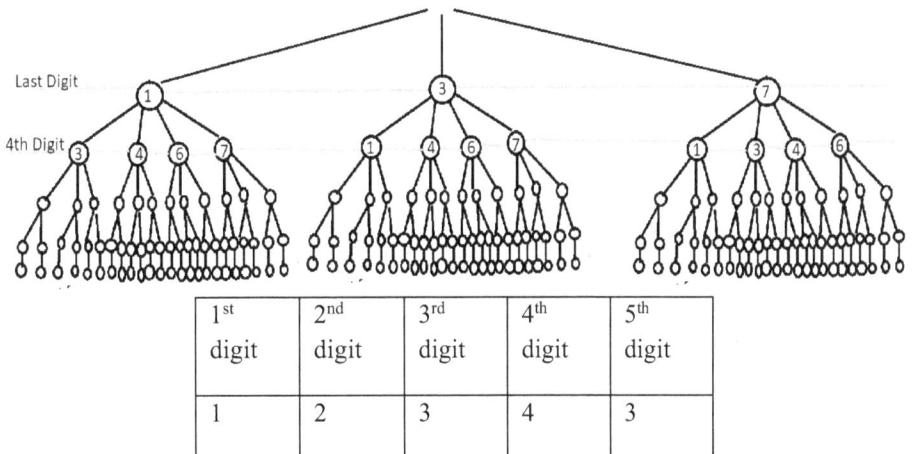

1st digit	2nd digit	3rd digit	4th digit	5th digit
1	2	3	4	3

v. For IGLOO there are 4 distinct letters to choose from, working forward with the restriction.

$3 \times 12 + 4! = 60$ ways.

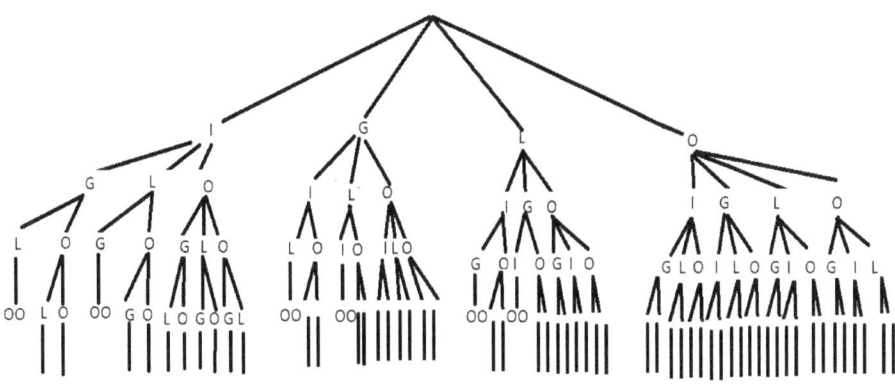

a) Roulette wheel

i. $\dfrac{1}{37}, \dfrac{36}{37}$

ii. $\dfrac{36}{37}, \dfrac{1}{37}$

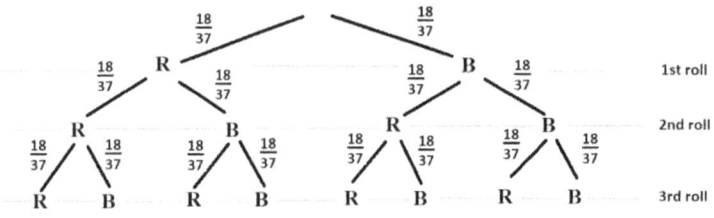

iii.

```
                              18                    18
                              37                    37
                    R                          B
                 18    18                   18    18
                 37    37                   37    37          1st roll
              R          B              R          B
           18   18    18   18        18   18    18   18
           37   37    37   37        37   37    37   37       2nd roll
           R    B     R    B         R    B     R    B        3rd roll
```

iv. {BRR, BRB, BBR, BBB}

v. One outcome , $\Pr(B_1 R_2 R_3) = \dfrac{18}{37} \times \dfrac{18}{37} \times \dfrac{18}{37}$ since the events are independent.

vi. Preceding events

$$\Pr(R_2/B_1) = \frac{\Pr\left(R_2 \cap B_1\right)}{\Pr\left(B_1\right)}$$

$$\Pr\left(R_2 \cap B_1\right) = \Pr(R_2/B_1)\Pr\left(B_1\right)$$

Since each roll is independent

$$\Pr(R_2/B_1) = \Pr\left(R_2\right)$$

$$\Pr\left(R_2 \cap B_1\right) = \Pr(R_2)\Pr\left(B_1\right)$$

$$\Pr\left(R_3 \cap R_2 \cap B_1\right) = \Pr\left(R_3\right) \times \Pr\left(R_2\right) \times \Pr\left(B_1\right)$$

Since each roll is independent

b) Chloe's salary, base salary $2000 + bonuses that are dependent on X

Linear Combination of random variable X: $salary = (1000X + 2000)$,

The expected value	$E(1000X + 2000) = 1000E(X) + 2000$
$E(1000X + 2000) = 1000E(X) + 2000$	$= 1000(3.04) + 2000$
$E(X) = \begin{pmatrix} (1 \times 0.22) + (2 \times 0.26) + (3 \times 0.18) \\ +(4 \times 0.14) + (5 \times 0.08) + (6 \times 0.06) \\ +(7 \times 0.04) + (8 \times 0.02) \end{pmatrix}$	$= 5040$
	The expected total salary for Chloe is $5040 per month.
$= 3.04$	
The standard deviation	$Var(X) = E\left(X^2\right) - \left[E\left(X\right)\right]^2$
$sD(1000X + 2000) = \sqrt{Var(1000X + 2000)}$	$= 12.52 - (3.04)^2$
$= \sqrt{1000^2 Var(X)}$	≈ 3.28
$= 1000\sqrt{Var(X)}$	$sD(1000X + 2000) = 1000\sqrt{Var(X)}$
$Var(X) = E\left(X^2\right) - \left[E\left(X\right)\right]^2$	$\approx 1000\sqrt{3.28}$
$E(X^2) = \begin{pmatrix} (1^2 \times 0.22) + (2^2 \times 0.26) + (3^2 \times 0.18) \\ +(4^2 \times 0.14) + (5^2 \times 0.08) + (6^2 \times 0.06) \\ +(7^2 \times 0.04) + (8^2 \times 0.02) \end{pmatrix}$	≈ 1811
	The standard deviation of the salary is $1811 per month.
$= 12.52$	

c) Nuclear family

i. What is the probability that both children are boys given that the first born child is a boy?

$$\Pr(BB) = \Pr(GG) = \frac{1}{4}, \Pr(BB \, or \, BG) = \frac{1}{2}$$

$$\Pr(BB / (BB \, or \, BG)) = \frac{\Pr(BB \cap (BB \, or \, BG))}{\Pr(BB \, or \, BG)}$$

$$= \frac{1/4}{2/4} = \frac{1}{2}$$

ii. We asked the mother if she had at least one son and she "Yes, I do.". With this extra information, what is the probability that both children are boys?

$$\Pr(BB) = \Pr(GG) = \frac{1}{4}, \Pr(at \, least \, one \, boy) = \frac{3}{4}$$

$$\Pr(BB / (at \, least \, one \, boy)) = \frac{\Pr(BB \cap (at \, least \, one \, boy))}{\Pr(at \, least \, one \, boy)}$$

$$= \frac{1/4}{3/4} = \frac{1}{3}$$

iii. If we randomly choose a child from this family and find out he's a boy. What is the probability that both children are boys?

$$\Pr(BB) = \Pr(GG) = \Pr(BG) = \Pr(GB) = \frac{1}{4} = 0.25$$

$$\Pr(Boy \, chosen / BB) = 1 \qquad\qquad \Pr(Boy \, chosen / GG) = 0$$

$$\Pr(Boy \, chosen / BG) = \frac{1}{2} = 0.5 \qquad\qquad \Pr(Boy \, chosen / GB) = \frac{1}{2} = 0.5$$

$$B_C = Boy \, chosen$$

$$\Pr(BB / B_C) = \frac{\Pr(BB \cap B_C)}{\Pr(B_C)}$$

$$= \frac{\Pr(B_C / BB)\Pr(BB)}{\Pr(B_C / BB)\Pr(BB) + \Pr(B_C / BG)\Pr(BG) + \Pr(B_C / GB)\Pr(GB) + \Pr(B_C / GG)\Pr(GG)}$$

$$= \frac{1 \times 0.25}{1 \times 0.25 + 0.5 \times 0.25 + 0.5 \times 0.25 + 0 \times 0.25}$$

$$= 0.5$$

Given a boy is chosen removes GG from the event space. The probability that a randomly chosen child from a family with BB is a boy is one, the probability for a family with BG is ½, or GB is ½, or GG is 0.

a) Binomial distribution, true or false?

i.   No. Exercise time is considered a continuous variable, which cannot be accounted for a binomial distribution. Another reason is that exercise time cannot be categorized into the probability of successes or failures.

ii.   Yes. A manufacturer of shirts picks a random sample of 20 shirts from the production of shirts each day for a detailed inspection. X is the number of shirts with a defect.

iii. Yes. X number of students eating five servings of fruits per day can be accounted as the probability of successes and those who do not eat these amount of fruits can be accounted as the probability of failures.

iv. X number of days when it is raining during Physical Eduction as the probability of failures and the probability of successes is the number of days where it doesn't rain during Physical Education classes. So in this case the binomial distribution is applied.

b) Faulty burglar alarm

i.

x	0	1	2	3	4	5
Pr(X=x)	0.81791	0.16523	0.01586	0.00096	0.00004	0

Tabulated using technology.

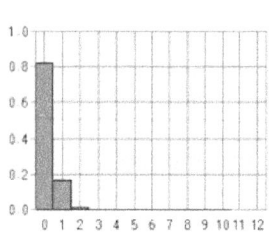

ii.

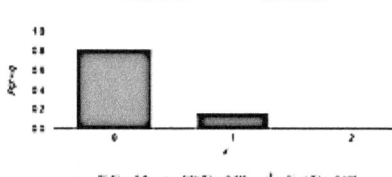

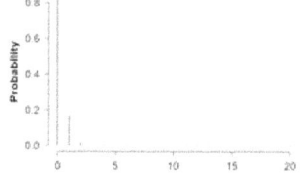

Binomial Distribution
n = 20 , p = 0.01

https://homepage.divms.uiowa.edu/~mbognar/applets/bin.html
https://keisan.casio.com/exec/system/1180573198
https://shiny.rit.albany.edu/stat/binomial/

iii. $$\Pr(X \geq 3) = 1 - \left[\Pr(X = 0) + \Pr(X = 1) + \Pr(X = 2)\right]$$
$$= 1 - \left[0.8179 + 0.1652 + 0.0159\right]$$

c) Multiple choice questions on exam.

i.
$$E(X) = np = 20 \times \frac{1}{5} = 4$$

ii.
$$\sigma = \sqrt{np(1-p)} = \sqrt{20 \times \frac{1}{5} \times \frac{4}{5}} = \frac{4}{\sqrt{5}}$$

iii.
$$X \sim Bi(n=20, p=\frac{1}{5})$$
$$Pr(X \geq 10) = 1 - Pr(X \leq 9)$$
$$\approx 1 - 0.9974$$
$$\approx 0.0026$$

iv.
$$Pr(X \geq 10 / X \geq 6) = \frac{Pr(X \geq 10 \cap X \geq 6)}{Pr(X \geq 6)}$$
$$= \frac{Pr(X \geq 10)}{Pr(X \geq 6)} \approx \frac{0.0026}{0.1958}$$
$$\approx 0.0133$$

d) 8 people board an empty bus.

i. Probability distribution having a birthday in May.

x	0	1	2	3	4	5	6	7	8
$Pr(X=x)$	0.4985	0.3626	0.1154	0.0210	0.0024	0.0002	0	0	0

Tabulated using technology.

ii. Probability that at least 3 people on the bus have a birthday in May.
$$Pr(X \geq 3) = 1 - \left(Pr(X=0) + Pr(X=1) + Pr(X=2)\right)$$
$$\approx 1 - \left(0.4985 + 0.3626 + 0.1154\right)$$
$$\approx 0.0235$$

iii. Probability that exactly 4 people on the bus have a birthday in May, given there is at least one person with a birthday in May.
$$Pr(X = 4 / X \geq 1) = \frac{Pr\left(X = 4 \cap X \geq 1\right)}{Pr(X \geq 1)}$$
$$= \frac{Pr\left(X = 4\right)}{Pr(X \geq 1)} = \frac{Pr\left(X = 4\right)}{1 - \left(Pr(X=0)\right)} \approx \frac{0.0024}{1 - \left(0.4985\right)} \approx 0.0048$$

iv. Least number of people that need to board the bus so that at least one person has a birthday in May is greater than 80% or 90% or 99%.

$Pr(X > 0) = 1 - Pr(X = 0)$	$Pr(X > 0) = 1 - Pr(X = 0)$	$Pr(X > 0) = 1 - Pr(X = 0)$
$1 - Pr(X = 0) > 0.8$	$1 - Pr(X = 0) > 0.9$	$1 - Pr(X = 0) > 0.99$
$1 - \binom{n}{0} p^0 (1-p)^n > 0.8$	$1 - \binom{n}{0} p^0 (1-p)^n > 0.9$	$1 - \binom{n}{0} p^0 (1-p)^n > 0.99$
$1 - \left(\frac{11}{12}\right)^n = 0.8$	$1 - \left(\frac{11}{12}\right)^n = 0.9$	$1 - \left(\frac{11}{12}\right)^n = 0.99$
$\left(\frac{11}{12}\right)^n = 0.2$	$\left(\frac{11}{12}\right)^n = 0.1$	$\left(\frac{11}{12}\right)^n = 0.01$
$n \approx \log_{\left(\frac{11}{12}\right)}(0.2)$	$n \approx \log_{\frac{11}{12}}(0.1)$	$n \approx \log_{\frac{11}{12}}(0.01)$
$n \approx 18.49$	$n \approx 26.47$	$n \approx 52.93$
19 people for 80% chance	27 people for 90% chance	53 people for 99% chance

e) *Itchy and Scratchery* lottery company claims that 20% of all tickets sold win a prize.

$p = 0.2, n = ?, x = ?$

$\Pr(X > 0) = 1 - \Pr(X = 0) = 0.5$

$1 - \binom{n}{0} p^0 (1-p)^n = 0.5$

$1 - (0.8)^n = 0.5$

$0.8^n = 0.5$

$n = \log_{0.8} 0.5 \approx 3.1063$

$n = 3$ is too low,

$n = 4$ meets the threshold

trials n	3	n=1,2,
probability of success p	0.2	0≦p≦1
[initial percentile	0	(success number)
increment	1	repetition :13]

Execute Clear Chart Store/Read Print

x	Binomial distribution
0	0.512
1	0.384
2	0.096
3	0.008
4	0
5	0
6	0
7	0
8	0
9	0
10	0
11	0
12	0

trials n	4	n=1,2,
probability of success p	0.2	0≦p≦1
[initial percentile	0	(success number)
increment	1	repetition :13]

Execute Clear Chart Store/Read Print

x	Binomial distribution
0	0.4096
1	0.4096
2	0.1536
3	0.0256
4	0.0016
5	0
6	0
7	0
8	0
9	0
10	0
11	0
12	0

Tabulated probability
https://keisan.casio.com/exec/system/1180573198

f) Match the graphs an distributions

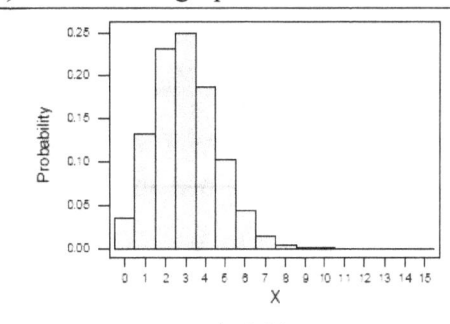

A (vi.)
$X = Bi(n = 15, p = 0.2)$

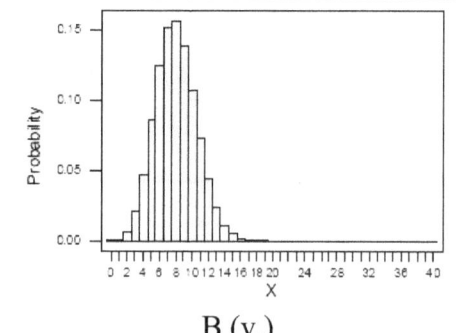

B (v.)
$X = Bi(n = 40, p = 0.2)$

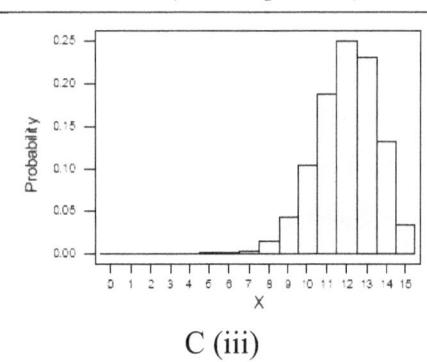

C (iii)
$X = Bi(n = 15, p = 0.8)$

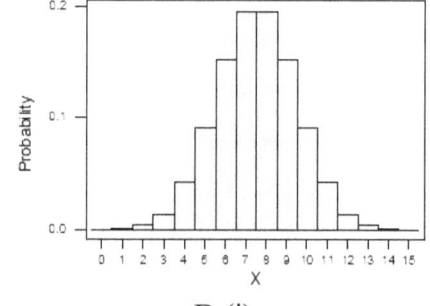

D (i)
$X = Bi(n = 15, p = 0.5)$

Continuous Probability

A random system exists if you do something the same way each time with a different outcome. The outcome of a random system can be a real number, random variables that can take any real number are called continuous random variables. Continuous random variables form the study of continuous probability which differs from discrete probability in the method of counting favourable events.

Real numbers are far more numerous than the natural numbers or discrete numbers and values. If you imagine the set of natural numbers as an infinite string of digits, growing positively larger and larger.

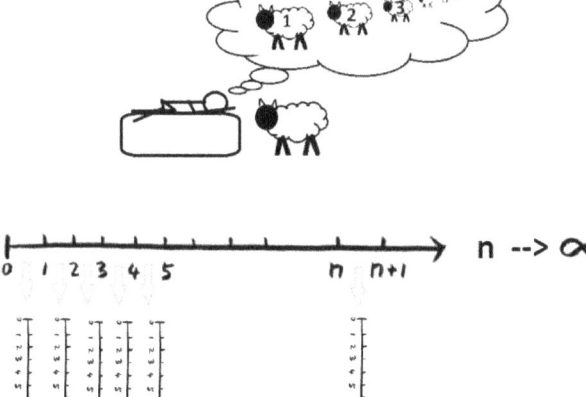

Then imagine the set of real numbers, there are an infinite number of real numbers between each consecutive natural number.

An "infinity of infinities" ($\infty \times \infty$) is a wondrous concept to contemplate.

Humans find ways of doing calculations with infinite real numbers that complete in finite time, using finite resources so as to be useful to humans, and to society by solving real world problems.

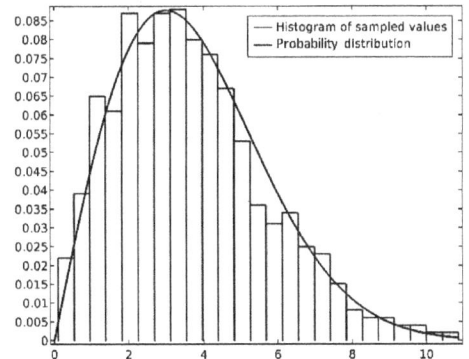

For calculations of probability and statistics, continuous and infinite random variables have their real values grouped into intervals defined by ranges of values, to allow them to be counted.

The counts or frequencies of real values within the intervals which are called groupings can be used for the creation of graphs, such as frequency histograms and frequency polygons.

Probability density functions of continuous random variables can be created from grouped values of the continuous random variable X.

Probability density function for continuous random variables.

The frequency histogram of the grouped values of the variable X can be transformed into a relative frequency histogram by dividing each bar's height divided by the total number of observations, creating a statistical model of the probability based on random sampling or whole population values.

The frequency polygon is derived from joining the tops of the bars of a frequency histogram or relative frequency histogram with a curve. The equation of the curve formed can be approximated by a function.

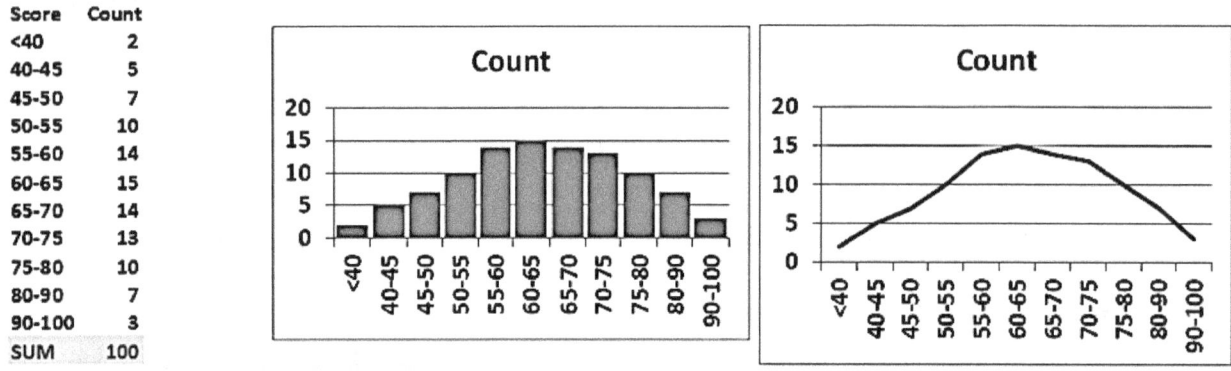

Score	Count
<40	2
40-45	5
45-50	7
50-55	10
55-60	14
60-65	15
65-70	14
70-75	13
75-80	10
80-90	7
90-100	3
SUM	100

This function is then scaled so that the total area under the frequency polygon is equal to 1. With the area under the function equal to 1, we have created a probability density function $f(x)$ where all the values of $f(x) \geq 0$ since they are derived from frequency counts for this continuous random variable. The probability density function derived using this method can then be used to calculate probabilities for that continuous random variable.

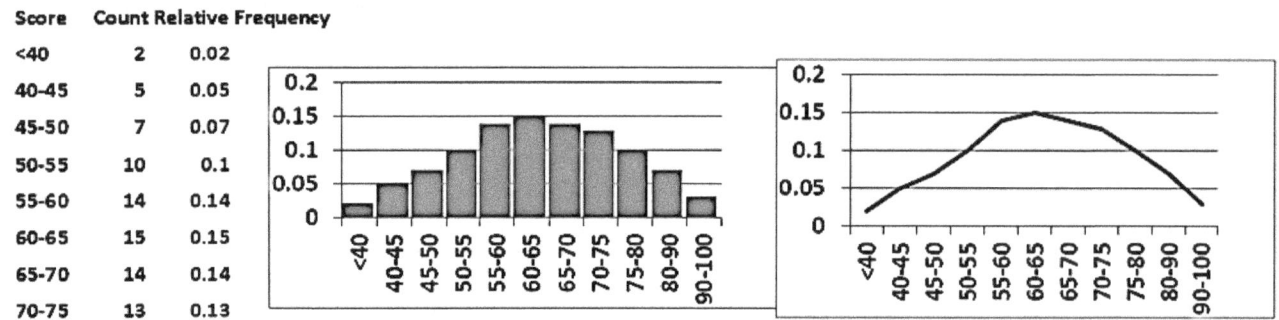

Score	Count	Relative Frequency
<40	2	0.02
40-45	5	0.05
45-50	7	0.07
50-55	10	0.1
55-60	14	0.14
60-65	15	0.15
65-70	14	0.14
70-75	13	0.13

The probability density function can take any shape with the restriction that the area under the function within the domain is equal to 1, and all the function values are greater than or equal to zero, since we cannot have negative probability values.

Probability density functions are used to find probabilities for continuous random variables. The word density is a reference to the number of individuals per unit of area in the creation of the function.

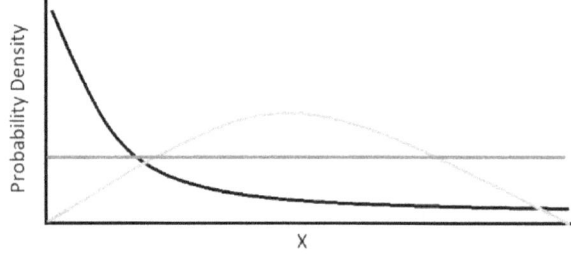

A certain brand of cocoa advertises that cans of cocoa weigh 250 grams. One randomly selected can of cocoa may weigh 255.1 grams, while another random selection may weigh 252.5 grams.

What is the probability that a randomly selected can of this cocoa weighs between 247.5 grams and 252.5 grams?

If X is the continuous random variable representing the weight of cans it can take any real number value.
How can we find the probability of $\Pr(247.5 \le X \le 252.5)$?

If we selected 100 cans of cocoa and weighed each one, we can create a frequency count of the weights by grouping the weights into intervals of say 4 grams.

Interval	Count
[245-249)g	23
[249-253)g	57
…………	……

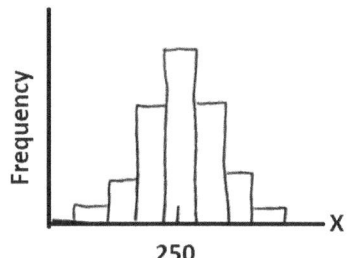

The resulting histogram should show that most of the sampled cans should be close to 250 grams, while some are heavier and some are lighter.

If the interval of the weight frequency count is made smaller say 2 grams for the 100 sampled cans then the bars of the frequency histogram become thinner, but the general shape of the distribution is similar to the general shape when the intervals were wider.

Interval	Count
[245-247)g	7
[247-249)g	16
[249-251)g	29
[251-253)g	28
…………	……

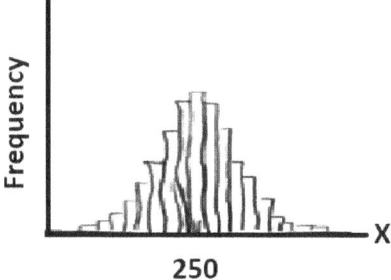

If the interval of the weight frequency count is made even smaller, say 1 gram for the 100 sampled cans, then the bars again become even thinner again.

If we join the top of the bars with a curve function $f(x)$ fitted to the tops of the very thin bars of the histogram, and then scaled the function $f(x)$ so that the area under the function was equal to 1, then that function $f(x)$ would represent the probability density function for the continuous variable X, the continuous random variable representing the weight of cans.

Interval	Count
[245-246)g	3
[246-247)g	4
[247-248)g	6
[248-249)g	10
[249-250)g	18
[250-251)g	21
[251-252)g	17
[252-253)g	11
…………	……

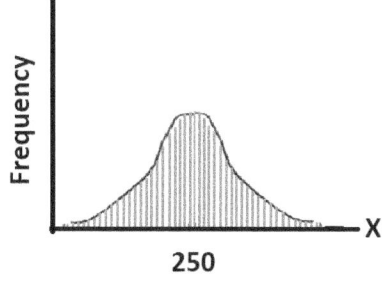

We find the probability of $\Pr(247.5 \le X \le 252.5)$ by

integration $\displaystyle\int_{247.5}^{252.5} f(x)dx$ of the

probability density function.

Comparing random variables with discrete and continuous distributions

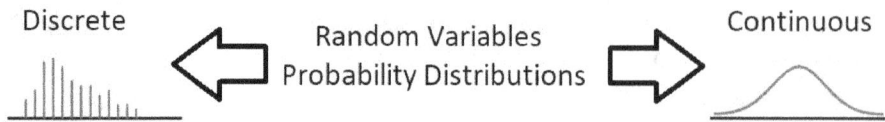

Probabilities for the discrete variables are found by finding the relative frequencies for a given value $\Pr(X=x)$ or by adding relative frequencies for a range of discrete values $\Pr\left(x_1 \le X \le x_k\right) = \sum_{i=1}^{k} c\left(x_i\right)$. Probabilities for the continuous random variables require finding the area under the probability density function in a given interval $X \in [a,b]$ $\Pr\left(a \le X \le b\right) = \int_a^b f(x)dx$. It should be noted that the probability cannot be determined for a unique value of a continuous random variable X although $f(x)$ can be evaluated.

Things to do with a probability density function $f(x)$ for a continuous random variable X, where the domain of the random variable is (min, max).

Show that all $f(x) \ge 1$ and the area under the function is equal to 1. $\int_{\text{min}}^{\text{max}} f(x)dx = 1$ 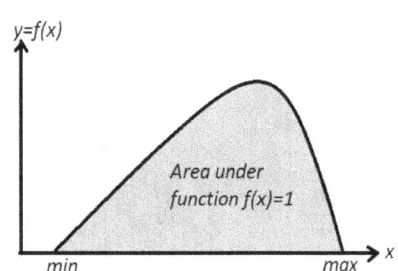	Find the probability that X is between two real values a and b which are in the domain for X. $\Pr\left(a \le X \le b\right) = \int_a^b f(x)dx$ 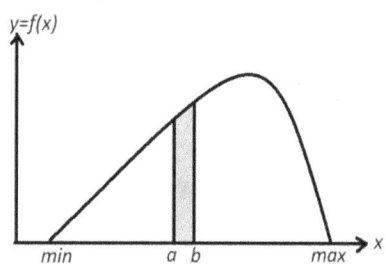

Find a central tendency measure.

Find the expected value $E(X)$, known as the mean μ of X. $E(X) = \mu = \int_{\text{min}}^{\text{max}} xf(x)dx$	Find the median value of X $\int_{\text{min}}^{\text{median}} f(x)dx = 0.5$ 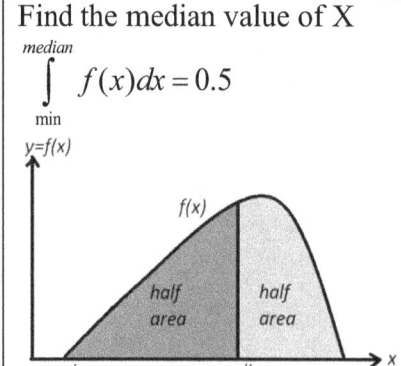	Find the mode value of X $f'(x) = 0$ solve for x to find the value with the highest probability. 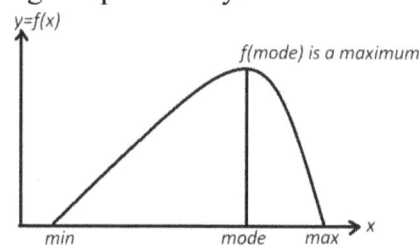

Find the spread of the distribution.

Find the variance of X, $Var(X)$, represented also by σ^2. $Var(X) = \sigma^2 = \int_{\text{min}}^{\text{max}} x^2 f(x)dx - \mu^2$	The standard deviation of X $sD(X)$, represented by σ where $\sigma = \sqrt{Var(X)}$

Example 8.1: Central tendency and spread for continuous probability distribution functions.

In the field of material science the Weibull probability density function of the time to failure

of materials is: $f(x) = \begin{cases} k\left(x^{k-1}\right)e^{-x^k}, x \geq 0 \\ 0, \quad x < 0 \end{cases}$, where

x is a measure of time.

Depending on the value of k a variety of probability density function curves can be generated.

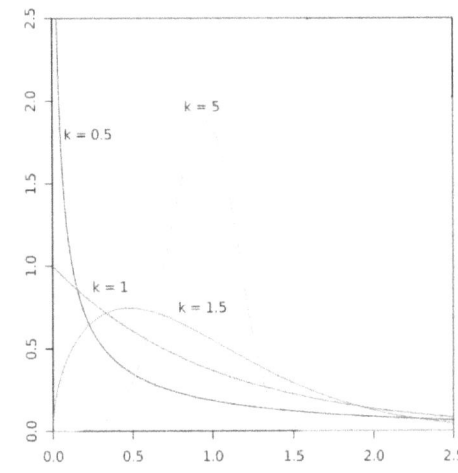

Source: Wikipedia, https://en.wikipedia.org/wiki/Weibull_distribution

For k=0.5, the Weibull probability density function

$$f(x) = \begin{cases} 0.5\left(x^{-0.5}\right)e^{-\sqrt{x}}, x \geq 0 \\ 0, \quad x < 0 \end{cases}$$

Area $\int_0^\infty f(x)dx = 1$, for $x \geq 0, f(x) \geq 0$

Central measures: mean, median, mode

Mean $E(X) = \mu = \int_0^\infty xf(x)dx = 2$

$\int_0^{median} f(x)dx \approx 0.4805$, median ≈ 1.337

Mode $f'(x) = 0$ cannot be solved for this function.

Measures of spread variance and standard deviation.

$$Var(X) = \sigma^2 = \int_0^\infty x^2 f(x)dx - \mu^2$$
$$= 24 - 2^2$$
$$= 20$$
$$\sigma = \sqrt{Var(X)} \approx 4.8990$$

The probability density function $f(r)$ for a neutral Hydrogen atom electrons in the 1s orbital with a radial distance of r units (1r $\approx 0.53 \times 10^{-8} metres$) from the nucleus is:

$$f(r) = 4r^2 e^{-2r}$$

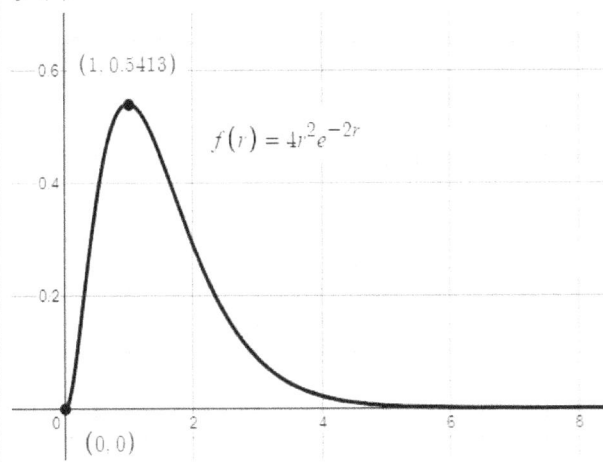

Area $\int_0^\infty f(r)dr = 1$, for $r \geq 0, f(r) \geq 0$

Central measures: mean, median, mode

Mean $E(R) = \mu = \int_0^\infty rf(r)dx \approx 1.5$

$\int_0^{median} f(r)dr = 0.5$, median ≈ 1.337

Mode found by solving the derivative function $f'(r) = 0, r \approx 0.54$

Measures of spread variance and standard deviation.

$$Var(R) = \sigma^2 = \int_0^\infty r^2 f(r)dr - \mu^2$$
$$= 2.99995 - \left(1.5\right)^2$$
$$\approx 0.7499$$
$$\sigma = \sqrt{Var(R)} \approx 0.8660$$

Application 8.2: Probability density functions and electron orbits.

Orbitals are mathematically derived regions of space around the nucleus of an atom with different probabilities of containing an electron. Orbitals can have various radii distances from the nucleus of an atom.

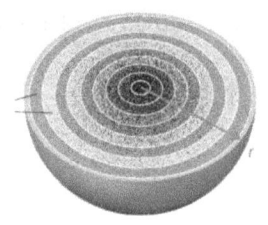

The radial probability, which is the probability of finding an electron at an "s Orbital" radius r from the nucleus, can be found by dividing the orbital space around the nucleus into very thin concentric shells, like the layers of an onion, and calculating the probability of finding an electron on each spherical shell.

Images:
https://chem.libretexts.org/Bookshelves/General_C hemistry/

---

s Orbitals are spherically symmetrical shells around the nucleus of atoms. The electron densities for the 1s, 2s, and 3s orbitals of the Hydrogen atom are shown in any cross-sectional plane that contains the nucleus.

Notice that there are some circular regions, where the electron density is zero.

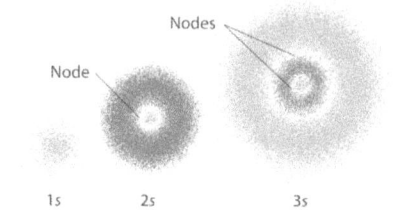

Image credit: UCDavis Chemwiki, CC BY-NC-SA 3.0 US

The corresponding probability density functions for each of the s Orbital shells, 1s, 2s, and 3s of the Hydrogen atom show the chance of finding electrons at that radial distance from the nucleus.

The uncharged Hydrogen atom is the simplest known atom with one proton and one electron bound to the nucleus by the Coulomb force, and it has the simplest probability density function for orbiting electrons.

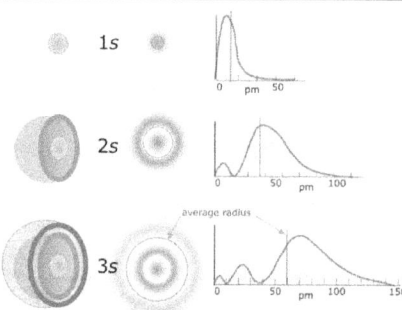

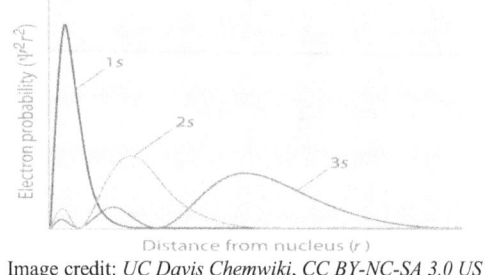

Image credit: UC Davis Chemwiki, CC BY-NC-SA 3.0 US

The probability density function for the 1s orbital of a neutral uncharged Hydrogen atom with radial distance r in some very tiny units ($1r \approx 0.53 \times 10^{-8}$ metres) is:

$$\psi_{1s}(r) = \frac{1}{\sqrt{\pi}} e^{-r} \text{ where}$$

$$\int_0^\infty [\psi_{1s}(r)]^2 \, 4\pi r^2 \, dr = 1$$

Source http://www.umich.edu/~chem461/QMChap7.pdf

Simplified the probability density function $f(r)$ for a neutral Hydrogen atom 1s orbital with a radial distance of r units ($1r \approx 0.53 \times 10^{-8}$ metres) is:

$f(r) = 4r^2 e^{-2r}$, refer to example 8.1

$f(r) = 4r^2 e^{-2r}$

The probability density functions for the position of electrons and other sub-atomic particles form the basis for the study of Quantum Physics and Quantum Computers.

Normal and Standard Normal Probability Density Functions (PDF).

The Normal distribution is the name given to a bell shaped symmetrical PDF described by the function

$f(x) = \dfrac{1}{\sigma\sqrt{2\pi}} e^{\frac{1}{2}\left(\frac{x-\mu}{\sigma}\right)^2}$, that has a mean of μ and a standard deviation of σ .

Many continuous random variables have a PDF defined as a Normal Distribution.

Properties of a Normal Distribution

- Symmetrical , bell shaped curve function, where the mean=median=mode, that is the highest point on the function.

- $X \sim N(\mu, \sigma)$ indicates the variable X has a normal distribution with a mean of μ and a standard deviation of σ . The continuous random variable X has an infinite theoretical range from $-\infty$ to ∞, the area under the PDF equals 1, $\displaystyle\int_{-\infty}^{\infty} f(x)dx = 1$

- $Z \sim N(\mu = 0, \sigma = 1)$. indicates the variable Z has a **Standard Normal Distribution** with a mean of $\mu = 0$ and a standard deviation of $\sigma = 1$. Any normal distribution can be transformed to a standard normal distribution by the application of translation and dilation to the data values of the function, $Z = \dfrac{X - \mu}{\sigma} \Leftrightarrow X = Z\sigma + \mu$. The Standard Probability Density function is $f(x) = \dfrac{1}{\sqrt{2\pi}} e^{\frac{1}{2}(x)^2}$.

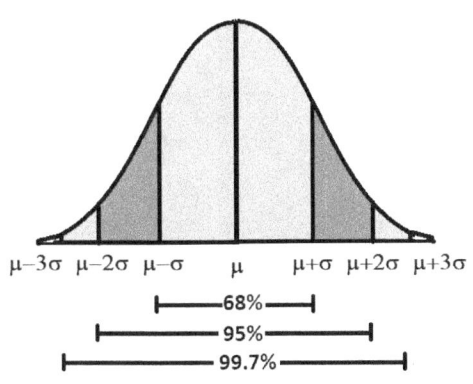

$\mu{-}3\sigma \;\; \mu{-}2\sigma \;\; \mu{-}\sigma \;\;\;\; \mu \;\;\;\; \mu{+}\sigma \;\; \mu{+}2\sigma \;\; \mu{+}3\sigma$

$\vdash\!\!-\!\!-\!\!-68\%\!\!-\!\!-\!\!-\dashv$
$\vdash\!\!-\!\!-\!\!-\!\!-95\%\!\!-\!\!-\!\!-\!\!-\dashv$
$\vdash\!\!-\!\!-\!\!-\!\!-99.7\%\!\!-\!\!-\!\!-\!\!-\dashv$

- Approximately 68% of the data values are within 1 standard deviations of the mean, $\Pr(\mu - 1\sigma \le X \le \mu + 1\sigma) \approx 0.68$

- Approximately 95% of the data values are within 2 standard deviations of the mean, $\Pr(\mu - 2\sigma \le X \le \mu + 2\sigma) \approx 0.95$

- Approximately 99.7% of the data values are within 3 standard deviations of the mean, $\Pr(\mu - 3\sigma \le X \le \mu + 3\sigma) \approx 0.997$

Example 8.2: Finding probabilities for Normally distributed continuous random variables.

The heights of women in Australia have a normal distribution. Let H be the continuous random variable representing the height: $H \sim N(\mu = 162.5cm, \sigma = 7.5cm)$ To find the probability that a randomly selected Australian women is taller than 175cm.	This probability could be found by integration $$\Pr(H \ge 175) = \int_{175}^{\infty} \dfrac{1}{7.5\sqrt{2\pi}} e^{\frac{1}{2}\left(\frac{x-162.5}{7.5}\right)^2}$$ However many technologies support finding the probabilities of Normally distributed variables using built in functions. $\Pr(H \ge 175) \approx 0.0478$
The variable H can be converted to a Standard Normal distribution variable Z by applying the transformation $Z = \dfrac{H - \mu}{\sigma}$ where $Z \sim N(\mu = 0, \sigma = 1)$.	$$\Pr(H \ge 175) = \Pr\left(Z \ge \dfrac{175 - 162.5}{7.5}\right)$$ $$= \Pr(Z \ge 1.6667)$$ $$\approx 0.0478$$

Symmetry of the Normal Distribution Probability Density Function (PDF)

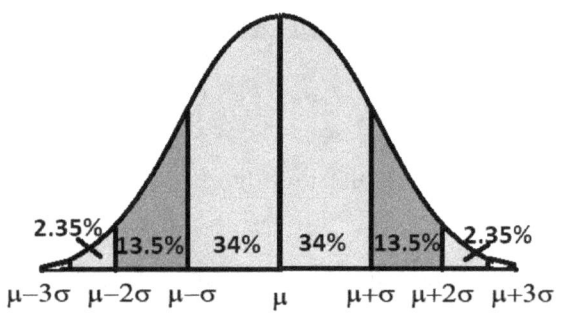

$$X \sim N(\mu, \sigma).$$

$$Z \sim N(\mu = 0, \sigma = 1). \text{ where } Z = \frac{X - \mu}{\sigma}$$

$$\Pr(X < C) = 1 - \Pr(X > C) \text{ by symmetry}$$

$$\Pr(Z < C) = 1 - \Pr(Z > C) \text{ by symmetry}$$

$$\Pr(X < \mu) = \Pr(X > \mu) = 0.5$$

$$\Pr(X < (\mu - \sigma)) = 1 - \Pr(X > (\mu - \sigma))$$

$$\approx 1 - (0.34 + 0.5)$$

$$\Pr(X > (\mu - \sigma)) = 1 - \Pr(X < (\mu - \sigma))$$

$$\approx 0.84$$

$$\Pr(Z < 0) = \Pr(Z > 0) = 0.5$$

$$\Pr(Z < -1) = 1 - \Pr(Z > -1)$$

$$\approx 1 - (0.34 + 0.5)$$

$$\Pr(Z > -1) = 1 - \Pr(Z < -1)$$

$$\approx 0.84$$

Example 8.3: Symmetry of Normally distributed continuous random variables.

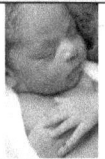	The distribution of newborns' weight is a normal distribution.	The average weight of a newborn child in Australia is 3.5 kg, with a standard deviation of 0.76 kg.

The chance that a healthy newborn Australian baby weighs 2.6 kg or less is 11.8%.

This can be calculated by finding the $\Pr(W \leq 2.6)$ where W is the continuous random variable representing the weights of newborns in Australia.

$$W \sim N(\mu = 3.5kg, \sigma = 0.76kg)$$

$$\Pr(W \leq 2.6kg) = 0.0118$$

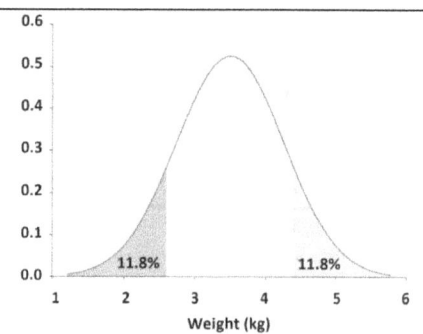

The variable W can be converted to a Standard Normal distribution PDF variable Z by applying the transformation $Z = \frac{W - \mu}{\sigma}$ where

$$Z \sim N(\mu = 0, \sigma = 1).$$

$$\Pr(W \leq 2.6) = \Pr(Z \leq \frac{2.6 - 3.5}{0.76})$$

$$= \Pr(Z \leq -1.1842)$$

$$\approx 0.0118$$

By the properties of symmetry

$$\Pr(Z > 1.1842) \approx 0.0118$$

Since $Z = \frac{W - \mu}{\sigma} \Rightarrow Z\sigma + \mu = W$

The variable Z can be transformed to W
$W = Z\sigma + \mu$

$$= (1.1842)(0.76) + 3.5$$

$$= 4.4000$$

$$\Pr(W > 4.4kg) \approx 0.0118$$

The shaded tails of the probability distribution are shown in the diagram above. Since the function is symmetrical, the areas shaded that represent the probability are equal..

The Normal Distribution in Real Life.

Many continuous random variables, especially those variables from nature that are derived from multiple sampling of real populations have a Normal Distribution.

"The mean gestational age at the onset of labour for women native to the area of study was 272 days with a standard deviation of 9 days"

A living histogram from the Connecticut State Agricultural College (1914).

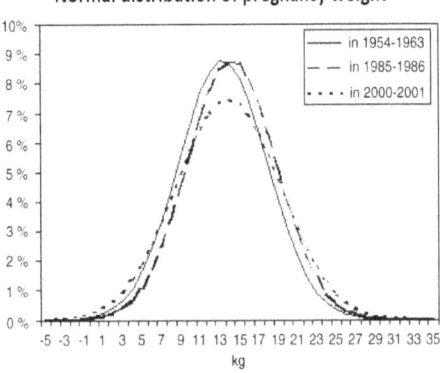

Normal distribution of pregnancy weight

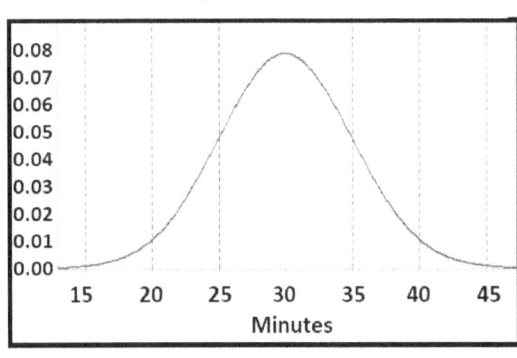

Distribution of Pizza Delivery Times, Normal Distribution
Mean=30 minutes, Standard Deviation=5 minutes

"the length of pregnancy in cats is approximately normal with a mean of 66 days and a standard deviation of 1 day."

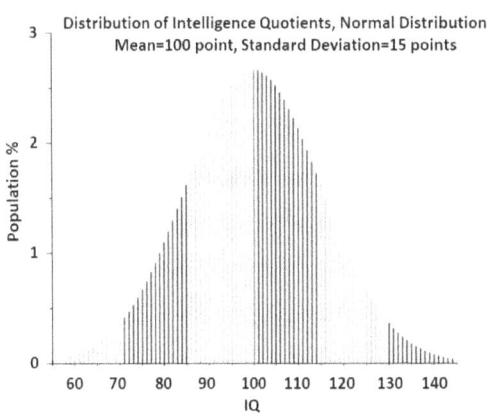

Distribution of Intelligence Quotients, Normal Distribution
Mean=100 point, Standard Deviation=15 points

What is so special about the Normal Distribution?

A theorem called the Central Limit Theorem proves that when we take the average of many (the convention is at least 30) randomly sampled data variables with replacement, the averaged values converge to the actual mean of the data variable, with a symmetric standard deviation. The distribution of the averaged data variables will have a normal distribution, bulging at the mean and decreasing exponentially at either side of the mean, in a symmetrical pattern, like a bell shaped curve.

The Central Limit Theorem holds true for the all averaged random sampling such as the sample mean $\overline{x}$ and the sample proportion of a particular attribute $\hat{p}$ (a special case of averaging the count of those with the certain attribute within the sample).

The Normal Distribution is supported by many technologies, apps and software.

Example 8.4: The Normal Distribution is supported by many technologies, apps and software.			
Finding probabilities for continuous random variables with Normal Distributions using technology. $X = N(\mu = 12, \sigma = 4)$	**Tinspire CAS**	**Casio Classpad**	**Mathematica**
$Pr(5 < X < 10)$ $Pr(X < 15)$ $Pr(X > 15)$	**Normal Cdf** Lower Bound: 5 Upper Bound: 10 μ: 12 σ: 4 normCdf(5,10,12,4) =0.2685 **Normal Cdf** Lower Bound: −∞ Upper Bound: 15 μ: 12 σ: 4 normCdf(-∞,15,12,4) =0.7734 normCdf(15,∞,12,4) =0.2266	**normCDf** Lower 5 Upper 10 σ 4 μ 12 lower boundary normCDf(5,10,4,12) =0.2685 **normCDf** Lower −∞ Upper 15 σ 4 μ 12 lower boundary normCDf(-∞,15,4,12) =0.7734 normCDf(15,∞,4,12) =0.2266	=CDF[NormalDistribution[12,4],10]-CDF[NormalDistribution[12,4],5] =0.2685 =CDF[NormalDistribution[12,4],15] 0.7734 =1-CDF[NormalDistribution[12,4],15] =0.2266
$Pr(X < C) = 0.37$ *find* C 0.37 C 12	**Inverse Normal** Area 0.37 μ 12 σ 4 invNorm(0.37,12,4) =10.6726	**invNormCDf** Tail setting Left prob 0.37 σ 4 μ 12 invNormCDf("L",0.37,4,12) =10.6726	InverseCDF[NormalDistribution[12,4],0.37] 10.6726
Use symmetry to find the upper tail probability $Pr(X > C) = 0.63$ $Pr(X > C) = 1 - Pr(X < C)$. $\qquad = 1 - 0.37$ $Pr(X < C) = 0.37$ *find* C	invNorm(0.37,12,4) =10.6726 invNormCDf("L",0.37,4,12) =10.6726 $C \approx 10.67$		0.37 0.63 C 12

Means, medians, quartiles and percentiles for Normal Distributions and their definitions:

Normal distributions have a theoretical domain of $(-\infty, \infty)$ for a continuous random variable X

- Q_1 is the lowest quartile, also called the 25th percentile where $\Pr(X < Q_1) = 0.25$

- $Q_2 = \mu = median$ is the middle quartile, also called the 50th percentile and is equal to the mean and median for a normal distribution where $\Pr(X < \mu) = \Pr(X < median) = \Pr(X < Q_2) = 0.5$

- Q_3 is the highest quartile, also called the 75th percentile where $\Pr(X < Q_3) = 0.75$

- The interquartile range is between the quartiles Q_1 and Q_3 where $\Pr(Q_1 < X < Q_3) = 0.5$

These quartiles can be found by integration for any PDF $f(x)$

- 25th percentile $\Pr(X < Q_1) = \int_{min}^{Q_1} f(x)dx = 0.25$

- 50th percentile $\Pr(X < median) = \int_{min}^{median} f(x)dx = 0.5$

- 75th percentile $\Pr(X < Q_3) = \int_{min}^{Q_3} f(x)dx = 0.75$

Example 8.5: Finding percentiles and quartiles using technologies, apps and software.

$X = N(\mu = 12, \sigma = 4)$

Tinspire CAS	Casio Classpad	Mathematica
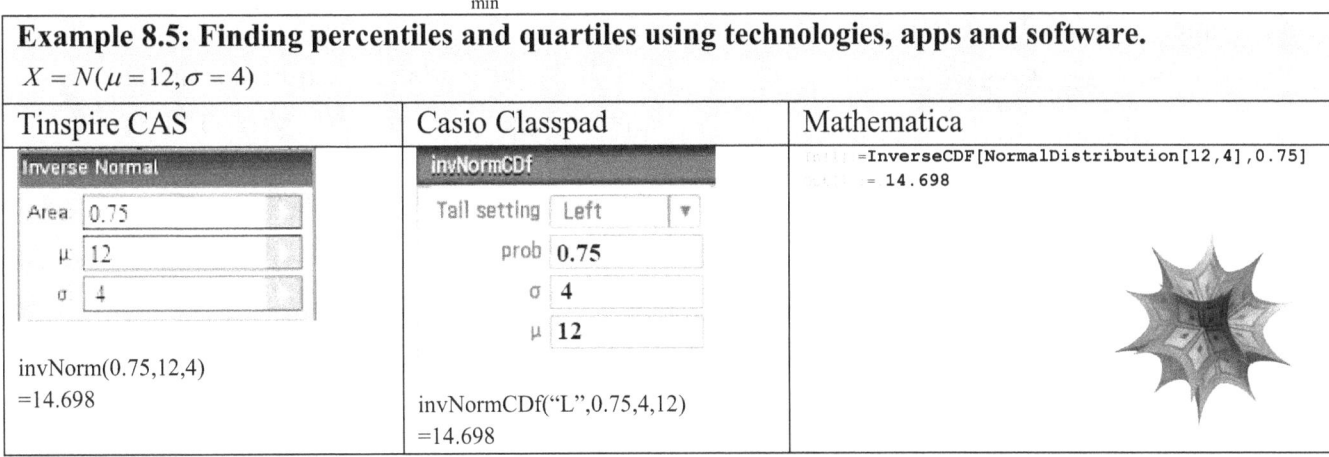 Inverse Normal — Area 0.75, μ 12, σ 4 invNorm(0.75,12,4) =14.698	invNormCDf — Tail setting Left, prob 0.75, σ 4, μ 12 invNormCDf("L",0.75,4,12) =14.698	=InverseCDF[NormalDistribution[12,4],0.75] = 14.698

Standard Normal Distribution, what is it good for?

$Z \sim N(\mu = 0, \sigma = 1)$ indicates the variable Z has a **Standard Normal Distribution** with a mean of $\mu = 0$ and a standard deviation of $\sigma = 1$. Any normal distribution can be transformed to a standard normal distribution by the application of translation and dilation to the data values of the function,

$$Z = \frac{X - \mu}{\sigma} \Leftrightarrow X = Z\sigma + \mu.$$

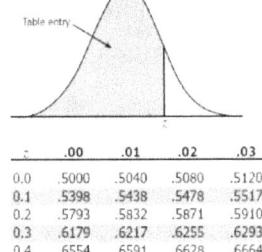

Before technology and computers people looked up Standard Normal Distribution probability values in a book of tables. To make the tables useful for any Normal distribution $X \sim N(\mu, \sigma)$, which could have any mean or standard deviation, the values were made standard by applying the transformation. $Z = \frac{X - \mu}{\sigma} \Leftrightarrow X = Z\sigma + \mu$

z	.00	.01	.02	.03
0.0	.5000	.5040	.5080	.5120
0.1	.5398	.5438	.5478	.5517
0.2	.5793	.5832	.5871	.5910
0.3	.6179	.6217	.6255	.6293
0.4	.6554	.6591	.6628	.6664
0.5	.6915	.6950	.6985	.7019

Symmetry Properties for Standard Normal Distributions.

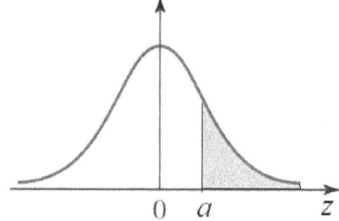
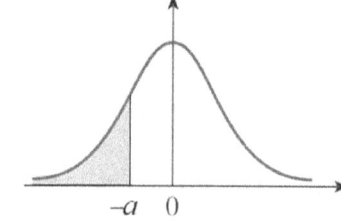

$$\Pr(Z > a) = \Pr(Z < -a)$$
$$\Pr(-a < Z < a) = \Pr(Z < a) - \Pr(Z < -a)$$
$$\Pr(Z < a) = 1 - \Pr(Z > a)$$
$$\Pr(Z > -a) = 1 - \Pr(Z < -a)$$

Example 8.6: Finding the unknown variables for a Normal Distribution using Standard Normal Distribution and Inverse standard normal distribution technology.

The time taken to complete a week's shopping in a supermarket is normally distributed with a mean of 30 minutes. The proportion of shoppers taking less than 17 minutes is 0.2. What proportion of shoppers take longer than an hour?

Let X be the randomly distributed variable representing the shopping time,

$$X \sim N(\mu = 30, \sigma = ?)$$

- Using $Z \sim N(\mu = 0, \sigma = 1)$, find

  $\Pr(Z < a) = 0.2$, using the Inverse Normal Distribution functionality we can find that $\Pr(Z < -0.8416) = 0.2$, so $a = -0.8416$, and $\Pr(X < 17) = 0.2$.

- Since $Z = \dfrac{X - \mu}{\sigma} \Rightarrow -0.8416 = \dfrac{17 - 30}{\sigma}$,

  solving for the standard deviation gives $\sigma = 15.45$ minutes.

- The full details of the normal distribution are now known, $X \sim N(\mu = 30, \sigma = 15.45)$ and can be used to find the required probability of $\Pr(X > 60) = 0.0261$ for the proportion of shoppers taking longer than one hour.

The average height of Peruvian women is 150cm. If the probability of being taller than 170cm is 6% find the standard deviation for this population?

Let H be the randomly distributed variable representing the height, $H \sim N(\mu = 150, \sigma = ?)$

We are given the information that $\Pr(H > 170) = 0.04$ which is equivalent to $\Pr(H < 170) = 0.94$ using symmetry.

- Using $Z \sim N(\mu = 0, \sigma = 1)$, find

  $\Pr(Z < b) = 0.94$, using the Inverse Normal Distribution $\Pr(Z < 1.5548) = 0.94$, so $b = 1.5548$, and $\Pr(H < 170) = 0.94$.

- Since $Z = \dfrac{H - \mu}{\sigma} \Rightarrow 1.5548 = \dfrac{170 - 150}{\sigma}$,

  solving for the standard deviation gives $\sigma = 12.8634$ centimetres.

- The full details of the normal distribution are now known, $H \sim N(\mu = 150, \sigma = 12.8634)$.

Conditional Probability Pr(A/B) applies to Continuous Random Variables also.

For dependent events, **Conditional Probability** is the probability that event A occurs given event B has already occurred. For continuous random variables discarding events outside of B, involves thinking about the restrictions imposed on the numeric values in an interval of X by the given condition B using a number line.

$$\Pr(A/B) = \frac{\Pr(A \cap B)}{\Pr(B)}$$

The only way that event A can happen is when the event belongs to the set A∩B.

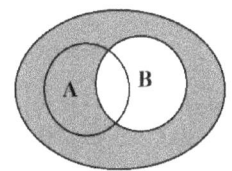

Discard events outside B.

For a continuous random variable X, this is an intersection of number regions A∩B, to determine the values of X required to calculate the probability.

Special cases of conditional probability, some thinking is required before applying the rule.

Intervals A and B are disjoint on a number line for random variable X, $A \cap B$ is empty.	Interval B is within interval A on a number line, $A \cap B = B$.	Interval A is within interval B on a number line, $A \cap B = A$.
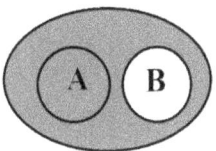	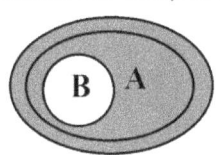	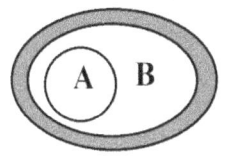
$\Pr(A/B) = \dfrac{\Pr(A \cap B)}{\Pr(B)}$ $= 0$ $\Pr(X \le 5 / X \ge 10)$ $= \dfrac{\Pr(X \le 5 \cap X \ge 10)}{\Pr(X \ge 10)}$ $= 0$	$\Pr(A/B) = \dfrac{\Pr(A \cap B)}{\Pr(B)}$ $= 1$ $\Pr(X \ge 0 / X \ge 1)$ $= \dfrac{\Pr(X \ge 0 \cap X \ge 1)}{\Pr(X \ge 1)}$ $= \dfrac{\Pr(X \ge 1)}{\Pr(X \ge 1)}$ $= 1$	$\Pr(A/B) = \dfrac{\Pr(A \cap B)}{\Pr(B)}$ $= \dfrac{\Pr(A)}{\Pr(B)}$ $\Pr(X \ge 1 / X \ge 0)$ $= \dfrac{\Pr(X \ge 1 \cap X \ge 0)}{\Pr(X \ge 0)}$ $= \dfrac{\Pr(X \ge 1)}{\Pr(X \ge 0)}$

Example 8.7: Conditional Probability for a continuous random variable.

| 

The length of farmed trout fish is normally distributed with a mean of 40cm and a standard deviation of 4cm. | Let T be the continuous variable of fish lengths: $T \sim N(\mu = 40, \sigma = 4)$.

Fish that are under 35 cm need to be thrown back into the water. | Given the length restrictions what is the probability that fish taken home are greater than 45 cm.

$\Pr(T > 45 / T > 35) = \dfrac{\Pr((T > 45) \cap (T > 35))}{\Pr(T > 35)}$

$= \dfrac{\Pr(T > 45)}{\Pr(T > 35)} \quad \dfrac{0.1057}{0.8944} \approx 0.1181$ |

Linear Transformation of Probability Density Functions.

For the linear transformation of a continuous random variable X where a is a constant dilation factor and b is a constant translation value, the mean and variance for the transformed variable $(aX+b)$ are :

$$E(aX+b) = aE(X)+b$$
$$Var(aX+b) = a^2 Var(X)$$

The probability density function $f(x)$ for X is transformed to $(aX+b)$ while the area under the transformed PDF is scaled in the height of the new PDF function to ensure that the area under the new transformed PDF is also equal to 1.

A transformation mapping of $(x,y) \rightarrow (ax+b, \frac{y}{a})$, which is equivalent to $\frac{1}{a} f\left(\frac{x-b}{a}\right)$.

Example 8.8: Linear transformation of a continuous random variable $X \rightarrow 2X+3$

Let $f(x) = \begin{cases} 5e^{-x}, & x \in [0, 0.2231) \\ 0, & elsewhere \end{cases}$ be a given PDF

If $y = f(x)$ the transformation is applied, where the y value is also scaled to keep the area under the new PDF equal to 1. $(x,y) \rightarrow (2x+3, \frac{y}{2})$

$x_{new} = 2x+3 \Rightarrow x = \dfrac{x_{new}-3}{2}$

$y_{new} = \dfrac{y}{2} \Rightarrow y = 2y_{new}$

We can find the transformed PDF function

$g(x) = \dfrac{1}{2} f\left(\dfrac{1}{2}(x-3)\right)$

$g(x) = \begin{cases} 2.5e^{-\left(\frac{x-3}{2}\right)}, & x \in [3, 3.4462) \\ 0, & elsewhere \end{cases}$

Define $f(x) = 5 \cdot e^{-x}$ *Done*

Define $g(x) = \dfrac{1}{2} \cdot f\left(\dfrac{1}{2} \cdot (x-3)\right)$ *Done*

$g(x)$ $\dfrac{5 \cdot e^{\frac{3}{2} - \frac{x}{2}}}{2}$

$\displaystyle\int_0^{0.2231} f(x)\, dx$ 0.999826

$\displaystyle\int_3^{3.4462} g(x)\, dx$ 0.999826

The expected value $E(2X+3) = 2E(X)+3$

The variance $Var(2X+3) = 2^2 Var(X)$

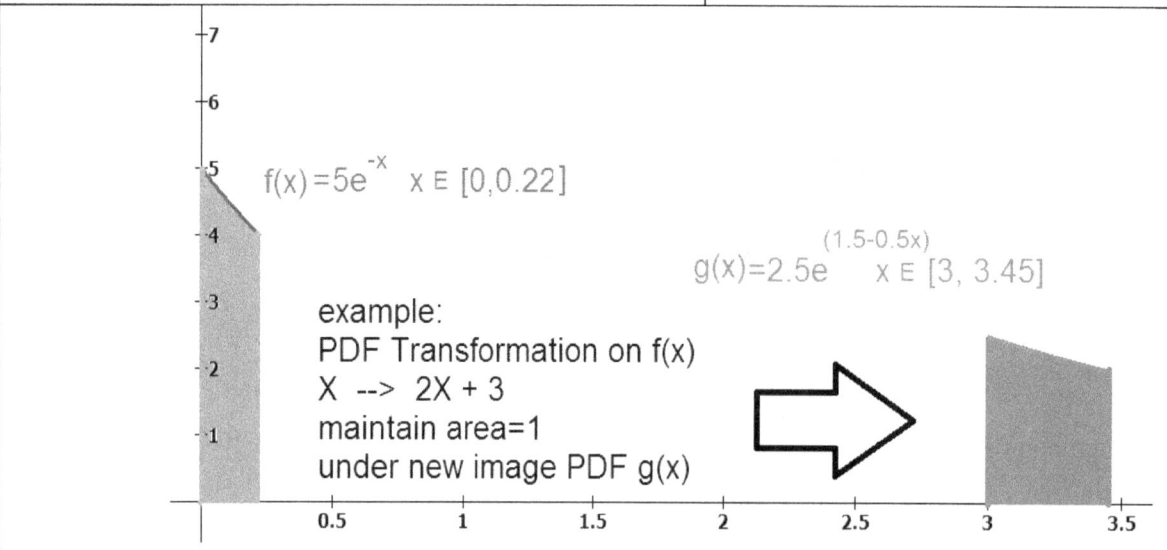

example:
PDF Transformation on f(x)
X --> 2X + 3
maintain area=1
under new image PDF g(x)

$f(x) = 5e^{-x}$ $x \in [0, 0.22]$

$g(x) = 2.5e^{(1.5-0.5x)}$ $x \in [3, 3.45]$

Combining Discrete and Continuous Probability

For a discrete random variable X with a **Binomial distribution** $X \sim Bi(n, p)$ as the number of trials n increases, the probability distribution of the random variable X becomes approximately symmetric, and the probabilities for X can be estimated using a **Normal Distribution** if $np(1-p) \geq 10$ by convention.

If $np(1-p) \geq 10$ then $X \sim Bi(n, p) \Rightarrow X \sim N\left(\mu = np, \sigma = \sqrt{np(1-p)}\right)$.

Example 8.9: Combination of Continuous and Discrete methods for finding probability.

The length of farmed trout fish is normally distributed with a mean of 40cm and a standard deviation of 4cm. Find the percentage of fish with length *>45 cm* to 4 decimal places.

Let L be the fish lengths $L \sim N(\mu = 40, \sigma = 4)$, we find $\Pr(L > 45) = 0.1057$ using technology,

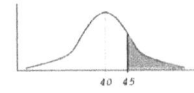

$\text{normCdf}(45, \infty, 40, 4) \qquad 0.10565$

which is 10.5650% of fish > 45 cm.

Let F be the event of catching a fish >45cm. The probability of binomial success $p = \Pr(L > 45 \mid L = N(40, 4)) \approx 0.1057$

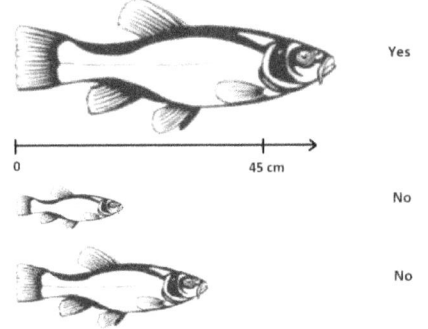

If 20 fish are caught the discrete random variable F has a binomial distribution where $n = 20, p = 0.1057$, $F \sim Bi(20, 0.1057)$

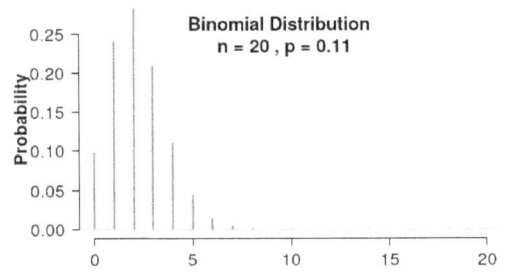

For $F \sim Bi(20, 0.1057)$, the probability that exactly 2 fish are longer than 45 cm to 4 decimal places.

$\Pr(F = 2) = \binom{20}{2} p^2 (1-p)^{18}, \Pr(F = 2) \approx 0.2842$

$\text{binomPdf}(20, 0.10565, 2) \qquad 0.284202$

For $F \sim Bi(20, 0.1057)$, the probability that more than 5 fish are longer than 45 cm to 4 decimal places. $\Pr(F \geq 5) \approx 0.0529$ $\text{binomCdf}(20, 0.1057, 5, 20)$

Notice that $np = 2.113$ for 20 fish. If the number of fish caught increases to 500, $F \sim Bi(500, 0.1057)$ and $np(1-p) = 52.85 \geq 10$, we can use the Normal Distribution to approximate the probabilities for F.

$F \sim Bi(n, p) \Rightarrow F \sim N\left(\mu = np, \sigma = \sqrt{np(1-p)}\right)$

$F \sim Bi(n = 500, p = 0.1057)$

$\Rightarrow F \sim N\left(\mu = 52.85, \sigma = \sqrt{47.26}\right)$

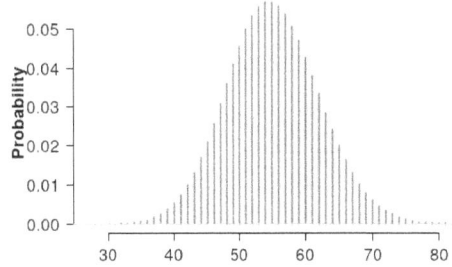

When catching 500 fish compare the probability of catching at least 60 fish that are longer than 45cm using the Binomial Method and the Normal Distribution method.

$F \sim Bi(500, 0.1057) \Rightarrow \Pr(X \geq 60) \approx 0.1663$

$F \sim N(\mu = 52.85, \sigma = 6.87) \Rightarrow \Pr(F \geq 60) \approx 0.1490$

Graphs created using app https://shiny.rit.albany.edu/stat/binomial/

CONTINUOUS PROBABILITY SUMMARY:

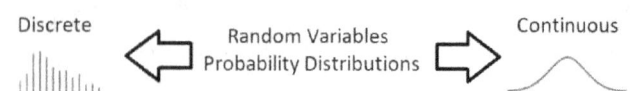

Discrete | Random Variables Probability Distributions | Continuous

A **PDF** (Probability Density Function) *f(x)* represent the probability for a continuous random variables *X*. The PDF can be any shape. Probability is given by finding the area under the PDF in a given interval

$$X \in [a,b] \quad \Pr(a \leq X \leq b) = \int_a^b f(x)dx$$

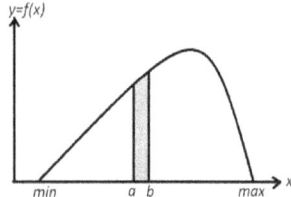

where $f(x) \geq 1$, area under PDF $\int_{min}^{max} f(x)dx = 1$

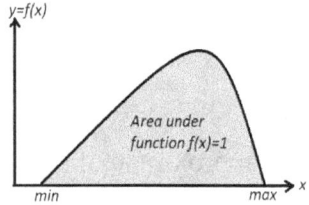

Mean $E(X) = \mu = \int_{min}^{max} xf(x)dx$

Median $\int_{min}^{median} f(x)dx = 0.5$

25th percentile $\Pr(X < Q_1) = \int_{min}^{Q_1} f(x)dx = 0.25$

50th percentile

$\Pr(X < median) = \int_{min}^{median} f(x)dx = 0.5$

75th percentile $\Pr(X < Q_3) = \int_{min}^{Q_3} f(x)dx = 0.75$

Mode, find the mode value of X, $f'(x) = 0$

Variance $Var(X) = \sigma^2 = \int_{min}^{max} x^2 f(x)dx - \mu^2$

Standard Deviation $\sigma = \sqrt{Var(X)}$

Linear Combinations of random variable X
$E(aX + b) = aE(X) + b$
$Var(aX + b) = a^2 Var(X)$

Normal Distribution PDF $X \sim N(\mu, \sigma)$

$$f(x) = \frac{1}{\sigma\sqrt{2\pi}} e^{\frac{1}{2}\left(\frac{x-\mu}{\sigma}\right)^2}$$

Bell shaped, Symmetrical
mean=median=mode

$\int_{-\infty}^{\infty} f(x)dx = 1$

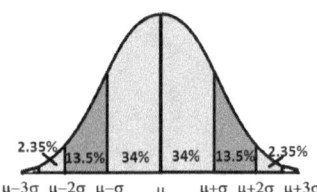

2.35% | 13.5% | 34% | 34% | 13.5% | 2.35%
$\mu-3\sigma$ $\mu-2\sigma$ $\mu-\sigma$ μ $\mu+\sigma$ $\mu+2\sigma$ $\mu+3\sigma$

Standard Normal Distribution $Z \sim N(\mu = 0, \sigma = 1)$

has the **PDF** $f(x) = \frac{1}{\sqrt{2\pi}} e^{\frac{1}{2}(x)^2}$

$Z = \frac{X - \mu}{\sigma} \Leftrightarrow X = Z\sigma + \mu$

Binomial Distribution $\Rightarrow$ **Normal Distribution**
if $np(1-p) \geq 10$ then by convention:

$X \sim Bi(n,p) \Rightarrow X \sim N\left(\mu = np, \sigma = \sqrt{np(1-p)}\right)$

Normal Distribution Symmetry

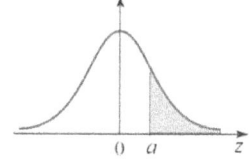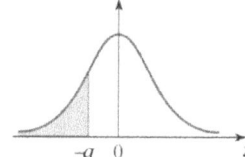

$\Pr(Z > a) = \Pr(Z < -a)$
$\Pr(-a < Z < a) = \Pr(Z < a) - \Pr(Z < -a)$
$\Pr(Z < a) = 1 - \Pr(Z > a)$
$\Pr(Z > -a) = 1 - \Pr(Z < -a)$

Conditional Probability

$\Pr(A/B) = \frac{\Pr(A \cap B)}{\Pr(B)}$

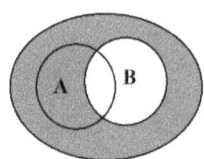

Discard events outside of B, involves thinking about the restrictions imposed on the interval of X by the given condition B using a number line.

CONTINUOUS PROBABILITY: Check your Understanding.

8.1 Probability Density Functions for continuous random variables..

a) A PDF is defined by the function $g(x) = kx(x-24), where\ x \in [0,24]$.

    i.   Find the value of k.

    ii.   Sketch the PDF function and shade the region where $X > 15$.

    iii.   Find the $\Pr(X > 15)$

    iv.   Find the Mean, Median and Variance of the PDF.

---

b) A PDF has the following general rule

$$f(x) = \begin{cases} \dfrac{1}{k^2}(x+k), & -k < x < 0 \\[2mm] \dfrac{1}{k^2}(k-x), & 0 \le x \le k \\[2mm] 0, & elsewhere \end{cases}$$

    i.   Show that the area under the PDF is equal to 1 for any real value of k.

If $k=2$

    ii.   Sketch the PDF

    iii.   Find the mean and standard deviation

    iv.   Find the 25th percentile of the PDF.

8.2 Normal Distributions.

a) IQ tests measure certain aspects of intelligence that are culture specific and not necessarily applicable to all people on our planet. If IQ is a continuous random variable with normal distribution, where $IQ \sim N(\mu = 100, \sigma = 16)$, the curve would look as follows.

IQ Normal Curve

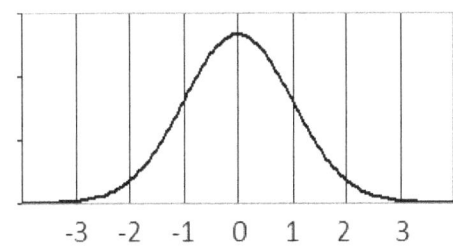

Standard Deviations	-3	-2	-1	0	1	2	3
Stanford-Binet IQ	36	52	68	84	100	116	132	148	164

Find the percentage of scores that are:

    i.   Less than 70

    ii.   Greater than 140

    iii.   Between 88 and 120

8.2 Normal Distributions. (Continued)

b)

In a population the height of adult men, M is a normally distributed random variable representing the height of men $M \sim N\left(\mu = 177cm, \sigma = 8cm\right)$, and the height of adult women, W is a normally distributed random variable representing the height of women $W \sim N\left(\mu = 162, \sigma = 7\right)$.

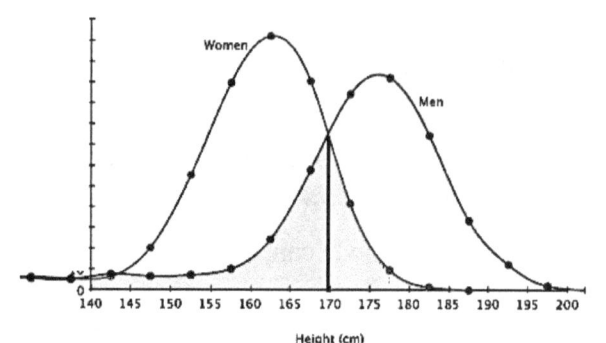

i. Find the probability than a man selected from this population is shorter than 170cm.
ii. Find the probability that a woman selected from this population is taller than 170cm.
iii. Find the probability that a woman selected from this population is between 160 and 170 cm tall.
iv. Given a woman is known to be between 160cm and 170cm tall, what is the probability that she is taller than 167cm?
v. Given a woman is known to be between 160cm and 170cm tall, what is the probability that she is between 162 and 167cm?
vi. Given a woman is known to be between 160cm and 170cm tall, what is the probability that she is between 155 and 159 cm?
vii. Given a man is shorter than 170cm, what is the probability he is shorter than 175cm?
viii. Given a man is shorter than 170cm, what is the probability he is taller than 175cm?

b) Find the mean and standard deviation of a normal distribution if
i. $\Pr(1.9 < X < 3.1) = 0.95$
ii. $\Pr(20 < Y < 32) = 0.997$
iii. $\Pr(6 < \left(2W + 1\right) < 10) = 0.68$
iv. Show the transformation required to turn each of the Normal Distributions into a Standard Normal Distribution and the reverse transformation to restore the original Normal Distribution.

c) The time spent waiting in a queue at airport security check-in is normally distributed.

If X is the time in minutes spent waiting in the line it is found that:
$$\Pr(4.25\,\text{min} \le X \le 5.75\,\text{min}) = 0.68$$

i. If $\mu + \sigma = 5.75$ minutes and $\mu - \sigma = 4.25$ minutes find the mean, and the standard deviation.
ii. What is the probability than a randomly selected person will get through the security check in 3 minutes or less?
iii. What is the probability that a randomly selected person will be in the queue for longer than 7 minutes>
iv. What percentage of people spend between 3 to 4 minutes waiting to pass through security?

8.3 Applications.

a) A local supermarket sells mangoes in 4 kg (4 kg=4000 g) boxes.

The weight of the boxes of mangoes is normally distributed with a mean of 4050 g and a standard deviation of 30 g.

Find the probability that a box of mangoes has a weight of:
  i.   less than 4 kg
  ii. more than 4020 g
  iii. between 4000 g and 4020 g
  iv. If 20% of the boxes are above the minimum weight, find the minimum weight.

b) A student applying for a scholarship has the following results across four subjects, where each subject's scores are normally distributed across the year level.

Subject	Student's Test Score	Subject Mean	Subject Standard Deviation
English	87	62	11
Mathematics	75	70	6
History	91	58	14
Geography	72	60	10

  i.   Find the standardized value for each subject's test score.
  ii. Show all the scores on a Standard Normal Distribution graph.
  iii. Which is this student's best subject using the test score information?

c) Find $\Pr(X \leq x)$ using the binomial probability $X \sim Bi(n,p)$. Then if the normal distribution can be used as an approximation $X \sim N\left(\mu = np, \sigma = \sqrt{np(1-p)}\right)$, find $\Pr(X \leq x)$ using the Normal Distribution and compare the results.
  i. $n = 60, p = 0.4, x \leq 20$
  ii. $n = 40, p = 0.25, x \leq 10$
  iii. $n = 850, p = 0.8, x \leq 700$

8.3 Applications. (Continued)

d) According to the Red Cross Blood Service, 7% of people in Australia have blood type A-.

  i. What is the probability that in a random sample of 600 people, 30 people or fewer have blood type A-?
 ii. Explain your method of find the probability for part i.
iii. Can we use the normal distribution to approximate the binomial distribution for this trial?
 iv. Compare the probability found using Normal Distribution methods for 30 people or fewer having blood type A-, and explain if this result can be used in terms of accuracy with reference to part iii.

e) Carbon rods are manufactured with an average length of 5 cm. Due to the variability in the production processes, the lengths of the carbon rods are approximately normally distributed, with a standard deviation of 0.02 cm.

  i. What percentage of Carbon rods have a length greater than 5.03 cm?

Carbon rods that have a length less than 4.95 cm or greater than 5.05 cm are rejected.

 ii. What percentage of carbon rods will be rejected?
iii. If 30,000 carbon rods are manufactured in a day, how many would be rejected?
 iv. If an order comes in for 50,000 carbon rods, how many rods should be manufactured if the order states that all rods must be between 4.97 cm and 5.03 cm in length?

CONTINUOUS PROBABILITY SOLUTIONS.

8.1 Probability Density Functions for continuous random variables. (SOLUTIONS)

a) PDF is defined by the function $g(x) = kx(x-24), where\ x \in [0,24]$.

i.

To be a PDF $\int_0^{24} g(x)\,dx = 1$

$$\int_0^{24} kx(x-24)\,dx = k\int_0^{24} x(x-24)\,dx$$

$$= k\left[\frac{x^3}{3} - 12x^2\right]_0^{24}$$

$$= k\left(\frac{(24)^3}{3} - 12(24)^2\right) - k\left(\frac{(0)^3}{3} - 12(0)^2\right)$$

$$= -2304k$$

$-2304k = 1$

$$k = \frac{-1}{2304}$$

$$\int_0^{24} \frac{-1}{2304} x(x-24)\,dx = 1$$

ii.

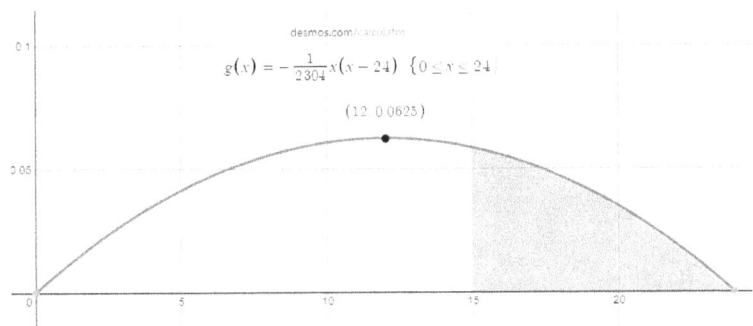

$g(x) = -\frac{1}{2304}x(x-24) \quad \{0 \le x \le 24\}$

$(12, 0.0625)$

iii.

$$\Pr(X > 15) = \int_{15}^{24} \frac{-1}{2304} x(x-24)\,dx$$

$$= \frac{-1}{2304} \int_{15}^{24} x(x-24)\,dx$$

$$= \frac{-1}{2304}\left[\frac{x^3}{3} - 12x^2\right]_{15}^{24}$$

$$= \frac{-1}{2304}\left(\frac{(24)^3}{3} - 12(24)^2\right) - \frac{-1}{2304}\left(\frac{(15)^3}{3} - 12(15)^2\right)$$

$$= 0.3164$$

iv.

$$E(X) = \int_0^{24} \frac{-1}{2304} x^2(x-24)\,dx$$

$$= 12$$

Median=E(X) for this PDF due to the symmetry of the quadratic.

$$Var(X) = E(X^2) - [E(X)]^2$$

$$= \int_0^{24} \frac{-1}{2304} x^3(x-24)\,dx - 12^2$$

$$= 172.8 - 144$$

$$= 28.8$$

8.1 Probability Density Functions for continuous random variables. (SOLUTIONS)

b)

i.

$$\int_{-k}^{0} \frac{1}{k^2}(x+k)\,dx = \frac{1}{k^2}\left[\frac{x^2}{2}+kx\right]_{-k}^{0} \qquad \int_{0}^{k} \frac{1}{k^2}(k-x)\,dx = \frac{1}{k^2}\left[kx-\frac{x^2}{2}\right]_{0}^{k}$$

$$=\frac{1}{k^2}\left[\left(\frac{(0)^2}{2}+k(0)\right)-\left(\frac{(-k)^2}{2}+k(-k)\right)\right] \qquad =\frac{1}{k^2}\left[\left(k(k)-\frac{(k)^2}{2}\right)-\left(k(0)-\frac{(0)^2}{2}\right)\right]$$

$$=\frac{1}{k^2}\left(-\frac{k^2}{2}+k^2\right) \qquad\qquad =\frac{1}{k^2}\left(k^2-\frac{k^2}{2}\right)$$

$$=\frac{1}{2} \qquad\qquad\qquad\qquad =\frac{1}{2}$$

Total area is 1.

ii.

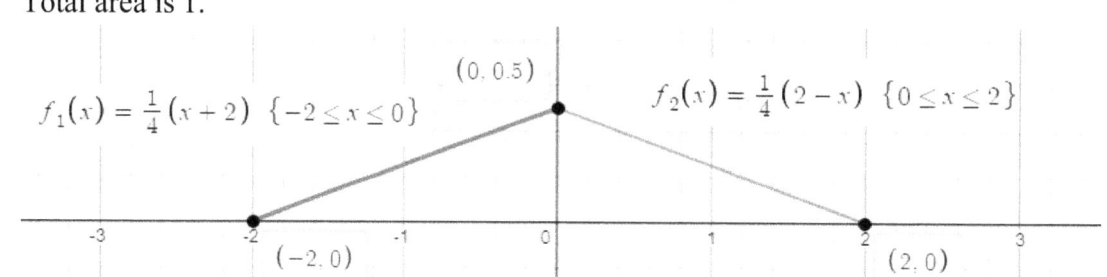

$f_1(x)=\tfrac{1}{4}(x+2)\ \ \{-2\le x\le 0\}$ (0, 0.5) $f_2(x)=\tfrac{1}{4}(2-x)\ \ \{0\le x\le 2\}$

$(-2,0)$ $(2,0)$

iii.

$$E(X)=\int_{-2}^{0}xf(x)\,dx+\int_{0}^{2}xf(x)\,dx$$

$$=\frac{1}{4}\int_{-2}^{0}x(x+2)\,dx+\frac{1}{4}\int_{0}^{2}x(2-x)\,dx$$

$$=\frac{1}{4}\int_{-2}^{0}\left(x^2+2x\right)dx+\frac{1}{4}\int_{0}^{2}\left(2x-x^2\right)dx$$

$$\frac{1}{4}\int_{-2}^{0}x(x+2)\,dx+\frac{1}{4}\int_{0}^{2}x(2-x)\,dx=\frac{-1}{3}+\frac{1}{3}$$

$$=0$$

$$E(X)=0$$

$$Var(X)=E(X^2)-\left[E(X)\right]^2$$

$$E(X^2)=\frac{1}{4}\int_{-2}^{0}x^2(x+2)\,dx+\frac{1}{4}\int_{0}^{2}x^2(2-x)\,dx$$

$$=\frac{1}{3}+\frac{1}{3}$$

$$=\frac{2}{3}$$

$$Var(X)=\frac{2}{3}$$

$$\sigma=\sqrt{Var(X)}=\sqrt{\frac{2}{3}}$$

iv.

$$\int_{-2}^{Q_1}\frac{1}{4}(x+2)\,dx=0.25$$

$$\frac{1}{4}\left[\frac{x^2}{2}+2x\right]_{-2}^{Q_1}=0.25$$

$$LHS=\frac{1}{4}\left[\left(\frac{Q_1^2}{2}+2(Q_1)\right)-\left(\frac{(-2)^2}{2}+2(-2)\right)\right]$$

$$=\frac{Q_1^2}{8}+\frac{Q_1}{2}+\frac{1}{2}$$

$$\frac{Q_1^2}{8}+\frac{Q_1}{2}+\frac{1}{2}=0.25$$

$$Q_1^2+4Q_1+4=2$$

$$Q_1^2+4Q_1+2=0$$

$$Q_1=-0.5858\ \ or\ \ Q_1=-3.4142$$

$$Q_1=-0.5858\ \text{is in interval}$$

8.2 Normal Distributions. (SOLUTIONS)

a) $IQ \sim N(\mu = 100, \sigma = 16)$

i.	$\Pr(IQ < 70) = 0.0304$	normCdf($-\infty$,70,100,16)	normCDf($-\infty$,70,16,100)
ii.	$\Pr(IQ > 140) = 0.0062$	normCdf(140,∞,100,16)	normCDf(140,∞,16,100)
iii.	$\Pr(88 < IQ < 120) = 0.6677$	normCdf(88,120,100,16)	normCDf(88,120,16,100)

b) Men: $M \sim N(\mu = 177cm, \sigma = 8cm)$, Women: $W \sim N(\mu = 162, \sigma = 7)$.

i.
$$\Pr(M < 170cm) = 0.1908$$

ii.
$$\Pr(W > 170cm) = 0.1265$$

iii.
$$\Pr(160 < W < 170) = 0.4859$$

iv.
$$\Pr(W > 167 / 160 < W < 170)$$

$$\Pr(A / B) = \frac{\Pr(A \cap B)}{\Pr(B)}$$

$$\Pr(W > 167cm / 160cm < W < 170cm) = \frac{\Pr([W > 167cm] \cap [160cm < W < 170cm])}{\Pr(160cm < W < 170cm)}$$

$$= \frac{\Pr([167cm < W < 170cm])}{\Pr(160cm < W < 170cm)} \approx \frac{0.1110}{0.4859} \approx 0.2284$$

v.
$$\Pr(162 < W < 167 / 160 < W < 170)$$

$$\Pr(A / B) = \frac{\Pr(A \cap B)}{\Pr(B)}$$

$$\Pr(162 < W < 167 / 160 < W < 170) = \frac{\Pr([162cm < W < 167cm])}{\Pr(160cm < W < 170cm)} \approx \frac{0.2625}{0.4859} \approx 0.5402$$

vi.
$$\Pr(155 < W < 159 / 160 < W < 170)$$

$$\Pr(A / B) = \frac{\Pr(A \cap B)}{\Pr(B)}$$

$$\Pr(155 < W < 159 / 160 < W < 170) = 0 \text{ no overlap of intervals}$$

vii.
$$\Pr(M < 175 / M < 170)$$

$$\Pr(A / B) = \frac{\Pr(A \cap B)}{\Pr(B)}$$

$$\Pr(M < 175 / M < 170) = \frac{\Pr([M < 175] \cap [M < 170cm])}{\Pr(M < 170cm)} = \frac{\Pr([M < 170cm])}{\Pr(M < 170cm)} = 1$$

viii.
$$\Pr(M > 175 / M < 170)$$

$$\Pr(M > 175 / M < 170) = 0$$

No overlapping intervals.

c) Find the mean and standard deviation of a normal distribution if

i. $\Pr(1.9 < X < 3.1) = 0.95$

$\Pr(\mu - 2\sigma < X < \mu + 2\sigma) = 0.95$

$\therefore$

$\mu - 2\sigma = 1.9$

$\mu + 2\sigma = 3.1$

Solving these equations simultaneously gives

$\mu = 2.5, \sigma = 0.3$

$X \sim N(\mu = 2.5, \sigma = 0.3)$

ii. $\Pr(20 < Y < 32) = 0.997$

$\Pr(\mu - 3\sigma < Y < \mu + 3\sigma) = 0.997$

$\therefore$

$\mu - 3\sigma = 20$

$\mu + 3\sigma = 32$

Solving these equations simultaneously gives

$\mu = 26, \sigma = 2$

$Y \sim N(\mu = 26, \sigma = 2)$

iii. $\Pr(6 < (2W + 1) < 10) = 0.68$

$\Pr(\mu - \sigma < (2W + 1) < \mu + \sigma) = 0.68$

$\therefore$

$\mu - \sigma = 6$

$\mu + \sigma = 10$

Solving these equations simultaneously gives

$\mu = 8, \sigma = 2$

$(2W + 1) \sim N(\mu = 8, \sigma = 2)$

iv. $Z = \dfrac{X - \mu}{\sigma} \Leftrightarrow Z\sigma + \mu = X$

i. $Z = \dfrac{X - 2.5}{0.3} \Leftrightarrow 0.3Z + 2.5 = X$

ii. $Z = \dfrac{Y - 26}{2} \Leftrightarrow 2Z + 26 = Y$

iii. $Z = \dfrac{(2W + 1) - 8}{2}$

$\Leftrightarrow 2Z + 8 = (2W + 1)$

d) The time spent waiting in a queue at an airport security check-in is normally distributed.

i. $\mu + \sigma = 5.75$ minutes and $\mu - \sigma = 4.25$ minutes , solving these equations simultaneously gives, $\mu = 5, \sigma = 0.75$.

If we let the W be the random variable representing the time waiting then

$W \sim N(\mu = 5, \sigma = 0.75)$

ii. $\Pr(W \le 3) \approx 0.0038$

iii. $\Pr(W > 7) \approx 0.0038$, could be found by symmetry part ii.

iv. $\Pr(3 \le W \le 4) \approx 0.0874$. Approximately 8.7% of people queue between 3 to 4 minutes.

8.3 Applications (SOLUTIONS)

a) A local supermarket sells mangoes in 4 kg (4kg=4000g) boxes.

Let W represent the weight of the boxes of mangoes: $W \sim N(\mu = 4050g, \sigma = 30g)$

   i. $\Pr(W < 4000g) \approx 0.0478$ ii. $\Pr(W > 4020g) \approx 0.8413$

   iii. $\Pr(4000g \leq W \leq 4020g)$

$$\approx \Pr(W < 4020g) - \Pr(W < 4000g) \approx 0.7935$$

   iv.   Use the Standard Normal Distribution $Z \sim N(\mu = 0, \sigma = 1)$ to solve.

$\Pr(Z > C) = 0.2$ by symmetry $\Pr(Z < -C) = 0.2$, using the inverse normal

functionality of our technology, $-C = -0.8414 \Rightarrow C = 0.8416$

Since $Z = \dfrac{W - \mu}{\sigma} = \dfrac{W - 4050}{30}$ $0.8416 = \dfrac{W - 4050}{30}$ $W \approx 4075.25g$

We can check this $\Pr(W > 4075.25g) \approx 0.2$

b) A student's test results.

Subject	Student's Test Score	Subject Mean	Subject Standard Deviation	i. Standardised Z-score equivalent
English	87	62	11	2.2727
Mathematics	75	70	6	0.8333
History	91	58	14	2.3571
Geography	72	60	10	1.2000

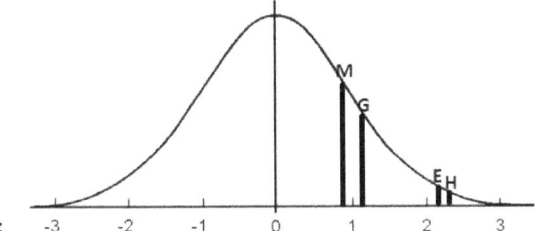

   ii.

   iii. History is the student's best subject.

c) Find $\Pr(X \leq x)$ using the Binomial and Normal Distribution and compare the results.

	Check $np(1-p) \geq 10$	Binomial $\Pr(X \leq x)$	Normal $\Pr(X \leq x)$
i.	$n = 60$, $p = 0.4$, $x \leq 20$	0.1786	$X \sim N\left(\mu = 24, \sigma = \sqrt{14.4}\right)$ $\Pr(X \leq 20) \approx 0.1459$
ii.	$n = 40, p = 0.25, x \leq 10$	0.5839	$np(1-p) = 7.5$ cannot be used
iii.	$n = 850$, $p = 0.8$, $x \leq 700$	0.9622	$X \sim N\left(\mu = 680, \sigma = \sqrt{136}\right)$ $\Pr(X \leq 700) \approx 0.9568$

d) According to the Red Cross Blood Service, 7% of people in Australia have blood type A-.

i. Let A^- represent people with blood type A-, then $A^- \sim Bi(n = 600, p = 0.07)$

$\Pr(A^- \leq 30) \approx 0.0285$

ii. There are 600 independent trials with each trial having a probability of success equal to 0.07. This is a binomial experiment.

iii. $np(1-p) = 600(0.07)(0.93) = 39.06 \geq 10$ Yes the Normal distribution can be used to approximate the probabilities.

iv. $A^- \sim N(\mu = 42, \sigma = \sqrt{39.06})$ giving $\Pr(A^- \leq 30) \approx 0.0274$

e) Carbon rods are manufactured with an average length of 5 cm, with a standard deviation of 0.02 cm.

Carbon rods that have a length less than 4.95 cm or greater than 5.05 cm are rejected.

i. Let C represent the length of the carbon rods: $C \sim N(\mu = 5, \sigma = 0.02)$

$\Pr(C > 5.03) \approx 0.0668$

ii. $1 - \Pr(4.95 \leq C \leq 5.05) \approx 0.0124$

iii. Amount rejected $30000 \times 0.0124 \approx 373$

iv. $\Pr(4.97 \leq C \leq 5.03) \approx 0.8664$, approximately 86.64% will be inside the required range. To find the amount to produce, $Amount \times 0.8664 \approx 50000$.
The $Amount \approx 57711$ Carbon rods need to be produced.

Population vs Sample

A population is a collection of people, items, or events about which you want to be able to understand enough to be able to make inferences and predictions. It is not always convenient or possible to examine every item, or survey every member of an entire population. The symbol N is used in statistics to represent the count of the whole population.

Examples of populations:

- all registered teachers in Victoria
- all Facebook users
- all houses on Phillip Island
- all recorded cyclones in Australia
- all B747 jets owned by Qantas
- all vehicles in a parking lot

A sample is a subset of people, items, or events from a larger population that you collect and analyse to make predictions and inferences. To represent the population fairly and well, a sample should be randomly collected and adequately large. The symbol n is used in statistics to represent the count of the sample items.

If the sample is random and large enough, you can use the information collected from the sample to make predictions and inferences about the whole population. For example, you could survey a randomly selected number of people about their favourite ice-cream flavour and then use the sample statistics to estimate the ice-cream flavour preferences of the entire population.

Population (N)

Sample (n)

When studying a population, there will be certain attributes or characteristics that we are trying to determine to understand the nature of some aspect of that population.

Based on a sample of size n drawn from a population of size N, the proportion $\hat{p}$ of individuals with the attribute in the sample can be used to infer the proportion p of that particular attribute in the entire population.

In real life it is very difficult to determine the parameter of population proportion p, and it is usually estimated using many samples and sample proportions $\hat{p}$.

Population (N)	Sample (n)
The **population proportion** p is the proportion of individuals in the entire population possessing a particular attribute, and is constant for a given population. $$p = \frac{\text{number with attribute in total population}}{\text{population size (N)}}$$ The **population mean** μ is the mean of all values of a measure in the entire population, and is constant for a given population. $$\mu = \frac{\text{sum of all data values in population}}{\text{size of population (N)}}$$	The **sample proportion** $\hat{p}$ is the proportion of individuals in a particular sample possessing this attribute, and varies from sample to sample. $$\hat{p} = \frac{\text{number with attribute in sample}}{\text{sample size (n)}}$$ The **sample mean** $\overline{x}$ is the mean of the values of this measure in a particular sample, and varies from sample to sample. $$\overline{x} = \frac{\text{sum of all data values in sample}}{\text{size of sample (n)}}$$

Example 9.1: Sample proportions, from multiple surveys.

20 students are randomly selected from a school population of 1400 students, and surveyed on what transport they use to get to and from school. This random survey is repeated 5 times on different days at the school. The attribute of main interest is public transport use by students.

- Sample proportion of first sample $\hat{p}_1 = 0.45$
- Sample proportion of second sample $\hat{p}_2 = 0.55$
- Sample proportion of third sample $\hat{p}_3 = 0.45$
- Sample proportion of fourth sample $\hat{p}_4 = 0.5$
- Sample proportion of fifth sample $\hat{p}_5 = 0.25$

Sample Number (n=20)	Car	Public Transport	Bicycle	Walking
1	5	9	2	4
2	2	11	4	3
3	3	9	5	3
4	4	10	4	2
5	4	5	6	5

For each sample taken, the sample proportion $\hat{p}$ is an estimate of the population proportion p of public transport users in the student population.

In the majority of cases we cannot examine every individual in a population to find the true proportion p for having a certain attribute..

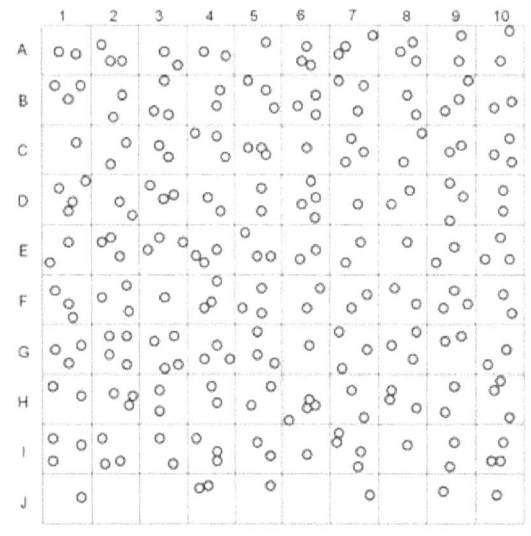

The grid shown is 100 cells. Each black circle represents one individual. If you randomly select several cells you can create a sample.

If you examine the individuals in the sample and record those that have the *certain attribute*, to find the sample proportion $\hat{p}$. The values of $\hat{p}$ are random variables represented by $\hat{P} = \dfrac{X}{n}$, where X is a binomial random variable of the number of items that have the certain attribute.

If we survey the entire population of a school and identify individuals that have the *certain attribute* of being "left handed", then we have found the proportion p of left-handers in the whole population.

If we take a sample from within the population of the same school, and identify individuals that have the *certain attribute* of being "left handed", then we have found the sample proportion $\hat{p}$. of left-handers in the sample (n). The values of $\hat{p}$ are random variables represented by $\hat{P} = \dfrac{X}{n}$, where X is a binomial random variable of the number of students that are left handed in a sample.

(N)

(n)

Using non-invasive aerial drones several random observations are made of Humpback whales, in Western Australia.

Images sourced from: Christiansen, Fredrik & Dujon, Antoine & Sprogis, Kate & Arnould, John & Bejder, Lars. (2016). Noninvasive unmanned aerial vehicle provides estimates of the energetic cost of reproduction in humpback whales. Ecosphere. 7. 10.1002/ecs2.1468.

Sampled individuals that have the *certain attribute* of reproductive health potential, representing the sample proportion $\hat{p}$. of Humpback whales with good reproductive health potential in the sample (n). The values of $\hat{p}$ are random variables represented by $\hat{P} = \dfrac{X}{n}$ based on random selections.

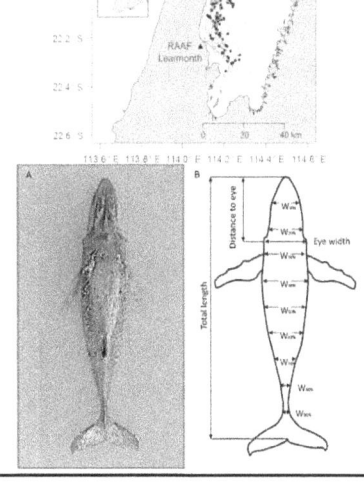

A summary of things you can do with statistical proportions and other univariate statistics.		
What is the population proportion of that attribute?	**Ask Question**	Attributes for human populations include gender, allergies, height, transport usage.
Survey Observation Experimental results	**Collect Data**	Random, independent sampling If population is known to have a normal distribution $n \geq 30$, otherwise perhaps $\sqrt{N}$ for N size population
Numeric Data Discrete (Integer, whole number) Continuous (Real, decimal values)	**Classify Data**	**Categorical Data** Nominal – selected from a list eg. Gender {M,F} Ordinal – selected from a list that has a ranking eg. School Grades {A+,A,B+,B,C+......}
Ungrouped data Grouped data – original data collected into class intervals Pie Charts Bar/Column Graphs Frequency Tables Frequency histogram Frequency polygon 	**Sort Data**	Stem-Leaf Plot | Stems	Leaves
---	---
4	3 6 6 7 9
5	0 0 1 1 1 2 4 6 8 8
6	1 2 3 3 5 6 6 7 9
7	2 5 7 8
8	0 1 2 2 4 4 5 7 8 9
9	1 3 3	 $4\|3 = 4.3$ Cumulative Frequency • Tables • Histogram • Polygon
Measures of central tendency Mean/Average $\bar{x} = \dfrac{\sum x}{n}$ Mode – most frequent value Median – middle value of an ordered data set.	**Analyse Data** Q_3 M Q_1	**Measures of spread** Range = max − min, 5 figure summary (min, Q_1, median, Q_3, max) Box Plots - Outliers Standard Error s – a measure of deviation from the mean value $$s = \sqrt{\sum \frac{(x - \bar{x})^2}{n - 1}}$$

If we take multiple samples and work out the sample proportion of a particular attribute for each sample we anticipate some variability due to the random selection of individuals from the total population.

| Example 9.2: Random variability of sample proportions, for particular population attributes. |

Sample 1	Sample 2	Sample 3	Sample 4

- Sample proportion of attribute 2 in the first sample $\hat{p}_1 = 0.45$
- Sample proportion of attribute 2 in the second sample $\hat{p}_2 = 0.46$
- Sample proportion of attribute 2 in the third sample $\hat{p}_3 = 0.48$
- Sample proportion of attribute 2 in the fourth sample $\hat{p}_4 = 0.46$

To analyse the behavior of any random variable such as a bunch of sample proportions we can try to determine its mean, median and mode, as well as its variance and standard deviation. We expect most sample proportions to be close to the values of the mean, median and mode, which intuitively we expect to be very close to the whole population proportion.

For large sample sizes (the convention is at least 30), we intuitively expect that spread around the population proportion will not be very great, with most values clustered around the true population proportion parameter p and fewer values either side, tapering away the further they are from p.

| Distribution of the sample proportions $\hat{p}$ |

If we take multiple random samples from a large population and find the sample proportions of the favourable attribute in each sample $\hat{p}_1, \hat{p}_2, \hat{p}_3, \hat{p}_4 \dots \hat{p}_k$, we know that each sample proportion is an estimate of the population proportion. While the sample mean is an average of a particular numeric attribute an individual in a population has, the sample proportion is an average of whether that attribute is present, it can be calculated by assigning a "1" to those individuals with the attribute, and "0" to those without it, and taking this average.

- Sample proportion (n=10) of first sample **[0,0,0,1,1,1,1,0,0,0]** $\hat{p}_1 = 0.4$ have the attribute.
- Sample proportion (n=10) of second sample **[1,0,1,0,1,0,0,0,0,0]** $\hat{p}_2 = 0.3$ have the attribute.
- Sample proportion (n=10) of third sample **[0,0,0,0,0,1,0,0,1,1]** $\hat{p}_3 = 0.3$ have the attribute.
- Sample proportion (n=10) of fourth sample **[1,1,0,0,1,0,0,1,0,1]** $\hat{p}_4 = 0.5$ have the attribute.
- Sample proportion (n=10) of kth sample **[0,0,0,1,0,1,0,0,1,0]** $\hat{p}_k = 0.3$ have the attribute.

The average of the sample proportions $\hat{p}_1, \hat{p}_2, \hat{p}_3, \hat{p}_4 \dots \hat{p}_k$ is normally distributed around the actual population proportion p. The more samples proportions we calculate, the higher the number that will have sample proportion values close to the actual population proportion, as stated by the **Central Limit Theorem.**

233

Act 1, Scene 1. Sample Proportions. The drama of random variable distributions.

If 30% of a population is known to have a certain attribute, the population proportion, p=0.3.

For random samples taken from this population, we would expect the sample proportion to be around 0.3.

When there is random selection, the sample proportion is a random variable since it can take many values which can be estimated using probability.

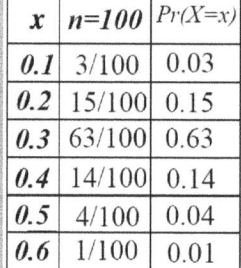

x	$n=100$	$Pr(X=x)$
0.1	3/100	0.03
0.2	15/100	0.15
0.3	63/100	0.63
0.4	14/100	0.14
0.5	4/100	0.04
0.6	1/100	0.01

Amazing

Algebrains.

For any random variable we try to determine the centre, spread and shape of the probability distribution by calculation of the expected value or mean, the standard deviation and graphing of the distribution.

The **Mean** is referred to as the expected value

$$E(X) = \mu = \sum_x x \times \Pr(X = x)$$

Variance $Var(X) = \sigma^2$

$$Var(X) = E\left[(X - \mu)^2\right]$$
$$= E(X^2) - \left[E(X)\right]^2$$

Standard Deviation

$$\sigma = \sqrt{Var(X)}$$

Act 1, Scene 2. Sample Proportions. The drama of random variable distributions.

For multiple random sampling we expect to see the values of the sample proportions to vary either side of true population proportion p.

We expect the averaged values of the sample proportions to become very close to the population proportion, which is 0.3 for this case.

Amazing

Algebrains.

Randomly selecting 50 individuals from a population is be more likely to be representative of the true probability distribution than selecting 5 individuals.

Sample size is important for measuring the spread.
We expect a smaller spread or standard deviations for larger samples, and larger standard deviation for smaller samples.

Sample proportions closest to 0.3 would be most common, and sample proportions far from 0.3 in either direction would be progressively less likely.

If the sample proportions are normally distributed, then the distribution of sample proportions will be centered at p.

Normal Distributions and the Central Limit Theorem.

Multiple Sample proportions observed

Population proportion = 0.45

Standard deviation of sample proportions=0.110

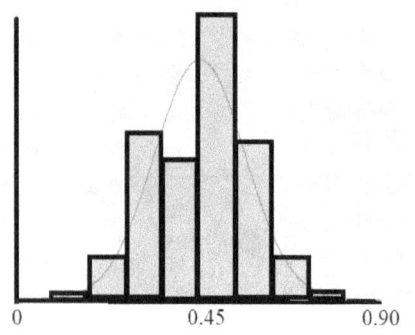

The **Central Limit Theorem** establishes that, in situations where multiple samples are randomly generated in a way that does not depend on the values of the other observations, the distribution of the sample proportions $\hat{p}_1, \hat{p}_2, \hat{p}_3, \hat{p}_4 .... \hat{p}_k$ will be closely approximated by a **Normal distribution** that is centred about the population proportion p .

If the sample size n, np and $n(1-p)$ are "large" the sampling distribution of the sample proportions $\hat{p}_1, \hat{p}_2, \hat{p}_3, \hat{p}_4 .... \hat{p}_k$ will be approximately normal with mean $\mu = p$ and standard deviation

$$\sigma = \sqrt{\frac{p(1-p)}{n}}$$

Another example of the Central Limit Theorem is observed with numerous tosses of a fair coin, the probability of getting a given number of heads in a series of tosses will approach a normal curve, as we have observed in Binomial Distribution graphs.

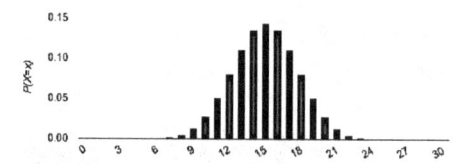

This Binomial distribution is symmetrical about the mean which is the expected value. We expect this because the chance of success and failure is equally likely. $X \sim Bi(n = 30, p = 0.5)$ $E(X) = np = 15$.

Due to the **Central Limit Theorem**, for large values of n the **Normal Distribution** may be used as an approximation to the **Binomial Distribution** if the distribution is symmetrical and not very skewed.

The probability density function for a random variable X with a Normal Distribution $X \sim N(\mu, \sigma)$ is symmetrical about the mean μ and the measure of spread is the standard deviation σ .

Theoretically the domain of the Normal distribution function is from $-\infty$ to ∞, $\int_{-\infty}^{\infty} \frac{1}{\sigma\sqrt{2\pi}} e^{\frac{1}{2}\left(\frac{x-\mu}{\sigma}\right)^2} dx = 1$.

The **Standard Normal Distribution** function $f(x) = \frac{1}{\sqrt{2\pi}} e^{-\frac{1}{2}x^2}$ is often represented by a Z variable,

$X \sim N(\mu = 0, \sigma = 1)$ is formed applying transformations of dilation and translation $Z = \frac{X - \mu}{\sigma}$ to the original probability density function $X \sim N(\mu, \sigma)$, so that its mean equals zero, and its standard deviation equals 1.

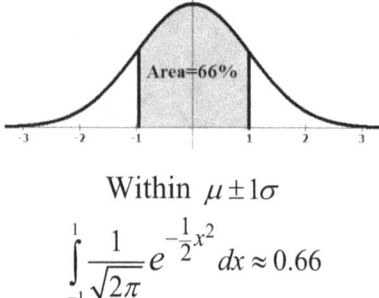

Within $\mu \pm 1\sigma$

$$\int_{-1}^{1} \frac{1}{\sqrt{2\pi}} e^{-\frac{1}{2}x^2} dx \approx 0.66$$

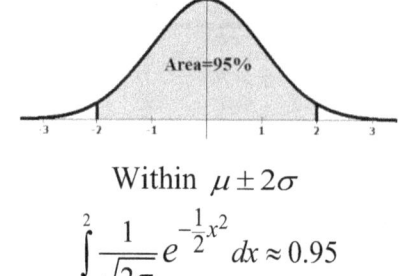

Within $\mu \pm 2\sigma$

$$\int_{-2}^{2} \frac{1}{\sqrt{2\pi}} e^{-\frac{1}{2}x^2} dx \approx 0.95$$

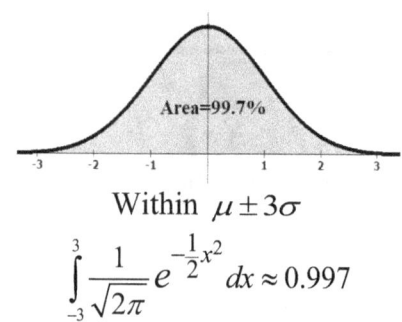

Within $\mu \pm 3\sigma$

$$\int_{-3}^{3} \frac{1}{\sqrt{2\pi}} e^{-\frac{1}{2}x^2} dx \approx 0.997$$

Scientists studying corals in a large coastal reef are looking for evidence of coral bleaching. Aerial photos of 30 adjacent regions are identified. Each area is mapped to a grid shown is 10 metres by 10 metres. Each cell of the grid is 1 m by 1 m. Each circle represents coral growth clumps.

For each of the 30 adjacent regions, 10 cells are randomly selected and the sample proportion $\hat{p}$ of coral bleaching is observed in each cell, making a total of 300 sample proportions. A histogram of the average value of the sample proportions in each region, indicates a normal distribution centred around a mean proportion of approximately 0.3. According to the Central Limit Theorem the population proportion is the mean $\mu = p$ and standard deviation

$\sigma = \sqrt{\dfrac{p(1-p)}{n}}$, the average of the sample

proportions $\hat{P}$ follow a normal distribution.

$$\hat{P} \sim N\left(\mu = 0.3, \sigma = \sqrt{\dfrac{0.3(0.7)}{300}}\right)$$

$$\hat{P} \sim N\left(\mu = 0.3, \sigma \approx 0.0265\right)$$

Region 22

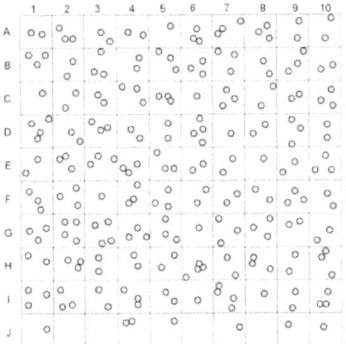

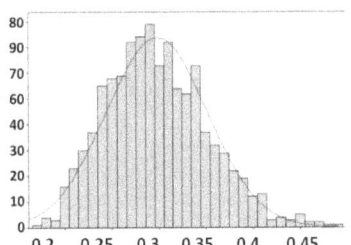

66% of the cells estimated to have a proportion of bleaching $\hat{P}$ between $\mu \pm 1\sigma$, 0.3 ± 0.0265.
95% of the cells estimated to have a proportion of bleaching $\hat{P}$ between $\mu \pm 2\sigma$, $0.3 \pm 2(0.0265)$.
99.7% of the cells estimated to have a proportion of bleaching $\hat{P}$ between $\mu \pm 3\sigma$, $0.3 \pm 3(0.0265)$.

The proportion p of left-handers in the whole population (N) of the world is thought to be approximately 12%. If L represents the random variable of a left handed person being selected from a population of 500.

$L \sim Bi(n = 500, p = 0.12)$

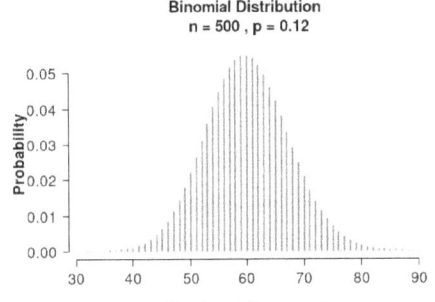

ROUGHLY 12% OF THE WORLD IS LEFT HANDED

For large values of n the **Normal Distribution** may be used as an approximation to the **Binomial Distribution** if the distribution is symmetrical and not very skewed.

$L \sim N\left(\mu = np, \sigma = \sqrt{np(1-p)}\right)$

$L \sim N\left(\mu = 60, \sigma \approx 7.2663\right)$

Binomial Distribution
n = 500 , p = 0.12

There is a 66% chance the number of left handers in this population is within $\mu \pm 1\sigma$, 60 ± 7.2663.
There is a 95% chance the number of left handers in this population is within $\mu \pm 2\sigma$, $60 \pm 2(7.2663)$.
There is a 99.7% chance the number of left handers in this population is within $\mu \pm 3\sigma$, $60 \pm 3(7.2663)$.

Act 2, Scene 1. Confidence intervals. The drama of finding how good the population proportion is.

Randomly sampling to find the population proportion of a certain attribute in the entire population. How good is that method?

Our intuition suggests and the Central Limit Theorem states, the more sample proportions we find randomly and independently, the distribution of the sample proportions will centre around the true population proportion.

The distribution of the sample proportions has Normal Distribution.

We can work out the probability the population proportion is in any interval.

Amazing

Algebrains.

$$\Pr\left(\hat{P} - 1.96\sqrt{\frac{p(1-p)}{n}} < p < \hat{P} + 1.96\sqrt{\frac{p(1-p)}{n}}\right) \approx 0.95$$

± 1.96 represents 95% using Standard Normal Distribution.

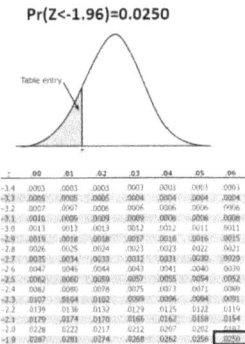

Pr(Z<-1.96)=0.0250

Pr(Z<1.96)=0.9750

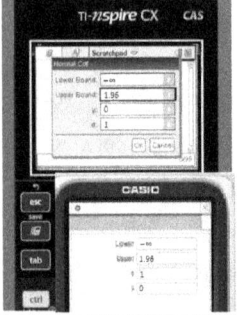

What are the chances that for a normally distributed random variable representing averaged sample proportions $\hat{P} = \dfrac{X}{n}$,

an interval will contain the true population proportion p ?

Commonly used intervals of **confidence** using Z scores where $Z \sim N(0,1)$ are:

- 90% chance which is $\mu \pm 1.645\sigma$
- 95% chance which is $\mu \pm 1.96\sigma$
- 99% chance which is $\mu \pm 2.58\sigma$

The interval gets bigger, as the required chance of confidence in the value of the population proportion p increases.

A confidence interval is an estimate of an interval which will contain the population proportion **p** with a given probability or chance.

Act 2, Scene 2. Confidence intervals. The drama of finding how good the population proportion is.

With the help of technology we can collect and calculate many averaged sample proportions. The statistics calculated for the distributions help us guess what the likely population proportion is for the attribute of interest.

Population proportion is inferred by multiple sampling.

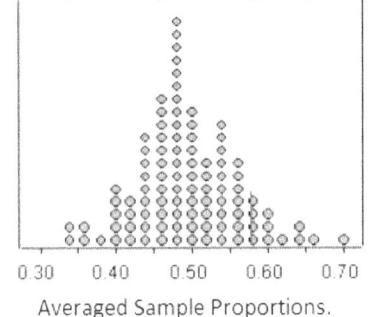

Averaged Sample Proportions.

Amazing

Algebrains

Random selection of samples from a population gives observations that are independent. They are like random variables, and we can use probability to find the likelihood of accuracy.

A sample proportion distribution $\hat{P}$ is shown in the table below, it is based on the frequency of the values of many random samples of sample proportions $\hat{p}_1, \hat{p}_2, \hat{p}_3, \hat{p}_4 \ldots \hat{p}_k$ recorded. $\hat{P}$ is a random measurement that can take many values and we use statistics to determine its probability.

Total samples taken = 365						
$\hat{P}$	0.5	0.55	0.63	0.67	0.71	0.75
Frequency Observed	11	50	130	120	45	9
ratio	$\dfrac{11}{365}$	$\dfrac{50}{365}$	$\dfrac{130}{365}$	$\dfrac{120}{365}$	$\dfrac{45}{365}$	$\dfrac{9}{365}$
Statistical Probability	0.03	0.14	0.36	0.33	0.12	0.02

According to the Central Limit Theorem as the number of samples gets very large the distribution of $\hat{P}$ is Normal.

The mean and standard deviation of $\hat{P}$ can be determined from the collected statistics.

$$p = E(\hat{P}) = 0.5(0.03) + 0.55(0.14) +$$
$$0.63(0.36) + 0.67(0.33) +$$
$$0.71(0.12) + 0.75(0.02)$$
$$= 0.64$$

The population proportion is the expected value, sometimes called the point estimate.

$$sd(\hat{P}) = \sqrt{\frac{p(1-p)}{n}} = \sqrt{\frac{0.64(0.36)}{365}}$$
$$= 0.025$$
$$\hat{P} \sim N(\mu = 0.64, \sigma = 0.025)$$

The interval estimate for p is called a confidence interval.

Parameters and Statistics

A parameter is a quantity that describes some aspect of the whole population. Most population parameters for large populations are not known since it is not practical to know all the attributes for every member in that population.

Statistics to estimate population parameters are derived from sampling parts of the entire population. Statistics are used to estimate and infer the value of a parameter using methods of probability.

After observing a large number of random samples, we use probability to determine the likelihood that a parameter equals a proposed statistically derived value using confidence intervals and margins of error.

	Population Parameter	Sampled Statistic
Proportion	p	$\hat{p}$
Mean	μ	$\overline{x}$
Variance	σ^2	s^2
Standard Deviation	σ	s

Example 9.3: Parameters of central tendency (mean) inferred from Statistics.

It is not possible to test all the groundwater in an area for acidity, population parameters for the groundwater in the area need to be estimated or inferred by statistical sampling.

The average acidity of groundwater in an area is to be determined by random sampling using a pH scale measurement. The pH is a measure of the hydrogen ion concentration in modes per litre.

We use the Central Limit Theorem to draw conclusions about a population mean by random sampling, even if the population distribution is not normal. The convention is that samples of size 30 or greater will have a fairly normal distribution of sample means $\overline{X}$

$\sim N\left(\mu, \dfrac{\sigma}{\sqrt{n}}\right)$ regardless of the shape of the distribution of the variable in the population.

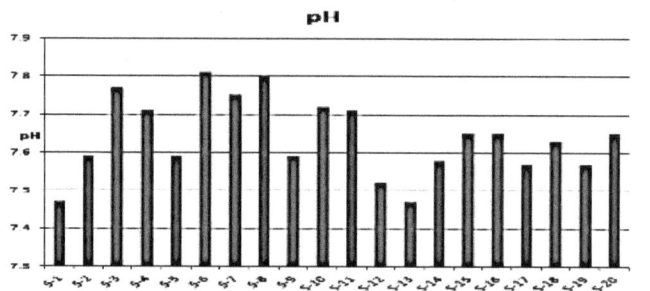

In the survey 20 samples are taken and tested and the pH is recorded .

Sampled pH values {7.48, 7.59, 7.77, 7.71, 7.59, 7.81, 7.75, 7.79, 7.59, 7.73, 7.71, 7.52, 7.47, 7.58, 7.65, 7.65, 7.57, 7.63, 7.57, 7.65}

The sample mean $\overline{x}$ of these values is a statistical estimate of the true mean parameter, in this case:

$$\overline{x} = \frac{152.81}{20} \approx 7.64$$

The Central Limit Theorem also informs that the sampling distribution of sample proportions is normally distributed with the expected value of p and a standard deviation of $\sqrt{\dfrac{p(1-p)}{n}}$.

The convention for application is if * $np \geq 10$ and * $n(1-p) \geq 10$ or the underlying population is known to follow a normal distribution.

*reference https://newonlinecourses.science.psu.edu/stat800/node/35/

Confidence Intervals and Margins of Error.

The estimated standard deviation of a sampled proportion $\hat{p}$ measures its accuracy. Using the sample proportion $\hat{p}$ and the standard deviation $sd\left(\hat{P}\right)=\sqrt{\dfrac{\hat{p}\left(1-\hat{p}\right)}{n}}$, and $Z\sim N(\mu=0,\sigma=1)$ to determine the

Confidence Interval $\hat{P}\pm Z\sqrt{\dfrac{p(1-p)}{n}}$, where $Z\sqrt{\dfrac{p(1-p)}{n}}$ is called the **Margin of Error**.

- 95% Confidence Interval $\Pr\left(\hat{P}-1.96\sqrt{\dfrac{p(1-p)}{n}}<p<\hat{P}+1.96\sqrt{\dfrac{p(1-p)}{n}}\right)\approx 0.95$

- 90% Confidence Interval $\Pr\left(\hat{P}-1.645\sqrt{\dfrac{p(1-p)}{n}}<p<\hat{P}+1.645\sqrt{\dfrac{p(1-p)}{n}}\right)\approx 0.90$

Where ± 1.96 represents 95%, ± 1.645 represents 90%, using Standard Normal Distribution Z scores.

For a parameter's estimation to be justifiable:
- The data's individual observation must be a random and independent sample of the population.
- Multiple random and independent samples are taken from the population.

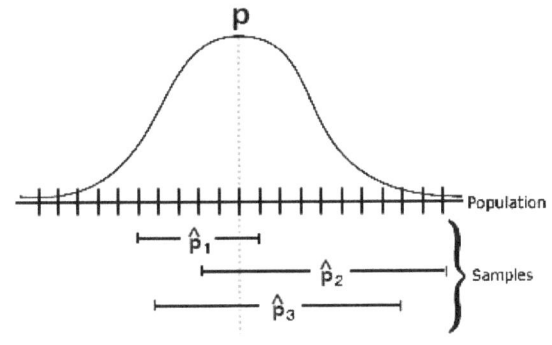

The line on either side of each of the sample proportions represents a confidence interval that has been constructed around the sample proportions $\hat{p}_1, \hat{p}_2, \hat{p}_3$, which identifies the range of values within which lies the true population proportion parameter.

The size of a confidence interval depends on the sample parameter's sample size and standard deviation. The bigger the interval the higher the confidence.

Example 9.4: Finding the Confidence Interval and Margin of Error for an estimated statistic.

1000 people were surveyed and 280 responded that they were vegetarians. Find the Margin of Error for a 90% confidence interval of this sample proportion.	Insert the values into the formula and solve: $$sd\left(\hat{P}\right)=\sqrt{\dfrac{\hat{p}\left(1-\hat{p}\right)}{n}}=\sqrt{\dfrac{0.28\left(0.72\right)}{1000}}$$ $$=0.014$$
Find sample proportion $\hat{p}$ by dividing the number of people who have the attribute of being a vegetarian in this sample. In this case, 280/1000 people (28%) responded positively to being a vegetarian, $\hat{p}=0.28$.	Margin of Error (M) $$M=1.645\sqrt{\dfrac{\hat{p}\left(1-\hat{p}\right)}{n}}=1.645\sqrt{\dfrac{0.28\left(0.72\right)}{1000}}$$ $$=1.645\times 0.014$$ $$=0.0233$$
Find the z-score that goes with the given confidence interval. A 90% confidence interval has a z-score of 1.645.	

SAMPLING STATISTICS SUMMARY:

Population parameters for large populations are not usually known, as it is not practical to count the attributes for every member in the population.

Population parameters are estimated from sampling parts of the entire population.

	Population Parameter	Sampled Statistic
Proportion	p	$\hat{p}$
Mean	μ	$\overline{x}$
Variance	σ^2	s^2

$$p = \frac{\text{number with attribute in total population}}{\text{population size (N)}}$$

$$\hat{p} = \frac{\text{number with attribute in sample}}{\text{sample size (n)}}$$

$\hat{p}$ is a random variable represented by $\hat{P} = \dfrac{X}{n}$, where X is a binomial random variable of the number of individuals in the population with the certain attribute.

Multiple random, independent samples are taken and the proportions of the certain attribute are found: $\hat{p}_1, \hat{p}_2, \hat{p}_3, \hat{p}_4 .... \hat{p}_k$

The sample proportions $\hat{p}_1, \hat{p}_2, \hat{p}_3, \hat{p}_4 .... \hat{p}_k$ are normally distributed around the actual population proportion p.

Population proportion is inferred by multiple sampling.

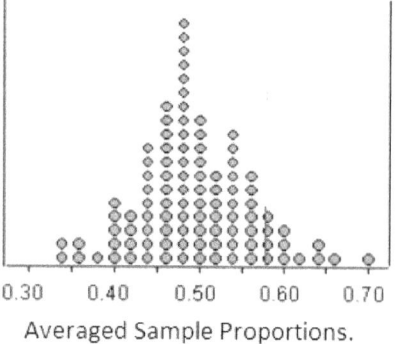

Averaged Sample Proportions.

The **Central Limit Theorem** informs that the sampling distribution of sample proportions is normally distributed with:

$$\mu = p \text{ and } \sigma = \sqrt{\frac{p(1-p)}{n}}.$$

The convention for application is if $* \, np \geq 10$ and $* \, n(1-p) \geq 10$ or the underlying population is known to follow a normal distribution.

For large values of n the **Normal Distribution** may be used as an approximation to the **Binomial Distribution** if the distribution is symmetrical and not very skewed

$\hat{P} = \dfrac{X}{n}$ is a random variable, $E(\hat{P}) = p$, with a standard deviation, or standard error given by:

$$sd(\hat{P}) = \sqrt{\frac{p(1-p)}{n}},$$

$$\hat{P} \sim N\left(\mu = \hat{p}, \sigma_{\hat{p}} = \sqrt{\frac{\hat{p}(1-\hat{p})}{n}}\right)$$

Confidence Interval $\hat{P} \pm Z\sqrt{\dfrac{p(1-p)}{n}}$, where

$Z\sqrt{\dfrac{p(1-p)}{n}}$ is called the **Margin of Error**.

The size of a confidence interval depends on the sample parameter's sample size and standard deviation.

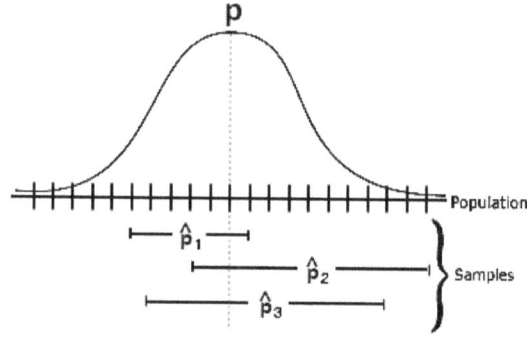

The bigger the interval the higher the confidence.

SAMPLING STATISTICS: Check your Understanding.

9.1 Sample proportions and standard deviations.

a) Can the following questions be answered using statistical methods? Explain your reasoning.

    i. Do English Literature novels use longer words than English novels?

    ii. Is ear size a useful predictor of height?

    iii. How does the level of carbon dioxide affect the growth of a plant?

    iv. How many people speak more than one language in Australia?

> Celebrate International Year of Indigenous Languages
>
> Rayakan Tahun Internasional Bahasa-bahasa Leluhu
>
> 庆祝国际土著语言年
>
> Feiern Sie das Internationale Jahr der einheimischen Sprache
>
> Εορτασμός του Διεθνούς Έτους Ιθαγενών Γλώσσων
>
> Celebrons l'année internationale des langues indigènes
>
> Comemore o Ano Internacional das Línguas Indígenas
>
> Отпразднуйте Международный Год Языков Коренных Народов

b) What is the expected value, and standard deviation of the sample proportions in the following situations?

    i. In a sample of 50 year 11 students, it is found that 23% of the students know the correct answer to a question. What is the expected value and the standard deviation of the sample proportion of year 11 students who know the correct answer?

    ii. Customers select pizza from the menu at a particular restaurant one-fifth of the time.

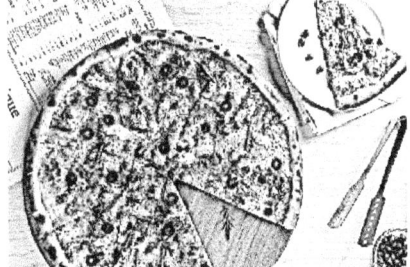

    40 menu selections are sampled. What is the expected value and the standard deviation of the sample proportion of pizza selection?

    iii. One in three of the people use Facebook regularly. If a samples of 70 people is taken, what is the expected value, and standard deviation of the sample proportion of regular Facebook users?

    iv. A master chef inspects meals coming out of a five star kitchen, and rejects sub-standard meals. One-tenth of the meals are rejected. 50 meals are inspected, what is the expected value and standard deviation of the sample proportion of rejected meals?

9.2 Normal approximation to the Binomial Distribution.

a) According to the taxation office, 80% of the population in Melbourne have full time employment. If 500 people from Melbourne are selected at random, find the probability that more than 85% of the people surveyed are in full time employment.

b) What is the approximate probability for 100 throws of a fair dice, the proportion of even numbers thrown will be between 0.45 and 0.55?

9.3 Confidence Intervals and Margins of error

a) A random sample of 100 public transport users is surveyed, and the proportion that own a car is 0.6.

    i.   There is a 95% chance that the sample proportion of car owning, public transport users falls between which two values?

    ii.   What is the probability that the sample proportion is less than or equal to 0.55?

b) In a survey of 100 university students, we found that 56 of them bring their lunch from home.
    i.   What is the estimate of p, the population proportion?
    ii.   What is the 90% confidence interval for p?

c) The Super CCCrunchy brand of corn chips was purchased by 1286 shoppers out of the 2010 that bought corn chips from a certain supermarket.
    i.   What is the estimate of p, the population proportion?
    ii.   Find the 90% confidence interval for p.
    iii.   Find the 95% confidence interval for p.
    iv.   Find the 99% confidence interval for p.

d) Media polls often use a sample size of 1200 to 1500 people.
    i.   How can a relatively small sample size relative to the whole population accurately predict an outcome?
    ii.   Complete the following table of possible p values, which value of p gives a maximum standard deviation?

p	$p(1-p)$
0.1	
0.2	
0.3	
0.4	
0.5	

    iii.   Hence, find the sample size n is needed if a margin of error of 0.025 will be tolerated?

SAMPLING STATISTICS SOLUTIONS.
9.1 Sample proportions and standard deviations. (SOLUTIONS)

a)

    i.   The sample mean word lengths can be collected from several English Literature novels and several English Novels and an estimate can be made about the differences in word lengths.

    ii.   Measurements of ear size and height can be randomly sampled from many individuals from the population and an estimate can be made of a correlation if one exists.

    iii.   Multiple plants can be put into different environments and their average growth statistics can be collected over time to make an estimate of which conditions are the most favourable.

    iv.   A survey could be carried out for many randomly selected individuals from the population and an estimate can be made about the population proportion that speak more than one language.

b) What is the expected value, and standard deviation of the sample proportions in the following situations?

    i.
$$n = 50, \hat{p} = 0.23, \sigma_{\hat{p}} = \sqrt{\frac{\hat{p}(1-\hat{p})}{n}} = \sqrt{\frac{0.23(0.77)}{50}} = 0.0595$$

    ii.
$$n = 40, \hat{p} = 0.2, \sigma_{\hat{p}} = \sqrt{\frac{\hat{p}(1-\hat{p})}{n}} = \sqrt{\frac{0.2(0.8)}{40}} = 0.0632$$

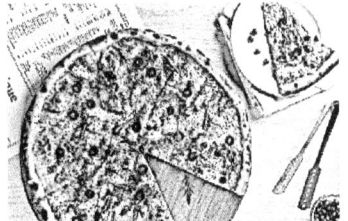

    iii.
$$n = 70, \hat{p} = \frac{1}{3}, \sigma_{\hat{p}} = \sqrt{\frac{\hat{p}(1-\hat{p})}{n}} = \sqrt{\frac{\frac{1}{3}\left(\frac{2}{3}\right)}{70}} = 0.0563$$

    iv.
$$n = 50, \hat{p} = 0.1, \sigma_{\hat{p}} = \sqrt{\frac{\hat{p}(1-\hat{p})}{n}} = \sqrt{\frac{0.1(0.9)}{50}} = 0.0424$$

9.2 Normal approximation to the Binomial Distribution. (SOLUTIONS)

a) 80% of the population in Melbourne have full time employment

Method 1.	Method 2.
Let X be the number of successes.	Let $\hat{P}$ be the proportion of successes.
$X \sim N\left(\mu = np, \sigma = \sqrt{np(1-p)}\right)$	$\hat{P} \sim N\left(\mu = \hat{p}, \sigma_{\hat{p}} = \sqrt{\dfrac{\hat{p}(1-\hat{p})}{n}}\right)$
$n = 500, p = 0.8$	
$X \sim N(\mu = 400, \sigma = 8.9443)$	$n = 500, p = 0.8$
$\Pr(X \geq (0.85 \times 500)) = \Pr(X \geq 425) \approx 0.0026$	$\hat{P} \sim N(\mu = 0.8, \sigma_{\hat{p}} = 0.0179)$
	$\Pr(\hat{P} \geq 0.85) \approx 0.0026$
	$\Pr(\hat{P} \geq 0.85) \approx 0.0026$

b) 100 throws of a fair dice

Method 1.	Method 2.
Let X be the number of successes.	Let $\hat{P}$ be the proportion of successes.
$X \sim N\left(\mu = np, \sigma = \sqrt{np(1-p)}\right)$	$\hat{P} \sim N\left(\mu = \hat{p}, \sigma_{\hat{p}} = \sqrt{\dfrac{\hat{p}(1-\hat{p})}{n}}\right)$
$n = 100, p = 0.5$	
$X \sim N(\mu = 50, \sigma = 5)$	
$\Pr(45 \leq X \leq 55) \approx 0.6827$	$n = 100, p = 0.5$
	$\hat{P} \sim N(\mu = 0.5, \sigma_{\hat{p}} = 0.05)$
	$\Pr(0.45 \leq \hat{P} \leq 0.55) \approx 0.6827$

9.3 Confidence Intervals and Margins of error. (SOLUTIONS)

a) A random sample of 100 public transport users is surveyed, and the proportion that own a car is 0.6.

  i. $n = 100, \hat{p} = 0.6, \sigma_{\hat{p}} = \sqrt{\dfrac{\hat{p}(1-\hat{p})}{n}} = \sqrt{\dfrac{0.6(0.4)}{100}} \approx 0.0490$

   The 95% Confidence Interval is given by

   $$\Pr\left(\hat{P} - 1.96\sqrt{\dfrac{p(1-p)}{n}} < p < \hat{P} + 1.96\sqrt{\dfrac{p(1-p)}{n}} \right) \approx 0.95$$

   $$\hat{P} - 1.96\sqrt{\dfrac{p(1-p)}{n}} \approx 0.6 - 1.96(0.0490) \approx 0.5039$$

   $$\hat{P} + 1.96\sqrt{\dfrac{p(1-p)}{n}} \approx 0.6 + 1.96(0.0490) \approx 0.6960$$

  ii. $\hat{P} \sim N\left(\mu = \hat{p}, \sigma_{\hat{p}} = \sqrt{\dfrac{\hat{p}(1-\hat{p})}{n}} \right)$

   $\hat{P} \sim N(\mu = 0.6, \sigma_{\hat{p}} = 0.0490)$

   $\Pr(\hat{P} \le 0.55) \approx 0.1538$

b) In a survey of 100 university students, we found that 56 of them bring their lunch from home.

  i. $n = 100, \hat{p} = \dfrac{56}{100}, \sigma_{\hat{p}} = \sqrt{\dfrac{\hat{p}(1-\hat{p})}{n}} = \sqrt{\dfrac{0.56(0.44)}{100}} \approx 0.0496$

  ii. The 90% Confidence Interval is given by

   $$\Pr\left(\hat{P} - 1.645\sqrt{\dfrac{p(1-p)}{n}} < p < \hat{P} + 1.645\sqrt{\dfrac{p(1-p)}{n}} \right) \approx 0.90$$

   $$\hat{P} - 1.645\sqrt{\dfrac{p(1-p)}{n}} \approx 0.56 - 1.645(0.0496) \approx 0.4783$$

   $$\hat{P} + 1.645\sqrt{\dfrac{p(1-p)}{n}} \approx 0.56 + 1.645(0.0496) \approx 0.6417$$

9.3 Confidence Intervals and Margins of error. (SOLUTIONS)

c) Super CCCrunchy corn chips were purchased by 1286 shoppers out of 2010.

i. $n = 2010, \hat{p} = \dfrac{1286}{2010}, \sigma_{\hat{p}} = \sqrt{\dfrac{\hat{p}(1-\hat{p})}{n}} = \sqrt{\dfrac{0.6399(0.3601)}{2010}} \approx 0.0107$

ii. The 90% Confidence Interval is given by

$\hat{P} \pm 1.645\sqrt{\dfrac{p(1-p)}{n}} \approx 0.6399 \pm 1.645(0.0107) \approx 0.6291, 0.6506$

iii. The 95% Confidence Interval is given by

$\hat{P} \pm 1.96\sqrt{\dfrac{p(1-p)}{n}} \approx 0.6399 \pm 1.96(0.0107) \approx 0.6190, 0.6609$

iv. The 99% Confidence Interval is given by

$\hat{P} \pm 2.58\sqrt{\dfrac{p(1-p)}{n}} \approx 0.6399 \pm 2.58(0.0107) \approx 0.6123, 0.6675$

d) Media polls often use a sample size of 1200 to 1500 people.

i. Large random independent sampling of individuals in a population can be used to estimate parameters of a population.

ii.

p	$p(1-p)$
0.1	0.09
0.2	0.16
0.3	0.21
0.4	0.24
0.5	0.25

$p=0.5$ gives the largest standard deviation

iii. The 95% Margin of error is equivalent to 0.025 difference is given by

$1.96\sqrt{\dfrac{p(1-p)}{n}} = 0.025$

$(1.96)^2\left(\dfrac{p(1-p)}{n}\right) = (0.025)^2$

$n = \dfrac{(1.96)^2\, p(1-p)}{(0.025)^2}$

$p=0.5$ gives the largest standard deviation from part ii., hence:

$p(1-p) = 0.5(0.5) = 0.25$

$n = \dfrac{(1.96)^2 (0.25)}{(0.025)^2} \approx 1536$

A

Abstraction	5
Addition of circular functions	97
Addition of functions	45
Addition rule of probability	177
Algebra	5
Algebra, Solutions for exercises	27
Algebra, Check your Understanding	25
Algebra & Algorithms Summary	24
Algorithms in pseudocode	16
Amplitude of Circular function	91
Antidifferentiation	149
Archimedes algorithm in pseudocode	18
Areas between functions	159
Areas between functions	160
Areas under functions	149
Areas under functions	150
Areas under functions	151
Asymptotes	45

B

Bernoulli sequence	179
Bernoulli sequence	180
binomCdf	189
Binomial coefficients	180
Binomial Distribution	179
Binomial Distribution	180
Binomial Distribution	183
Binomial Distribution	217
Binomial probability bar graph	184
Binomial probability bar graph	189
Biology derivatives	136
Bisection method f(x)=0	72
Bounded areas	156
Bounded areas	158
Business derivatives	136

C

Central Limit Theorem	211
Central Limit Theorem	236
Central Limit Theorem	240
Chain rule for differentiation	127
Change of logarithmic base	105
Changing p for Binomial Distributions	184
Chemistry derivatives	136
Circular functions	35
Circular functions	89
Combinations	190
Composite functions	38
Composite functions	41
Conditional structures in algorithms	17
Conditional probability	177
Conditional probability	185
Conditional probability	186
Conditional probability	187
Conditional probability	215
Confidence interval for sample proportions	238
Confidence interval for sample proportions	241
Consistent linear systems	19
Consistent linear systems	20
Constant of integration	153
Continuous function	151
Continuous Probability, Solutions for Exercises	223
Continuous random variables	203
Continuous random variables	205

D

Decibel scale	107
Decomposing functions	39
Definite Integrals	149
Definite Integrals	151
Definite Integrals	152
Definite Integrals	158
Definite Integrals	161
Degrees	90
Dependent events	187
Derivative function	128
Derivatives	121
Derivatives, Solutions for exercises	143
Derivatives, summary	139
Derivatives, Check your Understanding	140
Differentiation	121
Differentiation	149
Differentiation	151
Differentiation by first principles	122
Dilation	44
Dilation	42
Discrete Events	175
Discrete Outcomes	175
Discrete Outcomes	190
Discrete Probability	175
Discrete probability bar graph	178
Discrete probability bar graph	179
Discrete probability bar graph	183
Discrete probability bar graph	184
Discrete probability bar graph	187
Discrete probability distributions	178
Discrete probability distributions	179
Discrete Probability, Solutions for Exercises	197
Discrete Probability, Summary	192
Discrete Probability, Check your Understanding	193

D

Division of functions	45
Domain	34
Domain	40
Domain	71
Domain	93
Domain	99
Domain	103
Domain	134
dy/dx	121
dy/dx	122

E

Elimination in algebra	6
Elimination in algebra	9
Endpoints of functions	134
Equations	8
Euler's number	101
Euler's number	102
Even functions	65
Event space	175
Exact values	92
expand using CAS technology	22
Expected value	182
Expected value	183
Expected value	206
Exponential decay	99
Exponential decay	100
Exponential decay	101
Exponential functions	35
Exponential functions	89
Exponential functions	99
Exponential growth	99
Exponential growth	100
Exponential growth	101

F

Fabulous Functions, Solutions for exercises	112
Fabulous Functions, Check your Understanding	109
Factorial function defined in pseudocode	20
Factorising polynomial functions	67
factor using CAS technology	22
factors function defined in pseudocode	20
First derivative	123
For loops in pseudocode	17
Function graphs	36
Functions	33
Functions y=f(x) summary	46
Functions, Solutions for exercises	53
Functions, summary	108
Functions, Check your Understanding	47
Fundamental theorem of calculus	152
Fundamental theorem of calculus	161
f'(x)	121
f'(x)	122
f'(x)=0	129

G

General solutions for circular equations	95
Gradient	89
Gradient function	121
Gradient function	122
Gradient function	128
Graphs of functions y=f(x)	36
Greatest Common Divisor algorithm	20
Grouped real frequency data	203
Grouped real frequency data	204
Grouped real frequency data	205

H

Hybrid function	128
Hybrid function	134
Hybrid functions	37
Hybrid functions	41
Hyperbola function	70

I

Identity matrix	13
If – then – endif pseudocode	17
If – then –else – endif pseudocode	17
Indefinite integrals	152
Indefinite integrals	157
Independent events	177
Independent events	185
Index/Exponent laws	104
Infinite solutions	19
Infinite solutions	23
Infinitesimal	151
Instantaneous rate of change	121
Integrals	149
Integrals, Solutions for Exercises	169
Integrals, Summary	163
Integrals, Check your Understanding	164
Integration	149
Integration	151
Integration properties	153
Integration properties	155
Integration rules	155
Intersections of functions	63
Inverse circular functions	93
Inverse exponential function	103
Inverse functions	40
Inverse functions	41

J

K

INDEX A-Z

L

Limits of integration	151
Linear equations	10
Linear equations	10
Linear functions	62
Logarithm laws	104
Logarithmic base	103
Logarithmic base	104
Logarithmic base	105
Logarithmic base	106
Logarithmic functions	89
Logarithmic functions	103

M

Margin of error for sample proportions	241
Maxima and minima f'(x)=0	129
Maxima and minima f'(x)=0	132
Maxima and minima f'(x)=0	134
Mean	182
Mean	206
Mean	207
Mean	234
Mean=median=mode	209
Median	206
Median	207
Median, probability density functions	213
Minimum function defined in pseudocode	20
Mode	206
Mode	207
Multiplication of functions	45

N

Nature of stationary points	129
Nature of stationary points	130
Negative areas below x-axis	152
Negative areas below x-axis	156
Negative power functions	61
Negative power functions	68
Negative power functions	70
Newton's method for solving f(x)=0	137
Newton's method in pseudocode	137
Non-functions	35
Normal distribution	209
Normal distribution	217
Normal Distribution	236
Normal Distribution/Binomial Distribution	190
Null factor theorem	66
Number sets	34

O

Odd functions	65

P

Parabola	9
Parameters of populations	240
Parametric solutions	23
Pascal's triangle	180
Pascal's triangle	183
Percentiles, probability density functions	213
Period of Circular function	91
Physics derivatives	136
Piecewise functions	37
Polynomial functions	62
Polynomial long division	66
Population	229
Population mean	230
Population proportion	230

P

Positive areas above x-axis	152
Positive areas above x-axis	156
Power functions	35
Power functions	61
Power functions, Solutions for exercises	79
Power functions, summary	74
Power functions, Check your Understanding	75
Power rule for differentiation	125
Power rule, antiderivatives	153
Probability by integration	206
Probability density function	203
Probability density function	204
Probability density function	205
Probability density function	206
Probability density function	207
Product rule for differentiation	125
Python code	21

Q

Quadrants	92
Quadratic formula	66
Quadratic formula in pseudocode	20
Quadratic functions	62
Quantity by integration	157
Quantity by integration	158
Quantity by integration	149
Quantity by integration	161
Quartic functions	62
Quartiles, probability density functions	213
Quotient power functions	61
Quotient power functions	71
Quotient rule for differentiation	125
Quadrants	92
Quadratic formula	66
Quadratic functions	62
Quantity by integration	157
Quantity by integration	158
Quantity by integration	149

R

Radians	90
Range	34
Range	40
Range	71
Rate of change	121
Rate of change	134
Rate of change	162
Rates of change	137
Rates of change	138
Real life, Derivatives and Rate of Change	136
Real life, Logarithms	107
Real life, Binomial Distributions	190
Real life, Normal Distributions	211
Real life, Probability density functions	208
Real life, Sample Proportions	231
Real life, Sample Proportions	237
Reciprocal	45
Reference Angle	92
Reference Angle	93
Reference Angle	96
Reflection	43
Reflection	42
Remainder theorem	66
Richter scale	107
Riddles	8
Rules for antiderivatives	154
Rules for antiderivatives	155

S

Sample	229
Sample mean	230
Sample proportion	230
Sample proportion as random variables	234
Sample proportion as random variables	238
Sample proportion by random sampling	233
Sample proportion by random sampling	233
Sample proportions, Binomial distribution	237
Sample proportions, Normal distribution	237
Sampling Statistics	229
Sampling Statistics, Solutions for Exercises	245
Sampling Statistics, Summary	242
Sampling Statistics, Check your Understanding	243
Scalar multiplication	11
Simultaneous Linear Equations	12
Sine = Opposite/Hypotenuse	90
Smooth functions	126
Smooth functions	128
Smooth functions	134
Social trends derivatives	136
SohCahToa	89
solve using CAS Technology	22
Solving circular function equations	93
Solving circular function equations	96
Solving $f(x)=0$	72
Solving for n, Binomial Distributions	188
Square root function	71
Standard deviation	182
Standard deviation	183
Standard deviation	206
Standard deviation	207
Standard deviation	209
Standard deviation	234
Standard Normal distribution	209
Standard Normal distribution	213
Standard Normal Distribution	236

S

Stationary points	129
Stationary points	134
Statistics of samples	240
Strictly decreasing functions	133
Strictly decreasing functions	135
Strictly increasing functions	133
Strictly increasing functions	135
Substitution in algebra	6
Substitution in algebra	9
Subtraction of functions	45
Summation of area strips	151
Summation of area strips	161
Symbols in algebra	8
Symmetry of circular functions	92
Symmetry of Normal Distributions	210
Symmetry of Normal Distributions	214

T

Tangent lines on functions	130
Tangent=Opposite/Adjacent	90
Technology, Binomial Distribution	191
Technology, Binomial Distribution	189
Technology, Normal Distribution	212
Technology, Normal Distribution	213
Theoretical probability	175
Transformation matrices	15
Transformation of functions	42
Transformation of functions matrix algebra	43
Transformation of functions matrix algebra	44
Transformation Probability Density $f(x)$	216
Translation	43
Trapezoidal Rule	150
Tree diagrams	176
Trigonometric functions	89
Truncus function	70
Turning points maxima, minima	63
Turning points of functions	63

U

Unique solution	19
Unit Circle	90
Unit Circle	91
Univariate statistics	232

V

Variance	182
Variance	183
Variance	206
Variance	207
Variance	234
Variables in algorithms	16
Venn diagrams	176
Venn diagrams	185
Venn diagrams	187
Venn diagrams	215
Vertical line test	34

W

While loops in pseudocode	17

X

Y

Z

Chapter 1. Algebra and Algorithms

Abstraction in algebra	5
Algebra	5
Algebra, Solutions for exercises	27
Algebra, Check your Understanding	25
Algebra and Algorithms, Summary	24
Algorithms	16
Archimedes algorithm in pseudocode	18
CAS Technology solving algebra	11
CAS Technology built in functions	22
Conditional structures in pseudocode	17
Elimination method for simultaneous eqns	10
Elimination method for simultaneous eqns	13
Equations	8
Factorial function defined in pseudocode	20
factorise function pseudocode	20
Greatest common divisor function pseudocode	20
For loop in pseudocode	17
Functions in algorithms	19
If – then -end if in pseudocode	17
If – then – else – end if in pseudocode	17
Inconsistent linear systems	12
Infinite solution system	12
Loops in algorithms	17
Inverse matrices	14
Linear coordinate geometry	10
Linear equations	17
Linear equations	19
Minimum function defined in pseudocode	20
Numerical methods in algorithms	19
No solution system	12
Parabola	9
Parametric solutions	15
Pronumerals in algebra	8
Python code	21
Riddles	8
Sequential actions in algorithms	16
Simultaneous Linear Equations	6
Simultaneous Linear Equations	10
Simultaneous Linear Equations	13
Substitution method for simultaneous eqns	10
Substitution method for simultaneous eqns	9
Symbols in algebra	8
Transformation matrices	15
Unique solution systems	19
Variables in algorithms	16
While loop in pseudocode	17

Addition of functions	45
Asymptotes	45
Circular functions	35
Composite functions	38
Composite functions	41
Decomposing functions	39
Dilation of functions	44
Dilation of functions	42
Division of functions	45
Domain of functions	34
Domain of functions	40
Exponential functions	35
Function graphs	36
Functions	33
Functions y=f(x), Summary	46
Functions, Solutions for exercises	53
Functions, Check your Understanding	47
Graphs of functions y=f(x)	36
Hybrid functions	37
Hybrid functions	41
Inverse functions	40
Inverse functions	41
Multiplication of functions	45
Non-functions	35
Number sets	34
Piecewise functions	37
Power functions	35
Range of functions	34
Range of functions	40
Reciprocal functions	45
Reflection of functions	42
Reflection of functions	43
Subtraction of functions	45
Transformation of functions	42
Transformation of functions matrix algebra	43
Transformation of functions matrix algebra	44
Translation of functions	43
Translation of functions	42
Vertical line test	34

Chapter 3. Super Power Functions

Asymptotes	70
Bisection method f(x)=0	72
Cubic functions	62
Cubic functions	64
Domain restrictions	71
Even functions	65
Extreme values	70
Factorising polynomial functions	67
Hyperbola function	70
Intersections of functions	63
Inversely proportional functions	68
Linear functions	62
Negative power functions	61
Negative power functions	68
Negative power functions	70
Null factor theorem	66
Odd functions	65
Polynomial functions	62
Polynomial long division	66
Power functions	61
Power functions, Solutions for exercises	79
Power functions, Summary	74
Power functions, Check your Understanding	75
Quadratic formula	66
Quadratic functions	62
Quartic functions	62
Quotient power functions	61
Quotient power functions	71
Range of functions	71
Remainder theorem	66
Solving f(x)=0	72
Square root function	71
Truncus function	70
Turning points maxima, minima	63
Turning points of functions	63

Addition of circular functions	97
Amplitude of Circular function	91
Asymptotes	103
Change of logarithmic base	105
Circular functions	89
Cosine = Adjacent/Hypotenuse	90
Decibel scale, logarithms in real life	107
Degrees	90
Domain	93
Domain	99
Domain	103
Euler's number	101
Euler's number	102
Exact values	92
Exponential decay	99
Exponential decay	100
Exponential decay	101
Exponential functions	89
Exponential functions	99
Exponential growth	99
Exponential growth	100
Exponential growth	101
Fabulous Functions, Solutions for exercises	112
Fabulous Functions, Summary	108
Fabulous Functions, Check your Understanding	109
General solutions circular equations	95
Gradient	89
Index/Exponent laws	104
Inverse circular functions	93
Inverse exponential function	103
Irrational number	101
Logarithm laws	104
Logarithmic base	103
Logarithmic base	104
Logarithmic base	105
Logarithmic base	106

Logarithmic functions	89
Logarithmic functions	103
Period of Circular function	91
Quadrants	92
Radians	90
Real life Logarithms	107
Reference Angle	92
Reference Angle	93
Reference Angle	96
Richter scale, logarithms in real life	107
Sine = Opposite/Hypotenuse	90
SohCahToa	89
Solving circular function equations	93
Solving circular function equations	96
Symmetry of circular functions	92
Tangent=Opposite/Adjacent	90
Trigonometric functions	89
Unit Circle	90
Unit Circle	91

Chapter 5. Deriving Derivatives

Average rate of change	121
Biology derivatives in real life	136
Business derivatives in real life	136
Business derivatives in real life	138
Chain rule for differentiation	127
Chemistry derivatives in real life	136
Continuous functions	126
Continuous functions	134
Continuous functions	128
Derivative function	128
Derivatives	121
Derivatives, Solutions for exercises	143
Derivatives, summary	139
Derivatives, Check your Understanding	140
Differentiation	121
Differentiation by first principles	122
Domain	134
dy/dx	121
dy/dx	122
Endpoints	134
f'(x)	121
f'(x)	122
f'(x)=0	129
First derivative	123
Gradient function	121
Gradient function	122
Gradient function	128
Hybrid function	128
Hybrid function	134
Instantaneous rate of change	121
Maxima and minima f'(x)=0	129
Maxima and minima f'(x)=0	132
Maxima and minima f'(x)=0	134
Nature of stationary points	129
Nature of stationary points	130
Newton's method for solving f(x)=0	137
Newton's method in pseudocode	137
Physics derivatives in real life	136
Power rule for differentiation	125
Product rule for differentiation	125
Quotient rule for differentiation	125
Rate of change	121
Rate of change	134
Rates of change	137
Rates of change	138
Real life derivatives	136
Rules for differentiation	127
Smooth functions	126
Smooth functions	128
Smooth functions	134
Social trends derivatives	136
Stationary points	129
Stationary points	134
Strictly decreasing functions	133
Strictly decreasing functions	135
Strictly increasing functions	133
Strictly increasing functions	135
Tangent lines on functions	130

Anti-differentiation	149
Areas between functions	159
Areas between functions	160
Areas under functions	149
Areas under functions	150
Areas under functions	151
Average value of a function	162
Bounded areas	156
Bounded areas	158
Constant of integration	153
Continuous function	151
Definite Integrals	149
Definite Integrals	151
Definite Integrals	152
Definite Integrals	158
Definite Integrals	161
Differentiation	149
Differentiation	151
Fundamental theorem of calculus	152
Fundamental theorem of calculus	161
Indefinite integrals	152
Indefinite integrals	157
Infinitesimal	151
Integrals	149
Integrating Integrals, Solutions for Exercises	169
Integrating Integrals, Summary	163
Integrals, Check your Understanding	164
Integration	149
Integration	151
Integration properties	153
Integration properties	155
Integration rules	155
Limits of integration	151
Negative areas below x-axis	152
Negative areas below x-axis	156
Positive areas above x-axis	152
Positive areas above x-axis	156
Power rule, antiderivatives	153
Quantity by integration	157
Quantity by integration	158
Quantity by integration	149
Quantity by integration	161
Rate of change	162
Rules for antiderivatives	154
Rules for antiderivatives	155
Summation of area strips	151
Summation of area strips	161
Trapezoidal rule derivation	150
Trapezoidal rule example	150

Chapter 7. Discrete Probability

Addition rule of probability	177	Mean	182
Bernoulli sequence	179	Mean	183
Bernoulli sequence	180	Multiplication rule of probability	177
binomCdf	189	Normal Distribution / Binomial Distribution	190
Binomial coefficients	180	Pascal's triangle	180
Binomial distribution	179	Pascal's triangle	183
Binomial distribution	180	Real life, Binomial Distribution	190
Binomial distribution	183	Solving for n, Binomial Distributions	188
Binomial probability bar graph	184	Standard deviation	182
Binomial probability bar graph	189	Standard deviation	183
Changing p for Binomial Distributions	184	Technology, Binomial Distribution	191
Combinations	190	Technology, Binomial Distribution	189
Conditional probability	177	Theoretical probability	175
Conditional probability	185	Tree diagrams	176
Conditional probability	186	Variance	182
Conditional probability	187	Variance	183
Counting favourable outcomes	176	Venn diagrams	176
Dependent events	187	Venn diagrams	185
Discrete Events	175	Venn diagrams	187
Discrete Outcomes	175		
Discrete Outcomes	190		
Discrete Probability	175		
Discrete probability bar graph	178		
Discrete probability bar graph	179		
Discrete probability bar graph	183		
Discrete probability bar graph	184		
Discrete probability bar graph	187		
Discrete probability distributions	178		
Discrete probability distributions	179		
Discrete Probability, Solutions for Exercises	197		
Discrete Probability, Summary	192		
Discrete Probability, Check your Understanding	193		
Event space	175		
Expected value	182		
Expected value	183		
Independent events	177		
Independent events	185		

Binomial Distribution	217
Central Limit Theorem	211
Conditional probability	215
Continuous Probability, Solutions Exercises	223
Continuous random variables	203
Continuous random variables	205
Continuous Probability, Summary	218
Continuous Probability, Check your Understanding	219
Expected value	206
Grouped real frequency data	203
Grouped real frequency data	204
Grouped real frequency data	205
Inverse Normal Distribution	213
Mean	206
Mean	207
Mean=median=mode	209
Median	206
Median	207
Median, probability density functions	213
Mode	206
Mode	207
Normal distribution	209
Normal distribution	217
Percentiles, probability density functions	213
Probability by integration	206
Probability density function	203
Probability density function	204
Probability density function	205
Probability density function	206
Probability density function	207
Quartiles, probability density functions	213
Real life, Normal Distributions	211
Real life, probability density functions	208
Standard deviation	206
Standard deviation	207
Standard deviation	209

Standard Normal distribution	209
Standard Normal distribution	213
Symmetry of Normal Distributions	210
Symmetry of Normal Distributions	214
Technology, Normal Distribution	212
Technology, Normal Distribution	213
Transformation of Probability Density $f(x)$	216
Variance	206
Variance	207
Venn diagrams	215

Chapter 9. Sampling Statistics

Central Limit Theorem	236
Central Limit Theorem	240
Confidence interval for sample proportions	238
Confidence interval for sample proportions	241
Margin of error for sample proportions	241
Mean	234
Normal Distribution	236
Parameters of populations	240
Population	229
Population mean	230
Population proportion	230
Real life, Sample Proportion	231
Real life, Sample Proportion	237
Sample	229
Sample mean	230
Sample proportion	230
Sample proportion as random variables	234
Sample proportion as random variables	238
Sample proportion by random sampling	233
Sample proportion by random sampling	233
Sample proportions, Binomial distribution	237
Sample proportions, Normal distribution	237
Sampling Statistics	229
Sampling Statistics, Solutions for Exercises	245
Sampling Statistics, Summary	242
Sampling Statistics, Check your Understanding	243
Standard deviation	234
Standard Normal Distribution	236
Statistics of samples	240
Univariate statistics	232
Variance	234

www.ingramcontent.com/pod-product-compliance
Lightning Source LLC
Chambersburg PA
CBHW080824220526
45467CB00008B/2191